TISSUE REMODELING

ANNALS OF THE NEW YORK ACADEMY OF SCIENCES
Volume 995

TISSUE REMODELING

Edited by Marit Nilsen-Hamilton, Zena Werb, and Eli Keshet

The New York Academy of Sciences
New York, New York
2003

Library of Congress Cataloging-in-Publication Data

Growth Factor and Signal Transduction Conference (11th: 2002: Ames, Iowa).
Tissue remodeling / edited by Marit Nilsen-Hamilton ... [*et al.*]
p.; cm. — (Annals of the New York Academy of Sciences; v. 995)
"This volume is the result of a symposium entitled Tissue Remodeling (Proceedings of the Eleventh Growth Factor and Signal Transduction Conference), cosponsored by the Department of Biochemistry, Biophysics, and Molecular Biology at Iowa State University and the Cell Biology Program at the Mayo Clinic Cancer Center, held August 1–4, 2002, in Ames, Iowa."
Includes bibliographical references and index.
ISBN 1-57331-416-1 (cloth: alk. paper) — ISBN 1-57331-417-X (paper: alk. paper)
1. Morphogenesis—Congresses. 2. Transforming growth factors-beta—Congresses. 3. Neovascularization—Congresses. 4. Wound healing—Congresses. 5. Growth factors—Congresses. 6. Cellular signal transduction—Congresses.
[DNLM: 1. Growth Substances—physiology—Congresses. 2. Angiogenesis Factor—Congresses. 3. Bone Remodeling—Congresses. 4. Tissues—growth & development—Congresses. 5. Wound Healing—Congresses. QU 107 G8834 2003] I. Nilsen-Hamilton, Marit. II. Title. III. Series.
Q11.N5 vol. 995
[QH491]
500 s—dc21
[571.8

2003007406

GYAT / PCP

Printed in the United States of America
ISBN 1-57331-416-1 (cloth)
ISBN 1-57331-417-X (paper)
ISSN 0077-8923

ANNALS OF THE NEW YORK ACADEMY OF SCIENCES

Volume 995
May 2003

TISSUE REMODELING

Editors
MARIT NILSEN-HAMILTON, ZENA WERB, AND ELI KESHET

This volume is the result of a symposium entitled **Tissue Remodeling** (Proceedings of the Eleventh Growth Factor and Signal Transduction Conference), cosponsored by the Department of Biochemistry, Biophysics, and Molecular Biology at Iowa State University and the Cell Biology Program at the Mayo Clinic Cancer Center, held August 1–4, 2002, in Ames, Iowa.

CONTENTS

Financial assistance was received from:

Principal Sponsors

- CELL BIOLOGY PROGRAM/MAYO CLINIC CANCER CENTER
- DEPARTMENT OF BIOCHEMISTRY, BIOPHYSICS, AND MOLECULAR BIOLOGY/IOWA STATE UNIVERSITY

Sponsors

- COLLEGE OF AGRICULTURE/IOWA STATE UNIVERSITY
- F. WENDELL MILLER TRUST
- IOWA STATE UNIVERSITY BIOTECHNOLOGY COUNCIL
- NATIONAL ANIMAL DISEASE CENTER (AMES, IOWA)
- NATIONAL INSTITUTES OF HEALTH (GRANT NO. 1R13 CA91998)

Preface

MARIT NILSEN-HAMILTON,[a] ZENA WERB,[b] AND ELI KESHET[c]

[a]*Department of Biochemistry, Biophysics, and Molecular Biology, Iowa State University, Ames, Iowa, USA*

[b]*Department of Anatomy, University of California, San Francisco, California, USA*

[c]*Department of Molecular Biology, Hebrew University–Hadassah Medical School, Jerusalem, Israel*

Held at Iowa State University in August 2002, this symposium on Tissue Remodeling revealed, in its inspiring presentations, how the application of current technology can tease out a mechanistic understanding of physiological processes as complicated as tissue remodeling. This symposium was also memorable because it was originally planned for September 13, 2001, but had to be rescheduled for the next year due to the horrific events of September 11, 2001. The symposium can be counted as one of the many examples to show that such dreadful acts may delay civilized activities, but do not quell them.

The term "tissue remodeling" encompasses processes in which an existing tissue is physically altered as part of a normal physiological course of events. Examples of tissue remodeling are wound healing, compensatory adrenal gland growth, placentation, and the growth and involution of the uterus, mammary gland, and ovaries during female reproduction. These complex processes require the coordinated action of many cell types in the context of the intact organism. Cellular events include differentiation, dedifferentiation, proliferation, directed movement, and apoptosis. Proliferation dominates in a growing tissue and apoptosis dominates during regression or involution of a tissue. In all cases, normal tissue remodeling is characterized by an integrated and orderly progression of events that results in new tissue organization. Sequential events are tightly coordinated. Integration is driven by communication amongst cells within the tissues and with new cells that invade the remodeling tissues.

Pattern formation, particularly branching morphogenesis, is a common feature of tissue remodeling. Epithelial and endothelial cells display branching morphogenetic development in tissues such as the mammary glands, lungs, and placentae. The mechanistic basis of pattern formation is communication by cell-cell contact, by cell–extracellular matrix contact, and by secreted factors such as growth factors.[1]

Angiogenesis, another form of branching morphogenesis, is essential for tissue survival. Even cancer cells, which are free of many of the normal regulatory regimes, require a blood supply before they can grow beyond about 1 mm in diameter. Con-

Address for correspondence: Marit Nilsen-Hamilton, Department of Biochemistry, Biophysics, and Molecular Biology, 3206 Molecular Biology Building, Iowa State University, Ames, IA 50011. Voice: 515-294-9996; fax: 515-294-0453.

marit@iastate.edu

Ann. N.Y. Acad. Sci. 995: ix–xii (2003). © 2003 New York Academy of Sciences.

sequently, angiogenesis is an integral process in remodeling events that involve tissue growth. Vascular endothelial growth factor (VEGF) and fibroblast growth factors (FGFs) are well-established angiogenic factors. However, other factors, including cytokines such as macrophage migration inhibitory factor (MIF),[2] also stimulate angiogenesis *in vivo*. Hypoxia is also a regulator of angiogenesis and tissue remodeling. Genetic approaches can be taken, with simple organisms such as *C. elegans*, to unravel the regulatory mechanisms downstream of HIF1α, which is the major hypoxia-driven transcription factor.[3]

Studies that use transgenic mice in which VEGF production can be experimentally manipulated in specific tissues provide excellent means of evaluating the angiogenic response to a VEGF signal.[4] Other interactions, such as cell-to-cell contact through interaction between the Notch receptor and its Delta ligand, might also be involved in determining arterial/venous endothelial cell specification.[5] Identified negative regulators of angiogenesis may prove useful in suppressing tumor angiogenesis.[6]

To perform its function, the placenta requires extensive integration with the maternal blood supply. Placentation involves invasion and pseudovasculogenesis in which the ectodermally derived cytotrophoblasts take on molecular and morphological characteristics of endothelial cells and extravasating leukocytes.[7] Some tumor cells also form pseudovascular structures such as seen in the vasculogenic mimicry of melanoma cells.[8]

The immune system is likely involved in many aspects of tissue remodeling, particularly during tissue involution when the tissue regresses and is resorbed by the body. An acute phase protein that characterizes the innate immune response is highly expressed during involution of the mammary gland and the uterus after pregnancy.[9]

Growth factors play a central role in regulating tissue remodeling. No symposium can hope to cover all the growth factors that are known to exist. However, some growth factors that are centrally important to tissue remodeling events were discussed here. One important group is the transforming growth factor-β (TGFβ) growth factor family. The SMADs are the major intracellular signaling pathway by which the TGFβ family members alter cellular behavior through their transmembrane serine kinase receptors.[10] Although two family members, activin and TGFβ, promote fibrosis and wound healing, they seem to stimulate distinct functions in wound repair and to use different SMADs for signaling.[11] In addition to SMADs, another potential downstream signaling molecule, the TGFβ1-regulated antiapoptotic factor-1 (TIAF1), was discussed.[12]

Many growth factors act through tyrosine kinase transmembrane receptors and, for some growth factors, such as FGF and VEGF, there are more than one receptor. As discussed for VEGF, different receptors signal differently and thus their activation results in different cellular responses.[13] As well as encoding a transmembrane receptor for activating cellular functions, some receptor genes include alternative splice sites that result in the production of secreted extracellular domains with the potential capability of regulating growth factor accessibility to cells in the tissue. An example is the epithelial growth factor (EGF) receptor gene.[14]

Although most intercellular communications in tissue remodeling processes are probably generated locally and act locally, there are also examples of endocrine regulation of tissue remodeling. The mechanisms by which development and compensatory growth of the adrenal glands are regulated include signals from the pituitary hormones as well as locally generated signals.[15]

Cell-to-cell interactions and interactions with the extracellular matrix also provide signals to cells for their integrated differentiation and proliferation. For example, mutant mice that do not express one of the two abundant cartilage extracellular matrix components, perlecan or collagen II, have severe chondrodysplasia.[16] Cell–to–extracellular matrix interactions are critically important for placental development in which chorioallantoic attachment is mediated by an interaction between VCAM1, expressed on the allantoic tip, with α4-integrin expressed on the surface of the chorion.[1]

Systematic cell movements are characteristic of most tissue remodeling. Positive and negative signals within a tissue can direct the movement of individual cells through that tissue.[17] The environment through which the cells move can also influence cellular differentiation and integration into the remodeling tissue.[18] Cell movements also require the action of proteases.

However, the biological functions of proteases are much more extensive than to create a "pathway" for cell passage through a tissue matrix. Specific proteases can create protein fragments that affect tissue remodeling. In some cases, the protein fragment is a recognized growth factor. Members of the EGF growth factor family are produced as transmembrane precursors capable of juxtamembrane signaling. The EGF moiety is cleaved to form the soluble growth factor by a specific metalloproteinase of the ADAM family, called TACE.[19]

As well as cleaving cell surface proteins, secreted proteases also cleave extracellular matrix proteins and thus release growth factors and other "soluble" signaling proteins from their matrix associations. Frequently, the responsible proteases are members of the large metalloproteinase (MMP) family. For example, fragments of extracellular matrix proteins such as laminin, released by MMPs, can regulate the movements and patterning of cells.[8] The significance of MMP action to tissue remodeling is evident in the profound alterations in bone morphogenesis observed in MMP9-null mice.[20]

An important goal of understanding tissue remodeling is to apply this knowledge in medicine to correct aberrant remodeling processes. Good *in vivo* models are essential for testing hypotheses to explain the fundamental tissue remodeling process and to progress to a medical application. A strength of the presentations in this symposium was the frequent identification and use of *in vivo* models.

CITATIONS FROM THIS VOLUME

1. CROSS, J.C. 2003. Chorioallantoic morphogenesis and formation of the placental villous tree.
2. NISHIHIRA, J., T. ISHIBASHI, T. FUKUSHIMA *et al.* 2003. Macrophage migration inhibitory factor (MIF): its potential role in tumor growth and tumor-associated angiogenesis.
3. SHEN, C. & J.A. POWELL-COFFMAN. 2003. Genetic analysis of hypoxia signaling and response in *C. elegans*.
4. DOR, Y., V. DJONOV & E. KESHET. 2003. Induction of vascular networks in adult organs: implications to proangiogenic therapy.
5. SHAWBER, C.J., I. DAS, E. FRANCISCO *et al.* 2003. Notch signaling in primary endothelial cells.
6. IRUELA-ARISPE, M.L., D. CARPIZO & A. LUQUE. 2003. ADAMTS1: a matrix metalloprotease with angioinhibitory properties.

7. ZHOU, Y., O. GENBACEV & S.J. FISHER. 2003. The human placenta remodels the uterus by using a combination of molecules that govern vasculogenesis or leukocyte extravasation.
8. HENDRIX, M.J.C., E.A. SEFTOR, D.A. KIRSCHMANN *et al.* 2003. Remodeling of the microenvironment by aggressive melanoma tumor cells.
9. NILSEN-HAMILTON, M., Q. LIU, J. RYON *et al.* 2003. Tissue involution and the acute phase response.
10. ROBERTS, A.B., A. RUSSO, A. FELICI *et al.* 2003. Smad3: a key player in pathogenetic mechanisms dependent on TGF-β.
11. WANKELL, M., S. WERNER, C. ALZHEIMER *et al.* 2003. The roles of activin in cytoprotection and tissue repair.
12. KHERA, S. & N-S. CHANG. 2003. TIAF1 participates in the transforming growth factor β1–mediated growth regulation.
13. MEYER, R.D. & N. RAHIMI. 2003. Comparative structure-function analysis of VEGFR-1 and VEGFR-2: what have we learned from chimeric systems?
14. REITER, J.L. & N.J. MAIHLE. 2003. Characterization and expression of novel 60-kDa and 110-kDa EGFR isoforms in human placenta.
15. BLAND, M.L., M. DESCLOZEAUX & H.A. INGRAHAM. 2003. Tissue growth and remodeling of the embryonic and adult adrenal gland.
16. GUSTAFSSON, E., A. ASZÓDI, N. ORTEGA *et al.* 2003. Role of collagen type II and perlecan in skeletal development.
17. COFFMAN, C.R. 2003. Cell migration and programmed cell death of Drosophila germ cells.
18. SAKAGUCHI, D.S., S.J. VAN HOFFELEN & M.J. YOUNG. 2003. Differentiation and morphological integration of neural progenitor cells transplanted into the developing mammalian eye.
19. LEE, D.C., S.W. SUNNARBORG, C.L. HINKLE *et al.* 2003. TACE/ADAM17 processing of EGFR ligands indicates a role as a physiological convertase.
20. ORTEGA, N., D. BEHONICK, D. STICKENS *et al.* 2003. How proteases regulate bone morphogenesis.

Smad3: A Key Player in Pathogenetic Mechanisms Dependent on TGF-β

ANITA B. ROBERTS,[a] ANGELO RUSSO,[b] ANGELINA FELICI,[a] AND KATHLEEN C. FLANDERS[a]

[a]*Laboratory of Cell Regulation and Carcinogenesis,* [b]*Radiation Biology Branch, National Cancer Institute, Bethesda, Maryland, USA*

ABSTRACT: Transforming growth factor-β (TGF-β), a key player in a large variety of physiological and disease processes, signals through transmembrane receptor serine/threonine kinases to activate novel signaling intermediates called Smad proteins, which then modulate transcription of target genes. We have utilized mice with a targeted deletion of Smad3, one of two homologous proteins involved in signaling from TGF-β/activin, to investigate the function of this particular pathway in transducing such effects of TGF-β. The dramatic results of the absence of Smad3 on parameters of healing of cutaneous wounds, such as reepithelialization and influx of inflammatory cells, as well as on fibrosis as modeled by radiation fibrosis of skin in mice, suggest that signaling flux through Smad3 is critical for chemotactic activity of TGF-β, inhibitory effects of TGF-β on keratinocyte proliferation and migration, and chemoattraction and elaboration of extracellular matrix by fibroblasts in fibrotic diseases. We recently identified a novel molecule, TLP for TRAP-1-like protein, which selectively interferes with Smad3 signaling, and are currently investigating whether levels of this protein might be altered in disease to change the relative flow of information from Smad2 and Smad3.

KEYWORDS: TGF-β; Smad2; Smad3; wound healing; fibrosis; signaling

Transforming growth factor-β1 (TGF-β1) was first characterized biochemically in 1983 based on its purification from tissue sources including human platelets and its activity in inducing growth of nontransformed cells in soft agar.[1] Since that time, many molecules have been identified that share structural features with TGF-β1, such that the TGF-β superfamily now comprises over 40 different proteins, most of which can be classified into three broad groupings of TGF-βs, bone morphogenetic proteins, and the activins/inhibins.[2] The first receptors belonging to this family were not cloned until almost 10 years later,[3] and it is now generally accepted that the entire TGF-β superfamily signals through transmembrane receptors having intrinsic serine/threonine kinase activity.[4] These receptors are the only known transmembrane

Address for correspondence: Anita B. Roberts, Ph.D., Chief, Laboratory of Cell Regulation and Carcinogenesis, National Cancer Institute, Building 41, Room C629, 41 Library Drive, MSC 5055, Bethesda, MD 20892-5055. Voice: 301-496-5391; fax: 301-496-8345.
robertsa@dce41.nci.nih.gov

Ann. N.Y. Acad. Sci. 995: 1–10 (2003).

receptors with serine/threonine kinase activity; all other known serine/threonine kinases are intracellular proteins.

Our laboratory has focused on understanding the role of TGF-β in wound healing, fibrosis, and carcinogenesis. Several activities of TGF-β are common to these pathological events. It profoundly inhibits the proliferation of cells, particularly epithelial and lymphoid cells, which are the basis of most human cancers.[5] It regulates apoptosis and cellular differentiation,[1] strongly suppresses immune cell functions,[6] and regulates the accumulation of extracellular matrix (ECM).[7] While the inhibition of cellular proliferation by TGF-β is important in its tumor suppressor activities, its ability to promote influx of inflammatory cells and to stimulate the production of ECM is key to its roles in wound healing and fibrosis.

SMAD PROTEINS IN TGF-β SIGNAL TRANSDUCTION

The signal transduction pathways for TGF-β were elucidated in a burst of activity between 1996 and 1999. The clues came initially from genetic studies of other family members in *Drosophila*[8] and in *C. elegans*.[9] The acronym "Smad" is a condensation of the names of the signaling molecules identified downstream of TGF-β-superfamily ligands in *Drosophila* (Mad)[8] and *C. elegans* (Sma).[9] There are now five receptor-activated Smads and the common mediator Smad4, which function in direct signal transduction from the TGF-β-superfamily receptor kinases to the nucleus, modulated by two inhibitory Smads.[10,11] In the canonical signaling pathway, receptor-activated Smads are phosphorylated on C-terminal serines by the type I

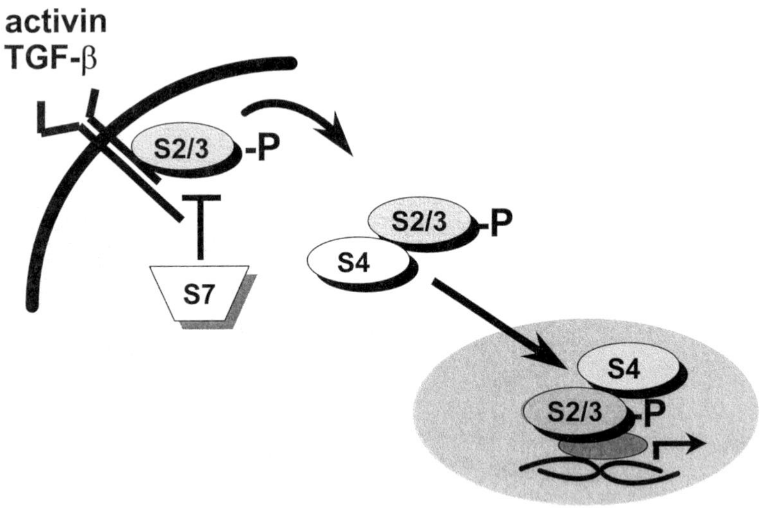

FIGURE 1. Basic features of the Smad signaling pathway downstream of TGF-β/activin receptors. Upon ligand binding, receptor-activated Smads2/3 are phosphorylated by the type I receptors, form a heteromeric complex with Smad4, and translocate to the nucleus where they regulate target gene expression both by direct DNA binding and through interaction with other transcription factors, coactivators, and corepressors.

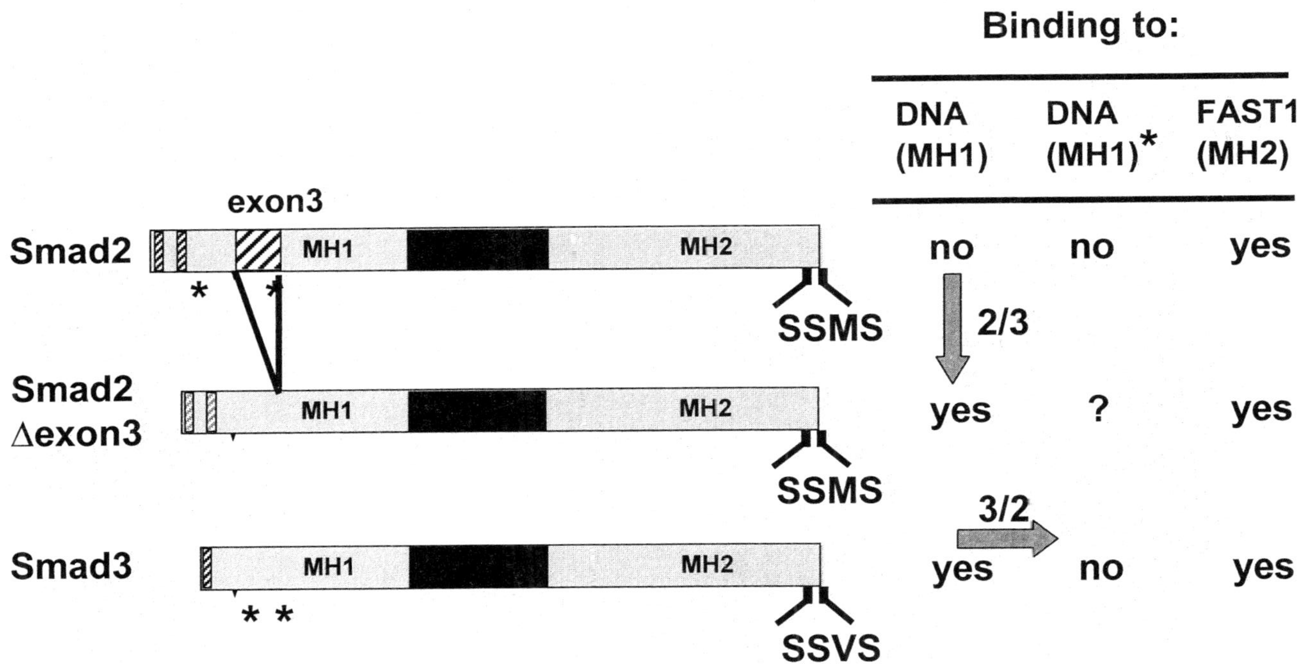

FIGURE 2. Smad2 and Smad3 differ in their ability to bind to DNA through their N-terminal MH1 domains, but share the ability to bind to transcription factors such as FAST1 through their C-terminal MH2 domains. Alternative splice forms of Smad2 lacking exon 3 acquire the ability to bind DNA directly through consensus GTCT sites, whereas phosphorylation of Smad3 on sites in its MH1 domain by protein kinase C interferes with its ability to bind DNA directly. Despite these mechanisms that modulate the distinguishing features of Smad2 and Smad3, their roles are not redundant.

TGF-β receptor, after which they partner with Smad4. The resulting complex translocates to the nucleus where it regulates the activity of target genes, either directly by binding to cognate DNA consensus sites or indirectly by interaction with other transcription factors (FIG. 1).

Of the five receptor-activated Smads, two of them, Smad2 and Smad3, are the principal transducers of signals from TGF-β receptors. Although highly homologous, Smad2 and Smad3 have many distinguishing features that contribute to unique patterns of gene activation[12] (FIG. 2). Whereas both Smad2 and Smad3 bind through an identical motif in their L3 loop to the L45 loop of the TGF-β type I receptor[13] and are phosphorylated with similar kinetics in most cells, Smad2 has an insertion of 30 amino acids in its N-terminal Mad-homology 1 (MH1) domain that precludes it from binding directly to DNA, as does Smad3.[14] Other pathways can also modulate the activity of Smad2 and 3. Phosphorylation of residues in the MH1 domain of Smad3 by protein kinase C interferes with the binding of Smad3 to DNA.[15] Mitogen-activated protein kinase (MAPK) phosphorylates serine/threonine residues in the middle linker region, which interferes with the nuclear translocation of Smad2/3 in certain instances.[16] Nuclear Smad3/4 complexes not only bind consensus GTCT motifs in the DNA directly, but they and Smad2/4 complexes also bind a variety of coactivators, corepressors, and transcription factors to bring about the multifaceted patterns of gene regulation attributed to TGF-β.[17,18]

Whereas the Smad-signaling pathway plays a prominent role in regulation of the transcriptome by TGF-β, other, more multipurpose signaling pathways are also activated directly by TGF-β. These include the phosphatidylinositol-3 kinase (PI3K) cascade; components of MAPK pathways including Ras, Raf, ERK, JNK, and p38; as well as the regulators of cadherin junctions, RhoA and Rac.[19–23] Genes are being identified that require cooperation between these other pathways and Smad pathways, such that the gene target or cellular process can be blocked by interruption of either pathway. For example, a recent study shows that epithelial to mesenchymal

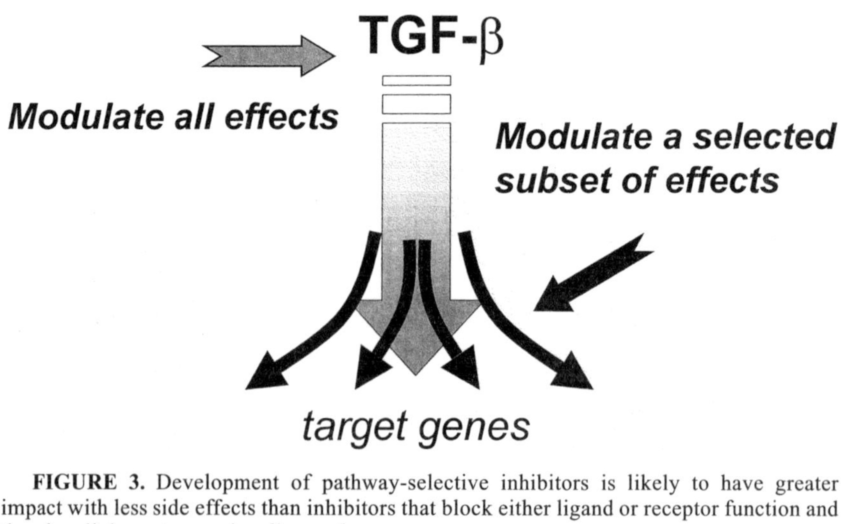

FIGURE 3. Development of pathway-selective inhibitors is likely to have greater impact with less side effects than inhibitors that block either ligand or receptor function and thereby all downstream signaling pathways.

transition (EMT) in mouse mammary epithelial NMuMg cells requires p38 and the Smad pathway, such that inhibition of either will block the transdifferentiation.[20,24] TGF-β-dependent EMT has, in some instances, also been shown to require activation of PI3K or RhoA.[19,23]

Given the multiple signaling pathways involved in mediating the complex, pleiotropic activities of TGF-β on cells, we have proposed that the development of pathway-selective inhibitors of these different pathways would serve two purposes: (1) it would allow for identification of the obligatory pathways involved in effecting various physiologic and pathophysiologic activities of TGF-β and (2) it would potentially enable a more specific and selective pharmacological approach to subtly control TGF-β effects, as compared to approaches involving interference at the level of either the ligand or its receptors (FIG. 3).

DISTINCT ROLES OF SMAD2 AND SMAD3

Both the TGF-βs and activins signal through Smad2 and Smad3[10,11] (FIG. 1). Clearly, there are nonredundant activities of these two Smad proteins since targeted deletion of Smad2 results in early embryonic lethality,[25] whereas the knockout of Smad3 results in a viable mouse that dies from defects in mucosal immunity at 1–6 months of age.[26,27] Smad3-null mice are characterized by the formation of multiple abscesses adjacent to mucosal surfaces that are filled with *Providencia*, bacteria that are not pathogenic except in an immune-compromised host, suggesting altered mucosal defense mechanisms.[26] Subsequent investigations have shown that TGF-β-mediated chemotaxis in a variety of cells, including neutrophils, macrophages, keratinocytes, and fibroblasts, requires Smad3. Importantly, the loss of Smad3 does not impair the ability of these cells to migrate in response to other stimuli, but selectively interferes with the recognition of signals from TGF-β.[26,28,29]

ROLE OF SMAD3 IN WOUND HEALING

Although it represents only one of many pathways activated by TGF-β in cells, a growing body of evidence suggests that Smad3 is required for many of the responses of cells to injury and in disease pathogenesis. The study of wound healing can be used to evaluate effects of modulated signal transduction in a setting in which many different cell types interact in a defined temporal pattern within a specific tissue architecture.[30] Unlike some other growth factors like KGF, which is produced by fibroblasts and acts in a very specific way on keratinocytes,[31] TGF-β is produced by all cells participating in the wound healing process, and all the cells can also respond to TGF-β.[7] In the initiation of wound healing, platelet alpha granules release TGF-β, which then activates production of TGF-β by other cells including keratinocytes and fibroblasts and both recruits macrophages and stimulates them to produce TGF-β.[7] To test whether certain effects of TGF-β on wound healing might be channeled through the Smad3 pathway, cutaneous wound healing was studied in the Smad3-null mice. Loss of Smad3 had a profound, dose-dependent effect on enhancing the rate of epithelialization of the wound to reestablish the cutaneous barrier, whereas Smad2 appeared to not be required in this process.[28] In this model and others, responses of Smad3 heterozygous mice were approximately midway between those of wild-type and null mice, suggesting that the level of Smad3 is limiting in certain responses. Loss of Smad3 also reduces the influx of macrophages into the wound bed and reduces the production of ECM by the granulation tissue.[28] While the decreased expression of collagen and certain other matrix proteins in wounds in Smad3-null mice may not be desirable in situations where wound strength is critical, this could likely be modulated by other factors in a clinical setting involving use of a putative Smad3 inhibitor.

ROLE OF SMAD3 IN MEDIATING DAMAGING EFFECTS OF IONIZING RADIATION

The mechanisms underlying the profibrotic effects of TGF-β are thought to involve its recruitment of macrophages to a site of injury and their subsequent production of TGF-β, which leads to recruitment of fibroblasts that then produce ECM proteins including collagen[32] (FIG. 4). Each of these aspects involves an obligatory role of Smad3. Many features of the role of TGF-β in radiation fibrosis are common to its roles in other models of fibrosis,[33] and we have chosen this important clinical problem to test the hypothesis that Smad3 is a key mediator of the pathogenic process. Mice irradiated locally with a single dose of 30 Gy to the skin show a severe thickening of the epidermis; a compacted, collagen-rich dermis; and hyperproliferation cells in the hair follicles at 5 weeks postirradiation.[34] Loss of Smad3 affords substantial protection from these unwanted effects of irradiation. As seen previously in wound healing, there is a gene-dosage effect of Smad3 in that the degree of

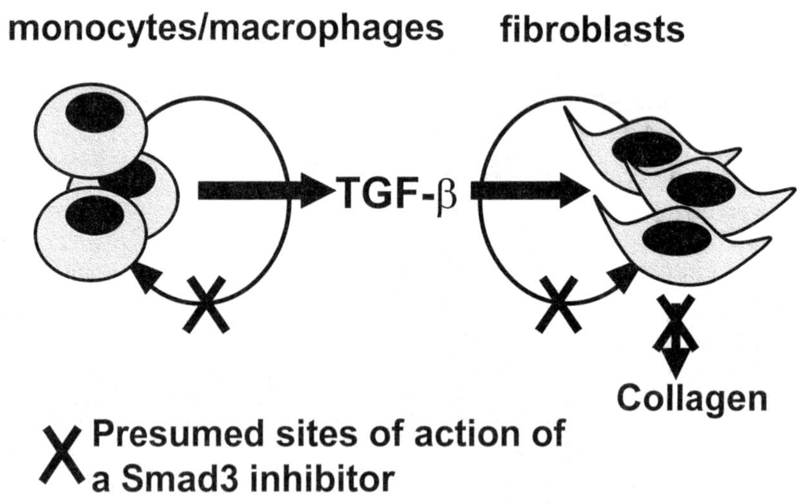

FIGURE 4. The role of TGF-β in fibrosis is thought to involve recruitment of monocytes/macrophages and their autoinduction of TGF-β, with subsequent recruitment of fibroblasts and their activation to express collagens and other matrix proteins. Each of the sites indicated by an "×" is a presumed site of interference by a putative inhibitor of Smad3.

thickening of the skin in Smad3 heterozygous mice is approximately midway between that of the wild-type and null genotypes.[34]

Since loss of Smad3 interferes with the chemotactic response of neutrophils, macrophages, keratinocytes, and fibroblasts to TGF-β,[26,28,29] examination of the differences in cellularity of sites of radiation in wild-type and Smad3-null mice can be used to infer which cell populations are likely recruited by TGF-β/Smad3 signals in the injury response. Quantitation of the numbers of inflammatory cells in skin lesions postirradiation shows that influx of neutrophils, and to a lesser extent mast cells and myofibroblasts, into irradiated sites is clearly TGF-β/activin/Smad3-dependent, whereas the recruitment of macrophages is less reliant on the Smad3 status and likely more a function of other signals.[34] Smad3 also has an important function in the upregulation of TGF-β at sites of injury. There is substantially stronger immunohistochemical staining for TGF-β in both the epidermis and dermis of irradiated skin in wild-type mice as compared to Smad3-null littermates.[34] Quantitative assessment of the collagen organization with picrosirius red staining under polarizing light shows that loss of Smad3 protects against the substantial scarring in the wild-type skin postirradiation, with Smad3-null skin having a "scar index" very close to that of nonirradiated control skin.[29] Overall, the data suggest that putative inhibitors of Smad3 might afford protection against many of the unwanted side effects of irradiation, including especially fibrosis.[32]

Wound healing in irradiated skin is important because surgeons are hesitant to operate on previously irradiated patients when the skin has become fibrotic and lost its pliability. Using as a model incisional wounds made in the skin of mice 5 weeks postirradiation with a single dose of 30 Gy, we have shown that loss of Smad3 still results in more rapid epithelialization of wounds than occurs in wild-type mice.[29] We observed profound decreases in the numbers of neutrophils and myofibroblasts in the wound beds associated with a decrease in the overall cellularity of the wound bed. Together, the data suggest that inhibition of Smad3, or even partial reduction in its activity, would be beneficial both to reduce the initial reaction of the skin to irradiation and to improve the parameters of wound closure postsurgery on irradiated tissue.[29]

ARE THERE ENDOGENOUS PROTEINS THAT CONTROL THE BALANCE BETWEEN SMAD2 AND SMAD3 SIGNALING?

Given what we know about the distinct roles of Smad2 and Smad3 in pathophysiological responses to TGF-β, it is difficult to understand why almost all cells express both of these signaling intermediates and why, in most cases, the kinetics of their phosphorylation appear to be nearly identical.[35] Despite the fact that Smad2 and Smad3 bind to the identical motif in the L45 loop of the TGF-β type I receptor through the identical domain in their L3 loops,[13] such that mutation in the Smad binding site of the receptor prevents binding of both Smad2 and Smad3,[20,24] provocative data obtained in our laboratory suggest that the binding and phosphorylation of Smad2 and Smad3 by the type I receptors are not competitive. Thus, we have reported that, in fibroblasts null for either Smad2 or Smad3, the levels of expression and, even more surprisingly, the levels of phosphorylation of the "other" Smad are unchanged from that in wild-type cells expressing both homologous Smads.[35] Moreover, in cells engineered to overexpress Smad2 or Smad3 up to 30-fold over baseline, no differences are seen in the level of phosphorylation of the "other" endogenous R-Smad compared to controls (Fang Tian and Stephen Yoo, unpublished). Such data suggest that Smad2 and Smad3 do not compete for receptor binding. Rather, they suggest either that the levels of receptors are in large excess compared to the R-Smads such that loss or overexpression of one R-Smad does not change the phosphorylation of the "other" R-Smad or, alternatively, that Smad2 and Smad3 might be phosphorylated by distinct receptor pools.

Recent data from other laboratories suggest that phosphorylation of Smad2 and Smad3 does not occur on receptors localized to the plasma membrane, but rather intracellularly in endosomes following clathrin-dependent endocytosis.[36,37] However, at present there is no evidence that Smad2 and Smad3 are phosphorylated in the same endocytic locale. Rather, the data demonstrating the independent nature of the phosphorylation of Smad2 and Smad3 are consistent with the hypothesis that these two Smad proteins are phosphorylated by distinct receptor pools possibly localized to different subcompartments or subdomains of endosomes, as have been shown to exist for different Rab GTPases[38] and even for the Smad2/3 binding protein SARA.[39] While SARA has been postulated to localize Smad2 to early endosomes,[39–41] at present little is known about the intracellular localization of Smad3 or the sites of Smad2/4 and Smad3/4 hetero-oligomerization.[42]

We have recently identified a novel molecule called TLP, for TRAP-1-like protein, that differentially regulates Smad2 and Smad3 activity.[43] Like its homologue TRAP-1,[44] TLP binds to TGF-β/activin receptors and to Smad4, but has the unique property of enhancing Smad2-dependent transcriptional responses while inhibiting Smad3-dependent transcriptional responses in the same cells. TLP is selective for the TGF-β/activin pathways and has no effects on BMP-dependent responses. Mechanistically, it exerts its selective effects subsequent to Smad phosphorylation at the level of formation of hetero-oligomeric complexes with Smad4, interfering selectively with formation of Smad3/4, but not Smad2/4 complexes.[43] We propose that TLP might be involved in localizing Smad4 to specific receptor/Smad complexes in different endosomal subcompartments such that it is available to activated Smad2, but not to activated Smad3. We are presently examining whether TLP expression might be modulated in disease states to alter the balance of signaling through the Smad2 and Smad3 pathways.

CONCLUSIONS

Although TGF-β activates both Smad2 and Smad3 in most cells, it is now clear that the two proteins mediate different transcriptional responses in cells and play distinct roles in pathophysiologic effects of TGF-β. Whether additional molecules like TLP will be identified that can control the balance of signaling between these two pathways, or whether expression of such molecules will be modulated in disease states, is not known. While our present data suggest that Smad3 and not Smad2 is the key mediator of pathogenic effects of TGF-β in fibrosis, increased knowledge of the unique roles and distinct transcriptional targets of Smad2 and Smad3 will ultimately allow design of more selective approaches to regulation of the activity of TGF-β in cells and tissues.

REFERENCES

1. ROBERTS, A.B. & M.B. SPORN. 1990. The transforming growth factors-β. *In* Handbook of Experimental Pharmacology: Peptide Growth Factors and Their Receptors, pp. 419–472. Springer-Verlag. New York/Berlin.
2. KINGSLEY, D.M. 1994. The TGF-beta superfamily: new members, new receptors, and new genetic tests of function in different organisms. Genes Dev. **8:** 133–146.
3. LIN, H.Y., X.F. WANG, E. NG-EATON *et al.* 1992. Expression cloning of the TGF-beta type II receptor, a functional transmembrane serine/threonine kinase. Cell **68:** 775–785.
4. DERYNCK, R. & X.H. FENG. 1997. TGF-beta receptor signaling. Biochim. Biophys. Acta **1333:** F105–F150.
5. WAKEFIELD, L.M. & A.B. ROBERTS. 2002. TGF-beta signaling: positive and negative effects on tumorigenesis. Curr. Opin. Genet. Dev. **12:** 22–29.
6. LETTERIO, J.J. & A.B. ROBERTS. 1998. Regulation of immune responses by TGF-beta. Annu. Rev. Immunol. **16:** 137–161.
7. ROBERTS, A.B. & M.B. SPORN. 1996. Transforming growth factor-β. *In* The Molecular and Cellular Biology of Wound Repair, pp. 275–308. Plenum. New York.
8. NEWFELD, S.J., E.H. CHARTOFF, J.M. GRAFF *et al.* 1996. Mothers against dpp encodes a conserved cytoplasmic protein required in DPP/TGF-beta responsive cells. Development **122:** 2099–2108.
9. SAVAGE, C., P. DAS, A.L. FINELLI *et al.* 1996. *Caenorhabditis elegans* genes sma-2, sma-3, and sma-4 define a conserved family of transforming growth factor beta pathway components. Proc. Natl. Acad. Sci. USA **93:** 790–794.
10. MASSAGUE, J. & D. WOTTON. 2000. Transcriptional control by the TGF-beta/Smad signaling system. EMBO J. **19:** 1745–1754.
11. ROBERTS, A.B. & R. DERYNCK. 2001. Meeting report: signaling schemes for TGF-beta. Sci. STKE **2001:** E43.
12. YANG, Y-C., E. PIEK, J. ZAVADIL *et al.* 2003. Hierarchical model of gene regulation by TGF-β. Submitted.
13. CHEN, Y.G., A. HATA, R.S. LO *et al.* 1998. Determinants of specificity in TGF-beta signal transduction. Genes Dev. **12:** 2144–2152.
14. YAGI, K., D. GOTO, T. HAMAMOTO *et al.* 1999. Alternatively spliced variant of smad2 lacking exon 3: comparison with wild-type smad2 and smad3. J. Biol. Chem. **274:** 703–709.
15. YAKYMOVYCH, I., P. TEN DIJKE, C.H. HELDIN *et al.* 2001. Regulation of Smad signaling by protein kinase C. FASEB J. **15:** 553–555.
16. KRETZSCHMAR, M., J. DOODY, I. TIMOKHINA *et al.* 1999. A mechanism of repression of TGFbeta/Smad signaling by oncogenic Ras. Genes Dev. **13:** 804–816.
17. MASSAGUE, J. & D. WOTTON. 2000. Transcriptional control by the TGF-beta/Smad signaling system. EMBO J. **19:** 1745–1754.
18. DERYNCK, R., Y. ZHANG & X.H. FENG. 1998. Smads: transcriptional activators of TGF-beta responses. Cell **95:** 737–740.
19. BHOWMICK, N.A., M. GHIASSI, A. BAKIN *et al.* 2001. Transforming growth factor-beta1 mediates epithelial to mesenchymal transdifferentiation through a RhoA-dependent mechanism. Mol. Biol. Cell **12:** 27–36.
20. YU, L., M.C. HEBERT & Y.E. ZHANG. 2002. TGF-beta receptor–activated p38 MAP kinase mediates Smad-independent TGF-beta responses. EMBO J. **21:** 3749–3759.
21. JANDA, E., K. LEHMANN, I. KILLISCH *et al.* 2002. Ras and TGF{beta} cooperatively regulate epithelial cell plasticity and metastasis: dissection of Ras signaling pathways. J. Cell Biol. **156:** 299–314.
22. MULDER, K.M. 2000. Role of Ras and Mapks in TGFbeta signaling. Cytokine Growth Factor Rev. **11:** 23–35.
23. BAKIN, A.V., A.K. TOMLINSON, N.A. BHOWMICK *et al.* 2000. Phosphatidylinositol 3-kinase function is required for transforming growth factor beta–mediated epithelial to mesenchymal transition and cell migration. J. Biol. Chem. **275:** 36803–36810.
24. ITOH, S., M. THORIKAY, M. KOWANETZ *et al.* 2002. Elucidation of Smad requirement in transforming growth factor-beta type I receptor–induced responses. J. Biol. Chem. In press.

25. Weinstein, M., X. Yang & C. Deng. 2000. Functions of mammalian Smad genes as revealed by targeted gene disruption in mice. Cytokine Growth Factor Rev. **11:** 49–58.
26. Yang, X., J.J. Letterio, R.J. Lechleider *et al.* 1999. Targeted disruption of SMAD3 results in impaired mucosal immunity and diminished T cell responsiveness to TGF-beta. EMBO J. **18:** 1280–1291.
27. Datto, M.B., J.P. Frederick, L. Pan *et al.* 1999. Targeted disruption of smad3 reveals an essential role in transforming growth factor beta–mediated signal transduction. Mol. Cell. Biol. **19:** 2495–2504.
28. Ashcroft, G.S., X. Yang, A.B. Glick *et al.* 1999. Mice lacking Smad3 show accelerated wound healing and an impaired local inflammatory response. Nat. Cell Biol. **1:** 260–266.
29. Flanders, K.C., C.D. Major, A. Arabshahi *et al.* 2003. Interference with TGF-β/Smad3 signaling results in accelerated healing of wounds compromised by irradiation. Submitted.
30. Singer, A.J. & R.A. Clark. 1999. Cutaneous wound healing. N. Engl. J. Med. **341:** 738–746.
31. Beer, H.D., M.G. Gassmann, B. Munz *et al.* 2000. Expression and function of keratinocyte growth factor and activin in skin morphogenesis and cutaneous wound repair. J. Invest. Dermatol. Symp. Proc. **5:** 34–39.
32. Roberts, A.B., E. Piek, E.P. Bottinger *et al.* 2001. Is Smad3 a major player in signal transduction pathways leading to fibrogenesis? Chest **120:** 43S–47S.
33. Martin, M., J. Lefaix & S. Delanian. 2000. TGF-beta1 and radiation fibrosis: a master switch and a specific therapeutic target? Int. J. Radiat. Oncol. Biol. Phys. **47:** 277–290.
34. Flanders, K.C., C.D. Sullivan, M. Fujii *et al.* 2002. Mice lacking Smad3 are protected against cutaneous injury induced by ionizing radiation. Am. J. Pathol. **160:** 1057–1068.
35. Piek, E., W.J. Ju, J. Heyer *et al.* 2001. Functional characterization of transforming growth factor beta signaling in Smad2- and Smad3-deficient fibroblasts. J. Biol. Chem. **276:** 19945–19953.
36. Penheiter, S.G., H. Mitchell, N. Garamszegi *et al.* 2002. Internalization-dependent and -independent requirements for transforming growth factor beta receptor signaling via the Smad pathway. Mol. Cell. Biol. **22:** 4750–4759.
37. Hayes, S., A. Chawla & S. Corvera. 2002. TGF beta receptor internalization into EEA1-enriched early endosomes: role in signaling to Smad2. J. Cell Biol. **158:** 1239–1249.
38. De Renzis, S., B. Sonnichsen & M. Zerial. 2002. Divalent Rab effectors regulate the sub-compartmental organization and sorting of early endosomes. Nat. Cell Biol. **4:** 124–133.
39. Hu, Y., J.Z. Chuang, K. Xu *et al.* 2002. Sara, a Fyve domain protein, affects Rab5-mediated endocytosis. J. Cell Sci. **115:** 4755–4763.
40. Panopoulou, E., D.J. Gillooly, J.L. Wrana *et al.* 2002. Early endosomal regulation of Smad-dependent signaling in endothelial cells. J. Biol. Chem. **277:** 18046–18052.
41. Itoh, F., N. Divecha, L. Brocks *et al.* 2002. The FYVE domain in Smad anchor for receptor activation (SARA) is sufficient for localization of SARA in early endosomes and regulates TGF-beta/Smad signalling. Genes Cells **7:** 321–331.
42. Moustakas, A., S. Souchelnytskyi & C.H. Heldin. 2001. Smad regulation in TGF-beta signal transduction. J. Cell Sci. **114:** 4359–4369.
43. Felici, A., J.U. Wurthner, W.T. Parks *et al.* 2003. TLP, a novel modulator of TGF-β signaling, has opposite effects on Smad2- and Smad3-dependent transcription. Submitted.
44. Wurthner, J.U., D.B. Frank, A. Felici *et al.* 2001. Transforming growth factor-beta receptor–associated protein 1 is a Smad4 chaperone. J. Biol. Chem. **276:** 19495–19502.

TIAF1 Participates in the Transforming Growth Factor β1–Mediated Growth Regulation

SMAIRA KHERA AND NAN-SHAN CHANG

Laboratory of Molecular Immunology, Guthrie Research Institute, Sayre, Pennsylvania 18840, USA

ABSTRACT: TGF-β induces growth suppression and apoptosis of various types of cells, but supports fibroblast growth. We previously isolated TIAF1 (TGF-β1-induced antiapoptotic factor 1), which protects murine L929 fibroblasts from TNF cytotoxicity. Here, we show that TIAF1 induced growth inhibition and apoptosis of monocytic U937 and other types of cells. In contrast, like TGF-β1, TIAF1 supported transforming growth of L929 fibroblasts. TIAF1 increased the expression of p53, Cip1/p21, and Smad proteins; suppressed ERK phosphorylation; and altered TGF-β1-mediated Smad2/3 phosphorylation in U937 cells. Antisense TIAF1 mRNA significantly enhanced the proliferation of mink lung Mv1Lu epithelial cells. Together, these observations indicate that TIAF1 participates in the TGF-β-mediated growth regulation.

KEYWORDS: TGF-β1; TIAF1; TNF; p53; Smad

INTRODUCTION

Transforming growth factor beta (TGF-β) mediates growth suppression and apoptosis of epithelial and liver cells and other types of cells, but supports the proliferation of fibroblasts. The mechanisms are unknown. TGF-β signaling utilizes three types of receptors and three intracellular signal-transducing proteins, namely, Smad2, 3, and 4, to regulate cell proliferation, extracellular matrix homeostasis, apoptosis, and others.[1,2] Immune defects and malignancy such as prostate and breast cancers are frequently associated with overexpression of TGF-β and excessive activation of latent TGF-β.[3–5] Also, TGF-β affects the development of immune cells, including lymphokine-activated killer cells, primitive human hematopoietic progenitor cells, and cytotoxic T cells, whereas tumor necrosis factor (TNF) reciprocally counteracts these TGF-β effects.[6–8]

TGF-β has been shown to inhibit TNF-mediated death of murine L929 fibroblasts and other types of cells.[9] We have identified several TGF-β1-induced proteins that induce TNF resistance in L929 cells.[10–13] One of these protective proteins is a 12-kDa TGF-β-induced antiapoptotic factor (TIAF1).[11] When stably expressed in L929

Address for correspondence: Nan-Shan Chang, Laboratory of Molecular Immunology, Guthrie Research Institute, 1 Guthrie Square, Sayre, PA 18840. Voice: 570-882-4632; fax: 570-882-4643. chang_nanshan@guthrie.org

Ann. N.Y. Acad. Sci. 995: 11–21 (2003).

cells, TIAF1 abolishes cell death by TNF, as well as by overexpressed TNF receptor adaptor proteins such as TRADD, FADD, and RIP.[11]

Here, we show that TIAF1 mimicked the functional activity of TGF-β in mediating growth inhibition, differentiation, and apoptosis of monocytic U937 cells.[14–16] TIAF1 increased the expression of p53, Cip1/p21, and Smad proteins; suppressed ERK phosphorylation; and altered TGF-β1-mediated Smad2/3 phosphorylation in U937 cells. Also, like TGF-β1, TIAF1 appeared to exert transforming growth activity when transiently overexpressed in fibroblasts. Antisense TIAF1 mRNA significantly enhanced proliferation of mink lung Mv1Lu epithelial cells, suggesting that TIAF1 is involved in the TGF-β-mediated growth regulation.

MATERIALS AND METHODS

Cell Lines and cDNA Expression Constructs

Cell lines used in the study were mink lung Mv1Lu epithelial cells, murine L929 fibroblasts, human neuroblastoma SK-N-SH, monocytic U937 and THP-1, and leukemia HL-60 cells (American Type Culture Collections, Manassas, VA). An N-terminal green fluorescent GFP-tagged TIAF1 construct, designated as GFP-TIAF1 (in pEGFP-C1, Clontech, Palo Alto, CA), was made as described previously.[11] Where indicated, GFP-TIAF1(Ala49) was used in transfection. This clone has a conservative transition of Ala-49 from Val-49 in the original clone (GenBank accession AF105277).[11]

Antibodies

Generation of polyclonal antibody against a synthetic peptide of TIAF1 was described previously.[11] The antibody interacts with the 12-kDa TIAF1.[11] Additional specific antibodies used in Western blotting were against the following proteins: p53, phosphorylated p42/44 mitogen-activated protein kinase or extracellular signal-regulated kinase (p-p42/44 MAPK or p-ERK), Smad2, Smad4, phospho-Smad2/3 (Santa Cruz Laboratory, Santa Cruz, CA), Cdk1/Cdc2, and p21/Cip1 (Transduction Laboratories, Lexington, KY). Where indicated, the NIH Image software was used to analyze the extent of protein expression in Western blots.

Stable Transfectants and Cell Cycle Analysis

U937 transfectants, which stably expressed GFP-TIAF1 or GFP alone, were established as described.[11] Briefly, U937 cells were electroporated with the purified TIAF1-pEGFP-C1 or pEGFP-C1 DNA (40 μg DNA/5 × 10^6 cells), followed by culturing in the continuous presence of G418 (up to 1 mg/mL) to select cell populations that stably expressed the desired protein. The percentages of cells that expressed GFP-TIAF1 or GFP were analyzed by flow cytometry (FACS, Becton Dickinson, Franklin Lakes, NJ). To determine the effect of TIAF1 on cell cycle progression, these cells were stained with propidium iodide,[17] and then the cellular DNA content was analyzed by flow cytometry. Similarly, stable transfectants of THP-1 and HL-60 cells expressing GFP or GFP-TIAF1 were also established.

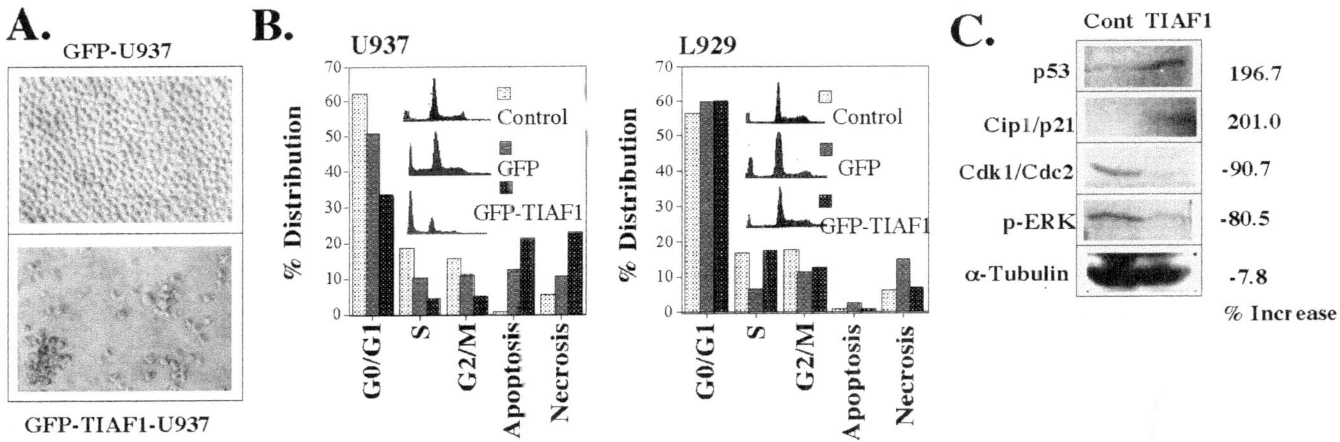

FIGURE 1. TIAF1 regulates cell cycle protein synthesis and mediates growth suppression and apoptosis in U937 cells. **(A)** Stable GFP-TIAF1-expressing U937 cells had a significantly retarded growth, and most of these cells adhered to the culture flasks and grew in aggregates. In contrast, control GFP-expressing cells grew normally in suspension, with no aggregation or no adherence to the culture flasks. **(B)** Cell cycle analysis showed a significantly reduced distribution of the GFP-TIAF1-expressing U937 cells in G0/G1, S, and G2/M phases, compared to nontransfected control and GFP-expressing cells. Also, there was an increased cell death, up to 45%, in the TIAF1-expressing cells (combining apoptosis and necrosis areas). In comparison, no growth suppression and apoptosis were observed in nontransfected L929 cells or in L929 stable transfectants expressing GFP-TIAF1 or GFP. **(C)** Analysis of the cell cycle protein expression showed that there were upregulation of p53 and Cip1/p21, as well as downregulation of Cdk1/Cdc2, which would promote G0/G1 arrest, in the GFP-TIAF1-expressing cells. Phosphorylation of ERK (pERK) was significantly reduced in the GFP-TIAF1-expressing cells, indicative of suppressed cell proliferation.

RESULTS

TIAF1 Mimics TGF-β1 Function in That It Mediates Growth Suppression and Apoptosis of Monocytic U937 Cells, but Promotes Transforming Growth of Fibroblasts

We determined whether TIAF1 is a downstream effector that contributes to the growth-inhibitory effect of TGF-β1 on U937 cells. U937 cells were used because TGF-β1 suppresses the proliferation, but induces differentiation of these cells.[14–16] Also, these TNF-sensitive cells do not express TIAF1 gene and protein.[11] U937 cells were electroporated with either a GFP-TIAF1 or a control GFP expression construct. Stable transfectants were selected by G418. As analyzed by flow cytometry, more than 90% of both stable transfectants were GFP-positive. The control GFP-expressing U937 cells grew normally, with little or no adherence to the culture flasks (FIG. 1A). In contrast, there was a significantly retarded proliferation of the GFP-TIAF1-expressing cells (FIG. 1A). In addition, most of the GFP-TIAF1-expressing cells adhered to the plastic flasks and grew in aggregates or colony-like (FIG. 1A), indicating that TIAF1 suppresses proliferation and induces differentiation of U937 cells.

By cell cycle analysis, it was determined that TIAF1 suppressed U937 cell growth and induced apoptosis. Compared to nontransfected controls or control GFP-expressing U937 cells, TIAF1-expressing cells had a significantly reduced distribution in the G0/G1, S, and G2/M phases, plus 45% dead cells (FIG. 1B). In comparison,

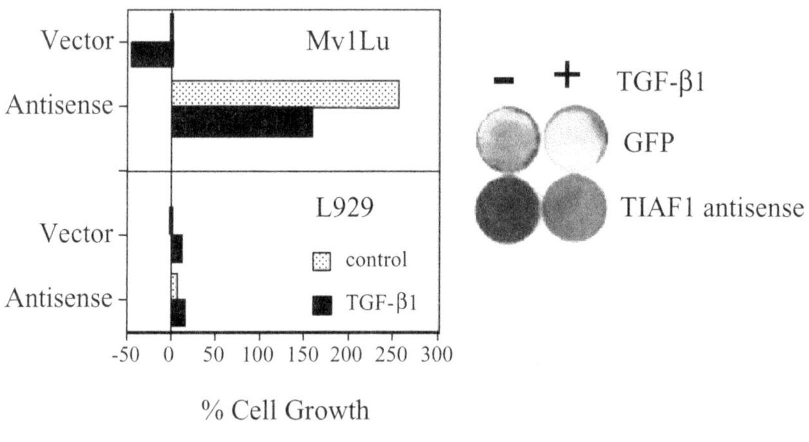

FIGURE 2. Antisense TIAF1 increases proliferation of Mv1Lu epithelial cells. Mv1Lu and L929 cells were transfected with GFP-TIAF1 antisense or control GFP construct by electroporation. After 24 h in culture, the cells were treated with TGF-β1 (2 ng/mL) for 48 h. Compared to control cells, antisense TIAF1 dramatically increased the growth of Mv1Lu cells. Normally, TGF-β1 increases L929 cell growth by 20–40% in 24–48 h.[11] Antisense TIAF1 had no significant effect on L929 cell growth. The results are an average of three independent experiments.

TIAF1 failed to induce growth arrest and apoptosis in L929 fibroblasts (FIG. 1B). These L929 cells were engineered to stably express GFP and GFP-TIAF1.[11] In these experiments, G418 was removed from the culture 24 h prior to cell cycle analysis. This was to prevent the possible toxic effect of G418 to these cells.

Expression of cell cycle proteins was altered in the TIAF1-expressing U937 cells (FIG. 1C). There was an increased expression of p53 and Cip1/p21, as well as a reduced expression of Cdk1/Cdc2, in the TIAF1-expressing cells. Cip1/p21 promotes G0/G1 arrest, while p53 may induce apoptosis. In addition, phosphorylation of ERK (or p42/44 MAPK) was significantly reduced in the GFP-TIAF1-expressing cells (FIG. 1C), indicating that cell growth is suppressed.[18–20]

To further demonstrate that TIAF1 is an effector of the TGF-β1-mediated growth suppression, we examined the effect of antisense TIAF1 mRNA on epithelial cell proliferation. TGF-β1-sensitive mink lung Mv1Lu epithelial cells were electroporated with an antisense TIAF1 construct or GFP alone, followed by culturing for 48 h in the presence of fresh culture media. The results showed that the antisense TIAF1 dramatically increased Mv1Lu cell proliferation up to ~255% (an average of three experiments) compared to GFP-expressing cells (FIG. 2). TGF-β1 inhibited ~50% growth of Mv1Lu cells in 48 h, and the growth inhibition was reduced by the antisense TIAF1 (down to ~30%) (FIG. 2). In comparison, TGF-β1 increases L929 cell growth by 20–40% in 24–48 h.[11] Antisense TIAF1 had no significant effect on L929 cell growth (FIG. 2).

TIAF1 Supports Transforming Growth of Fibroblasts

TGF-β1 mediates adherence-independent growth of fibroblasts, the so-called transforming growth.[21] Like TGF-β1, TIAF1 appeared to exert the transforming growth activity. L929 fibroblasts were electroporated with TIAF1 expression constructs (in pcDNA3, without GFP tag) and cultured for 48 h. More than 80% of the TIAF1-expressing cells lost adherence and grew in suspension as aggregates or colony-like (FIG. 3A). In controls, null or empty vector–electroporated cells adhered to the flasks and grew as a monolayer (FIG. 3A). Under similar experimental conditions, TIAF1 also stimulated adherence-independent growth of neuroblastoma SK-N-SH cells (FIG. 3B). Compared to control cells, TIAF1 failed to limit both L929 and SK-N-SH cell growth and induce apoptosis (data not shown). Similar results were observed using the GFP-TIAF1 and control GFP constructs.

In comparison, nonadherent monocytic THP-1 cells were tested. These cells are more differentiated and refractory to TNF apoptosis than U937 cells. THP-1 cells were transfected with the GFP-TIAF1 or GFP construct by electroporation. After 48 h in culture, the resulting transfectants also showed aggregate (or colony-like) formation, whereas no aggregates were observed in cells electroporated with nothing or GFP only. Selection of stable transfectants using G418 also revealed the continual presence of aggregates or colonies in TIAF1-expressing cells (FIG. 3C). THP-1 cells have endogenous TIAF1.[11] Ectopic TIAF1 could not induce apoptosis in these cells. Compared to the original clone (GenBank accession AF105277), a homologue GFP-TIAF1(Ala49) construct yielded similar results (FIG. 3C). This cDNA encodes an alanine-49, a conserved transition from valine-49. Similar results were observed when testing leukemia HL-60 cells (data not shown). TGF-β1 is known to stimulate the differentiation of HL-60 cells.

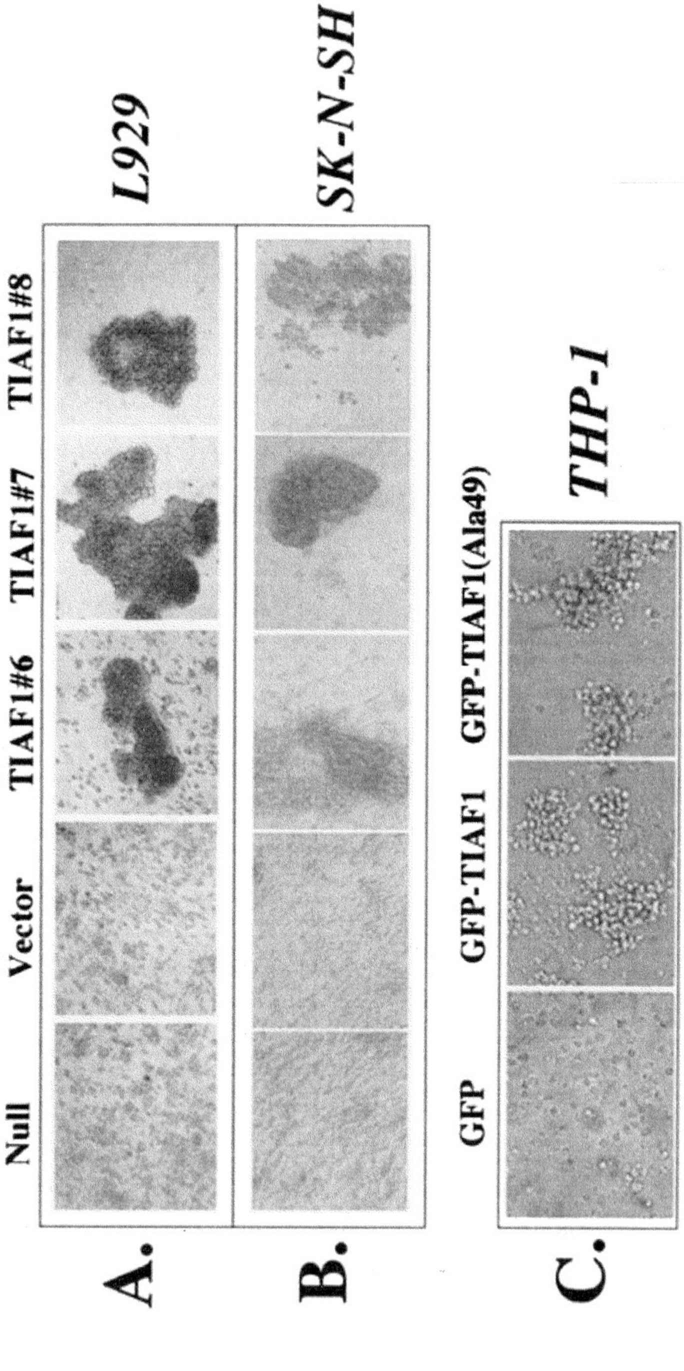

FIGURE 3. *See following page for legend.*

TIAF1 Upregulates Smad2/4 and Alters the TGF-β/Smad Signaling Pathway

The ability of TIAF1 in affecting the TGF-β/Smad signaling pathway was examined. U937, L929, and SK-N-SH cells were electroporated with the GFP-TIAF1 or GFP construct, cultured for 48 h (in the presence of 10% fetal bovine serum), and then exposed to TGF-β1 for 1 h. TIAF1 upregulated the expression of Smad2 and Smad4 in U937 cells (FIG. 4A). TGF-β1 increased the nuclear expression of Smad2, as well as nuclear Smad2/3 phosphorylation, in U937 cells expressing GFP, but not expressing GFP-TIAF1 (FIG. 4A). The increased nuclear Smad2 was probably due to TGF-β1-induced protein translocation from the cytoplasm. In contrast, TGF-β1 reduced the phosphorylation of cytosolic Smad2/3 in GFP-expressing cells, but increased the phosphorylation of Smad2/3 in GFP-TIAF1-expressing cells (FIG. 4A), suggesting that TIAF1 alters the TGF-β signaling pathway in U937 cells.

TIAF1-mediated cytosolic sequestration of Smad proteins was shown in HL-60 cells. Three stable transfectants of HL-60 cells expressing GFP, GFP-TIAF1, and GFP-TIAF1(Ala49) were established. Unlike U937, TIAF1 failed to induce apoptosis of HL-60 cells. Smad2 and Smad4 were mainly present in the nuclei of HL-60 cells expressing GFP, when cultured in the presence of serum (FIG. 4B). In contrast, Smad2 and Smad4 were mostly sequestered in the cytoplasm of HL-60 cells expressing GFP-TIAF1 or GFP-TIAF1(Ala49) (FIG. 4B).

In contrast to the above cells, Smad2 and Smad4 were present mainly in the nuclei of the serum-stimulated L929 and SK-N-SH cells, and TIAF1 upregulated the expression of these proteins (FIGS. 4C and 4D). TGF-β1 increased Smad2 nuclear translocation, as well as Smad2/3 phosphorylation, in L929 expressing GFP and GFP-TIAF1 (FIG. 4C). There was also an increased nuclear Smad2, as well as phosphorylated Smad2/3, in the TGF-β1-treated SK-N-SH cells expressing GFP, but not expressing GFP-TIAF1 (FIG. 4D). Smad2 and Smad4 in L929 and SK-N-SH cells migrated at ~65 kDa in SDS-PAGE under reducing conditions, suggesting that these proteins are phosphorylated. In contrast, these Smad proteins were found at the predicted ~45 kDa in U937 and HL-60 cells.

FIGURE 3. TIAF1 mediates adherence-independent growth of L929 and SK-N-SH cells. **(A)** L929 fibroblasts were electroporated with the TIAF1-pcDNA3 constructs (nos. 6, 7, and 8 with an identical DNA sequence) or an empty vector, and cultured for 48 h. Compared to null or empty vector-electroporated cells, TIAF1 failed to suppress cell growth and induce apoptosis, as determined by MTS and DNA fragmentation assays. Also, more than 80% of the TIAF1-transfected cells lost adherence and grew in suspension as aggregates (or colony-like). **(B)** Under similar experimental conditions, TIAF1 also supported the adherence-independent growth of neuronal SK-N-SH cells. **(C)** Similarly, monocytic THP-1 cells were electroporated with the GFP-TIAF1, GFP-TIAF1(Ala49), or GFP construct. TIAF1 mediated cellular aggregation in 48 h (data not shown). These cells were treated with G418 for 2–3 weeks for selection of stable transfectants. TIAF1 continued to mediate cellular aggregation, whereas control cells failed to form aggregates. Similar results were observed by testing the homologue GFP-TIAF1(Ala49) construct. TIAF1(Ala49) has an alanine-49 residue, a conserved transition from valine-49. Unlike U937, THP-1 cells have endogenous TIAF1 protein.[11] Ectopic TIAF1 could not induce THP-1 cell death.

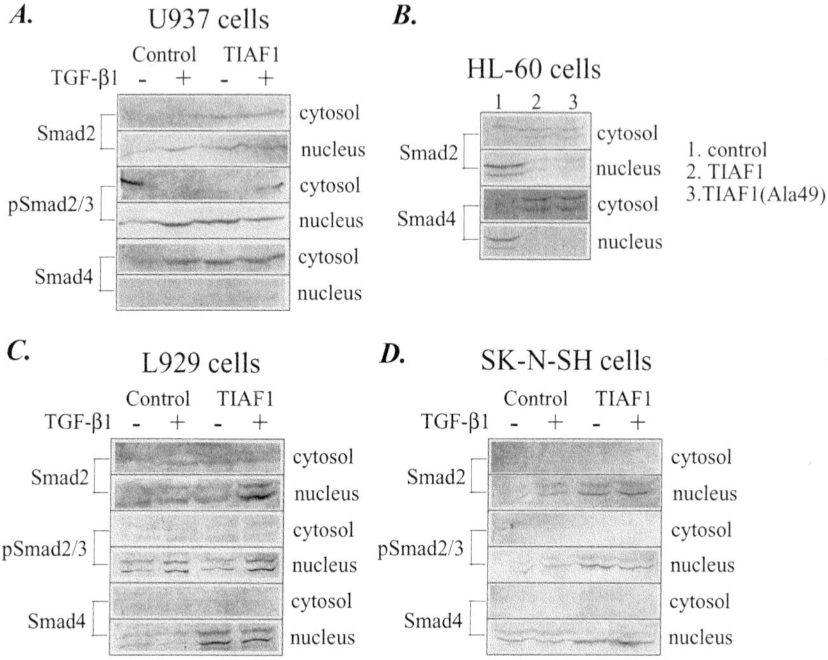

FIGURE 4. TIAF1 upregulates Smad2/4 and alters the TGF-β/Smad signaling pathway in U937 cells. U937, L929, and SK-N-SH cells were electroporated with the GFP-TIAF1 or GFP construct, cultured for 48 h, and exposed to TGF-β1 (2 ng/mL) for 1 h. **(A)** TIAF1 upregulated the expression of Smad2 and Smad4 in U937 cells. TGF-β1 increased the nuclear translocation of Smad2, as well as nuclear Smad2/3 phosphorylation, in U937 cells expressing GFP, but not expressing GFP-TIAF1. In contrast, TGF-β1 reduced the phosphorylation of cytosolic Smad2/3 in GFP-expressing cells, but increased the phosphorylation of Smad2/3 in GFP-TIAF1-expressing cells. **(B)** Three stable transfectants of HL-60 cells expressing GFP, GFP-TIAF1, and GFP-TIAF1(Ala49) were established. Unlike U937 cells, TIAF1 failed to mediate apoptosis of HL-60 cells. Smad2 and Smad4 were mainly present in the nuclei of HL-60 cells expressing GFP. In contrast, Smad2 and Smad4 were sequestered in the cytoplasm of HL-60 cells expressing GFP-TIAF1 or GFP-TIAF1(Ala49). **(C)** TGF-β1 increased Smad2 nuclear translocation, as well as Smad2/3 phosphorylation, in L929 expressing GFP or GFP-TIAF1. **(D)** There was also an increased nuclear Smad2, as well as phosphorylated Smad2/3, in the TGF-β1-treated SK-N-SH cells expressing GFP, but not expressing GFP-TIAF1. The figure shows a set of representative results from three independent experiments.

DISCUSSION

TIAF1 was originally shown to protect L929 fibroblasts from TNF cytotoxicity.[11] In this study, we demonstrate that TIAF1 mimics the functional activity of TGF-β1 in growth regulation. When transiently overexpressed, TIAF1 induces apoptosis of U937 monocytes and various nonfibroblast cells such as MCF-7 breast cancer cells,

Du145 prostate cancer cells, and Molt-4 T lymphocytes (data not shown). Like TGF-β1, the overexpressed TIAF1 fails to induce apoptosis, but supports the transforming growth of fibroblasts. Blocking of TIAF1 expression by antisense mRNA dramatically enhances the proliferation of TGF-β-sensitive Mv1Lu epithelial cells. These observations suggest that TIAF1 is a downstream effector of the TGF-β signaling pathway to mediate growth inhibition and promotion in different types of cells.

Like TGF-β1, TIAF1 upregulated the expression of Smad proteins. The increased nuclear Smad2/3 were indeed phosphorylated. TIAF1-mediated Smad2/3 phosphorylation appears to correlate positively with its induction of U937 cell death. Whether the phosphorylated Smad2/3 induce the expression of p53 and other apoptogenic proteins in U937 cells remains to be established.

In contrast, TIAF1 may alter the TGF-β/Smad signaling pathway in U937 cells. A reduced phosphorylation of cytosolic Smad2/3 was observed in the control U937 cells expressing GFP when exposed to TGF-β1 for 1 h. In contrast, the cytosolic Smad2/3 phosphorylation was increased in the TGF-β1-treated U937 cells expressing GFP-TIAF1. Conversely, TGF-β1 mediated phosphorylation of nuclear Smad2/3 in the GFP-expressing U937 cells, while the phosphorylation was reduced in the GFP-TIAF1-expressing cells. Intriguingly, TIAF1 sequestered Smad2 and Smad4 in the cytoplasm of HL-60 cells. The mechanism is unknown. Whether TIAF1 physically interacts with Smad proteins *in vivo* remains to be established.

We have shown that TIAF1 is located in the perinuclear and nuclear areas.[11] Further studies showed that a portion of TIAF1 is present in the mitochondria and migrates to the nuclei during cell death (Chang *et al.*, unpublished). We have identified two critical amino acid residues in TIAF1 that participate in TIAF1 nuclear translocation (data not shown). Also, two different amino acid residues are involved in the apoptosis-inducing activity of TIAF1 in nonfibroblast cells. Yeast two-hybrid analysis revealed that TIAF1 physically interacts with several mitochondrial proteins *in vivo*. Nonetheless, the functional role of TIAF1 in mitochondria is largely unknown.

A recent study showed that both TIAF1 mRNA and protein are upregulated in patients with kidney and liver allografts with chronic rejection.[22] TIAF1 is also upregulated in activated Th2 helper T cells in these allografts, suggesting a potential role of TIAF1 in regulating Th2 cell function.

Together, TIAF1 is a downstream effector of the TGF-β-mediated growth regulation. TIAF1 upregulates Smad expression and increases their phosphorylation. The mechanisms remain to be established.

ACKNOWLEDGMENTS

Smaira Khera was a surgical resident of the Robert Packer Hospital of the Guthrie Healthcare System.

Research was supported by the American Heart Association and the Guthrie Foundation for Education and Research. We thank Jeffery Mattison for DNA sequencing analysis and SK-N-SH cells, and John Heath and Lori Schultz for technical assistance.

REFERENCES

1. TEN DIJKE, P., K. MIYAZONO & C.H. HELDIN. 2000. Signaling inputs converge on nuclear effectors in TGF-beta signaling. Trends Biochem. Sci. **25:** 64–70.
2. LAWRENCE, D.A. 1996. Transforming growth factor-beta: a general review. Eur. Cytokine Netw. **7:** 363–374.
3. REISS, M. & M.H. BARCELLOS-HOFF. 1997. Transforming growth factor-beta in breast cancer: a working hypothesis. Breast Cancer Res. Treat. **45:** 81–95.
4. BARRACK, E.R. 1997. TGF beta in prostate cancer: a growth inhibitor that can enhance tumorigenicity. Prostate **31:** 61–70.
5. LETTERIO, J.J. & A.B. ROBERTS. 1998. Regulation of immune responses by TGF-beta. Annu. Rev. Immunol. **16:** 137–161.
6. ESPEVIK, T., I.S. FIGARI, G.E. RANGES & M.A. PALLADINO, JR. 1988. Transforming growth factor-beta 1 (TGF-β1) and recombinant human tumor necrosis factor-alpha reciprocally regulate the generation of lymphokine-activated killer cell activity. J. Immunol. **140:** 2312–2316.
7. RANGES, G.E., I.S. FIGARI, T. ESPEVIK & M.A. PALLADINO, JR. 1987. Inhibition of cytotoxic T cell development by transforming growth factor beta and reversal by recombinant tumor necrosis factor alpha. J. Exp. Med. **166:** 991–998.
8. SNOECK, H.W., S. WEEKX, A. MOULIJN *et al.* 1996. Tumor necrosis factor alpha is a potent synergistic factor for the proliferation of primitive human hematopoietic progenitor cells and induces resistance to transforming growth factor beta, but not to interferon gamma. J. Exp. Med. **183:** 705–710.
9. CHANG, N-S. 1998. Transforming growth factor-β protection of cancer cells against tumor necrosis factor cytotoxicity is counteracted by hyaluronidase [review]. Int. J. Mol. Med. **2:** 653–659.
10. CHANG, N-S. 1995. Transforming growth factor-beta 1 induction of novel extracellular matrix proteins that trigger resistance to tumor necrosis factor cytotoxicity in murine L929 fibroblasts. J. Biol. Chem. **270:** 7765–7772.
11. CHANG, N-S., J. MATTISON, H. CAO *et al.* 1998. Cloning and characterization of a novel transforming growth factor–beta1–induced TIAF1 protein that inhibits tumor necrosis factor cytotoxicity. Biochem. Biophys. Res. Commun. **253:** 743–749.
12. CAREY, G.B. & N-S. CHANG. 1998. Cloning and characterization of a transforming growth factor beta1–induced anti-apoptotic adhesion protein TIF2. Biochem. Biophys. Res. Commun. **249:** 283–286.
13. CHANG, N-S. 2000. TGF-β-induced matrix proteins inhibit p42/44 MAPK and SAPK/JNK activation and suppress TNF-mediated IκBα degradation and NF-κB nuclear translocation in L929 fibroblasts. Biochem. Biophys. Res. Commun. **267:** 194–200.
14. PIACIBELLO, W., A. SEVERINO, A. STACCHINI & M. AGLIETTA. 1991. Differential effect of transforming growth factor beta 1 on the proliferation of human lymphoid and myeloid leukemia cells. Haematologica **76:** 460–466.
15. BOMBARA, C. & R.A. IGNOTZ. 1992. TGF-beta inhibits proliferation of and promotes differentiation of human promonocytic leukemia cells. J. Cell. Physiol. **153:** 30–37.
16. DEFACQUE, H., D. PIQUEMAL, A. BASSET *et al.* 1999. Transforming growth factor-beta1 is an autocrine mediator of U937 cell growth arrest and differentiation induced by vitamin D3 and retinoids. J. Cell. Physiol. **178:** 109–119.
17. GIVAN, A.L. 1992. Cells from within: DNA and molecular biology. *In* Flow Cytometry: First Principles, pp. 103–134. Wiley–Liss. New York.
18. RICE, P.L., R.J. GOLDBERG, E.C. RAY *et al.* 2001. Inhibition of extracellular signal-regulated kinase 1/2 phosphorylation and induction of apoptosis by sulindac metabolites. Cancer Res. **61:** 1541–1547.
19. WENG, L.P., W.M. SMITH, J.L. BROWN & C. ENG. 2001. PTEN inhibits insulin-stimulated MEK/MAPK activation and cell growth by blocking IRS-1 phosphorylation and IRS-1/Grb-2/Sos complex formation in a breast cancer model. Hum. Mol. Genet. **10:** 605–616.
20. HOSHINO, R., S. TANIMURA, K. WATANABE *et al.* 2000. Blockade of the extracellular signal-regulated kinase pathway induces marked G1 cell cycle arrest and apoptosis in

tumor cells in which the pathway is constitutively activated: up-regulation of p27(Kip1). J. Biol. Chem. **276:** 2686–2692.

21. Roberts, A.B., L.C. Lamb, D.L. Newton *et al.* 1980. Transforming growth factors: isolation of polypeptides from virally and chemically transformed cells by acid/ethanol extraction. Proc. Natl. Acad. Sci. USA **77:** 3494–3498.
22. van der Leij, J., A. van den Berg, E.W.J.A. Albrecht *et al.* 2002. High expression of *tiaf-1* in chronic kidney and liver allograft rejection and in activated t helper cells. Transplantation. In press.

TACE/ADAM17 Processing of EGFR Ligands Indicates a Role as a Physiological Convertase

DAVID C. LEE,[a,b] SUSAN WOHLER SUNNARBORG,[a] C. LEANN HINKLE,[a] TIMOTHY J. MYERS,[a] MARY STEVENSON,[c] WILLIAM E. RUSSELL,[c] BEVERLY J. CASTNER,[d] MARY J. GERHART,[d] RAYMOND J. PAXTON,[d] ROY A. BLACK,[d] AILEEN CHANG,[a] AND LESLIE F. JACKSON[a]

[a]Department of Biochemistry and Biophysics, [b]UNC-Lineberger Comprehensive Cancer Center, School of Medicine, University of North Carolina at Chapel Hill, Chapel Hill, North Carolina 27599-7260, USA

[c]Departments of Pediatrics and Cell Biology, Vanderbilt University, Nashville, Tennessee 37232, USA

[d]Amgen Incorporated, Seattle, Washington 98101, USA

ABSTRACT: EGF family growth factors, including transforming growth factor-alpha (TGFα), amphiregulin (AR), and heparin-binding EGF (HB-EGF), are invariably expressed as transmembrane precursors that are cleaved at one or more sites in the extracellular domain to release soluble growth factor. Considerable attention has focused on the identification of proteases responsible for these processing events. We previously implicated tumor necrosis factor-alpha converting enzyme (TACE/ADAM17) in the generation of soluble TGFα from its transmembrane precursor, proTGFα. Here, we review our findings that primary keratinocytes from Tace$^{\Delta Zn/\Delta Zn}$ mice, which express a nonfunctional TACE, released dramatically lower levels of soluble TGFα compared to their normal counterparts, even though TGFα mRNA and cell-associated protein levels were similar in the two cell populations. Restoration of TACE activity in Tace$^{\Delta Zn/\Delta Zn}$ cells increased shedding of TGFα species, including the mature, 6-kDa protein. Further, exogenous TACE enzyme accurately cleaved the N-terminal processing site of proTGFα in cell lysates, as well as both physiologic sites of a soluble proTGFα ectodomain. TACE also accurately cleaved peptide substrates corresponding to the processing sites of several additional EGF family members, and restoration of TACE activity enhanced the shedding of soluble AR and HB-EGF proteins from Tace$^{\Delta Zn/\Delta Zn}$ cells. Finally, reduction of functional TACE gene dosage greatly exacerbated the open-eye defect of Egfr$^{wa\text{-}2/wa\text{-}2}$ newborns, which is regulated by redundant actions of several EGF family ligands. The implications of these results for the biology of the EGF family and TACE are discussed.

KEYWORDS: TACE/ADAM17; ADAM; ERBB; TGFα; EGF family; EGF receptor

Address for correspondence: David C. Lee, Department of Biochemistry and Biophysics, 405 Mary Ellen Jones Building, Campus Box 7260, School of Medicine, University of North Carolina at Chapel Hill, Chapel Hill, NC 27599-7260. Voice: 919-966-5912; fax: 919-966-2852.
dclee@med.unc.edu

Ann. N.Y. Acad. Sci. 995: 22–38 (2003).

INTRODUCTION

Tyrosine kinase receptors belonging to the ERBB family are essential for normal development and are prominent targets of dysregulation in cancer. The family includes epidermal growth factor receptor (ERBB1) and the related ERBB2/c-neu, ERBB3, and ERBB4.[1,2] These receptors are differentially bound and activated by polypeptide ligands from the epidermal growth factor (EGF)/neuregulin (NRG) superfamily.[3,4] Members of this superfamily are defined by a conserved 36–40 amino acid sequence (the EGF-like motif), which forms a disulfide-bonded, three-loop structure that is required for receptor binding. Besides its namesake, the EGF family also includes transforming growth factor-alpha (TGFα), amphiregulin (AR), heparin-binding EGF (HB-EGF), betacellulin (BTC), epiregulin (EPR), and the recently described epigen.[5] These ligands all bind EGFR with relatively high affinity, while a subset (HB-EGF, BTC, EPR) additionally binds ERBB4.[6] On the other hand, multiple NRGs derived from four genes by alternate splicing differentially activate ERBB3 and ERBB4.[7–9] The complexity of the ERBB signaling network is further enhanced by ligand-dependent hetero- as well as homodimerization, with the orphan ERBB2 receptor serving as a preferred dimerization partner.[1,2]

ERBB ligands are generally expressed as integral membrane precursors that contain a single bioactive EGF-like sequence in their ectodomain.[5,10] Soluble growth factor is derived in each case by proteolytic cleavage of a site within the precursor's juxtamembrane stalk region. Precursors may be cleaved upstream of the EGF-like motif as well, but the site and extent of N-terminal cleavage varies. Thus, some EGF family members are shed as larger proteins with N-terminal leaders that may confer heparin binding (e.g., HB-EGF, AR), nuclear localization (e.g., AR), or other properties. However, the contribution of these variable N-terminal sequences to the bioactivity of ERBB ligands is poorly understood. Cleavage in the stalk region occurs at a similar distance from the membrane in each case, but the sequences flanking the cleavage sites vary. Intact EGF family precursors can also activate EGFR, raising the possibility of juxtacrine in addition to auto- and paracrine signaling.[10] However, similarities of the phenotypes of knockout mice lacking EGF family ligands and a protease implicated in their processing (see below) thus far underscore the importance of soluble forms.[11]

Considerable attention has focused on the identity of the protease(s) responsible for the generation of soluble ERBB ligands, especially TGFα. The 159/160-amino acid TGFα precursor (proTGFα) is cleaved at two Ala-Val bonds to release fully processed, mature 50-amino-acid growth factor.[12,13] However, larger bioactive forms resulting from cleavage only within the juxtamembrane stalk are frequently observed,[14–16] even though the membrane-distal site is more rapidly cleaved in pulse chase analyses.[17] This apparent contradiction may be explained by the proximity of an N-linked glycosylation site to the N-terminal processing site. Release of TGFα from cells, which is prevented by metalloproteinase inhibitors,[18] occurs slowly under nonstimulated conditions, but is rapidly induced by phorbol esters, growth factors, calcium ionophores, and phosphatase inhibitors.[19,20] These are characteristics of ectodomain shedding, a global phenomenon that results in the rapid release of extracellular regions of various cell surface proteins, including receptors and cell recognition molecules. Moreover, release of soluble TGFα is inhibited in a mutant CHO cell clone that is defective in other ectodomain shedding events, indicating the existence of a shared, critical component.[18] Thus, understanding the regulation of TGFα release in the context of ectodomain shedding is a major challenge.

The discovery of tumor necrosis factor-alpha (TNFα) converting enzyme (TACE/ADAM17) was a watershed event in studies of growth factor processing.[21,22] TACE is an integral membrane, Zn^{2+}-dependent metalloprotease belonging to the burgeoning ADAM (a disintegrin and metalloproteinase) subfamily of metzincin proteases. Approximately 30 different ADAMs have been characterized in the past several years, and these have been implicated in either proteolytic or cell-cell recognition events.[23] Although proTNFα is a particularly efficient substrate, accumulating evidence has implicated TACE as a TGFα convertase. Hence, TGFα$^{-/-}$ mice and mice homozygous for an inactivating TACE mutation (Tace$^{\Delta Zn}$) display an identical wavy hair and open-eye-at-birth phenotype, immortalized fibroblasts from Tace$^{\Delta Zn/\Delta Zn}$ mice shed dramatically reduced levels of TGFα, and TACE correctly cleaves peptide substrates corresponding to the two proTGFα cleavage sites.[11] Thus, combined genetic and functional evidence supports the hypothesis that TACE is a key regulator of TGFα release. Since Tace$^{\Delta Zn/\Delta Zn}$ mice die in late gestation or shortly after birth and

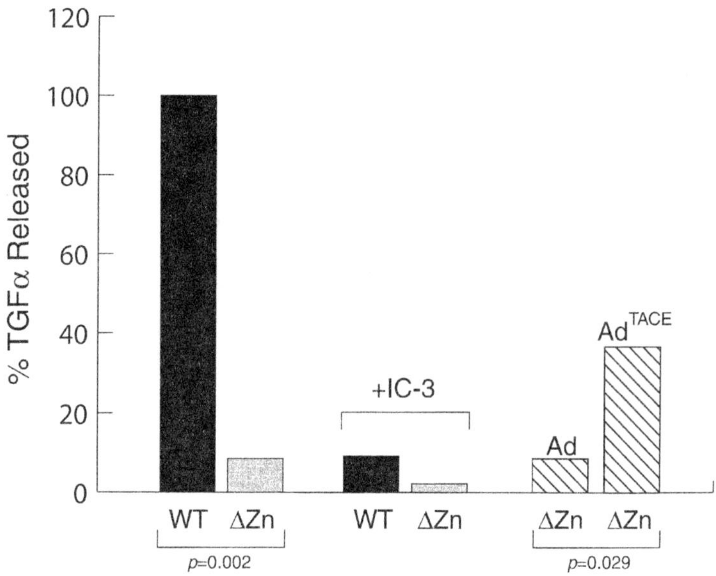

FIGURE 1. TACE is essential for efficient shedding of TGFα from primary epidermal keratinocytes. Primary keratinocytes from Tace$^{+/+}$ or Tace$^{\Delta Zn/\Delta Zn}$ newborn littermates were cultured to ~90% confluence and then shifted to conditioning media. Media and cell lysates were harvested after 48 h and their TGFα levels determined using a specific RIA. Results are expressed as the ratio of TGFα in the medium vs. lysate and are normalized relative to the wild-type ratio. The hydroxamic acid inhibitor, IC-3 (50 μM), was included as indicated. TACE restoration was achieved by infecting Tace$^{\Delta Zn/\Delta Zn}$ cultures with adenovirus (Ad) encoding TACE (Ad^{TACE}) at a multiplicity of infection of 4–10; comparison with control Ad is shown. Statistical significance determined by the Wilcoxon rank sums test: for WT vs. ΔZn, $p = 0.002$; for IC-3 treatment ($n = 2$); for virus reconstitution, $p = 0.029$. [With permission of ref. 24.]

display a spectrum of epithelial defects reminiscent of those observed in $Egfr^{-/-}$ mice, it is possible that TACE functions more broadly to regulate shedding of EGF-like growth factors.[11] Here, we review additional evidence that TACE is a direct mediator of TGFα release.[24]

RESULTS

TACE Is Required for Conversion of proTGFα in Primary Cells

To further assess the impact of the $Tace^{\Delta Zn}$ mutation on the conversion of membrane-anchored to soluble TGFα, we used a specific radioimmunoassay (RIA) to quantitate the levels of TGFα protein in media conditioned for 48 h by primary epidermal keratinocytes from $Tace^{+/+}$ versus $Tace^{\Delta Zn/\Delta Zn}$ newborn mice. We found that keratinocytes released ~12-fold less when they lacked functional TACE, a highly significant result (FIG. 1). This difference was not attributable to altered expression of the precursor since TGFα mRNA levels were comparable in the two populations of keratinocytes, and the two keratinocyte populations were similarly differentiated as reflected by keratin 5 levels. Consistent with this result, release of TGFα by $Tace^{+/+}$ keratinocytes was impaired ~10-fold when the cells were treated with a hydroxamate-based inhibitor, IC-3, that is effective against TACE. Residual release of TGFα by $Tace^{\Delta Zn/\Delta Zn}$ keratinocytes is presumably due to functionally redundant metalloproteinases since this activity was further reduced by IC-3 treatment. Media TGFα levels were restored to ~35% of control values when $Tace^{\Delta Zn/\Delta Zn}$ keratinocytes were infected with an adenovirus encoding the wild-type enzyme (FIG. 1). We attributed the incomplete restoration to partial infection of $Tace^{\Delta Zn/\Delta Zn}$ cells; this was confirmed by monitoring parallel cultures infected with adenovirus encoding green fluorescent protein.

Restoration of TACE Produces Mature TGFα

Endogenous TGFα is not reliably detected by Western blot in media from normal mouse cells. Hence, to visualize the forms of TGFα produced upon restoration of TACE activity, we transfected immortalized clonal $Tace^{\Delta Zn/\Delta Zn}$ fibroblasts with epitope-tagged proTGFα (see FIG. 2A) alone or in the presence of TACE, and then examined lysate and media forms of TGFα by Western blot. RIA of conditioned media from these cell lines confirmed our previous observation[11] that polyclonal $Tace^{\Delta Zn/\Delta Zn}$ fibroblasts released ~20-fold less TGFα than $Tace^{+/+}$ fibroblasts (FIG. 2B). Western blot analysis using an antibody directed against a FLAG epitope at the C-terminus of proTGFα revealed several species of growth factor in the absence of functional TACE. These included intact, immature glycoprotein precursors of 22 and 25 kDa, a mature glycoprotein of 36 kDa, and a stable 16-kDa "tail" that is generated upon release of soluble growth factor and corresponds to the membrane and cytoplasmic domains of the precursor.[25] As shown in FIGURE 2C, cotransfection of TACE largely eliminated the intact precursor species and correspondingly increased the levels of the cell-associated, 16-kDa cleavage product. Analysis of the media using an antibody directed against an HA epitope in the EGF-like sequence showed that cotransfection of TACE resulted in the appearance of a prominent 6-kDa

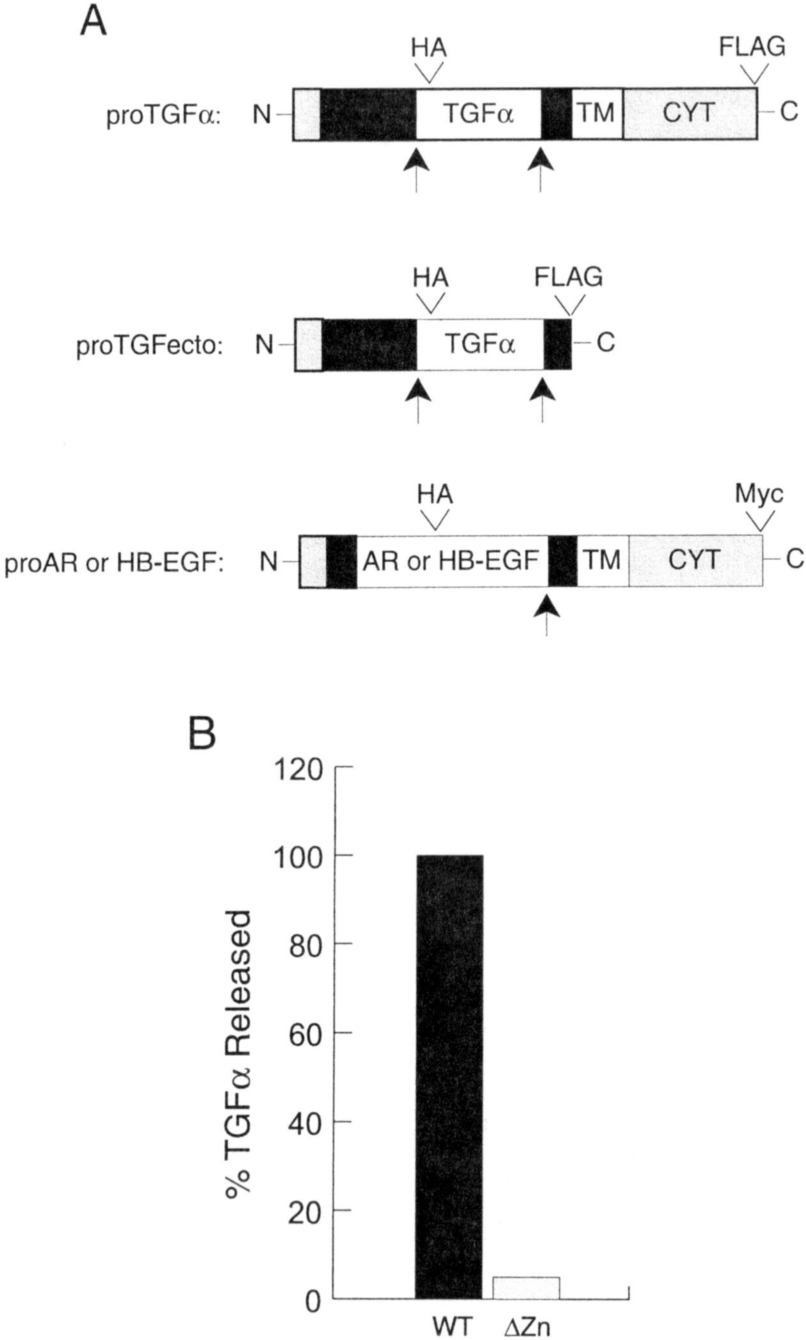

FIGURE 2. *See following page for legend.*

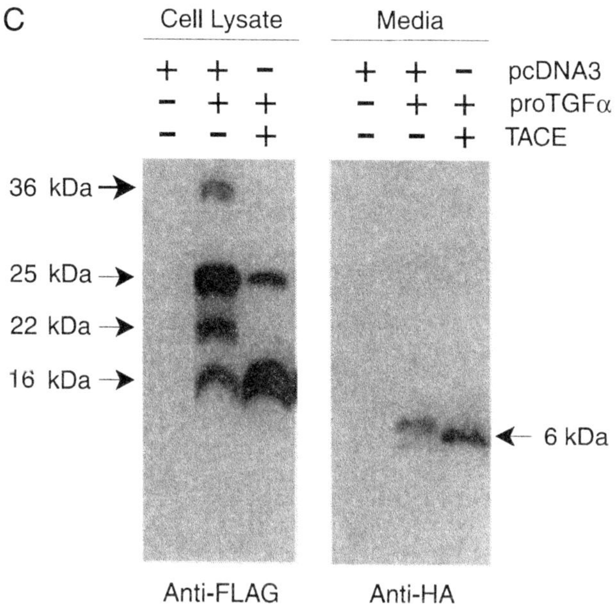

FIGURE 2. Restoration of TACE activity promotes processing of proTGFα by Tace$^{\Delta Zn/\Delta Zn}$ fibroblasts. **(A)** Diagrams of growth factor pro-forms used in this study showing the locations of HA, FLAG, or Myc epitope tags. Abbreviations: TM, transmembrane domain; CYT, cytoplasmic domain. Sites cleaved to release mature growth factors are indicated by arrows. **(B)** TGFα release by Myc/Ras-transformed Tace$^{\Delta Zn/\Delta Zn}$ and Tace$^{+/+}$ fibroblasts was assayed as described in FIGURE 1 (n = 2). **(C)** Effects of restoring TACE activity on TGFα processing as assessed by Western blot. Tace$^{\Delta Zn/\Delta Zn}$ fibroblasts were transfected with epitope-tagged proTGFα in combination with pcDNA3 or pcDNA3 containing the murine TACE cDNA,[45] and, after 24 h, shifted to conditioning media. Cell lysates and media were harvested 24 h later, and media concentrated using Sep Pak C-18 cartridges. TGFα species of the indicated molecular weights were then detected by Western blot using FLAG or HA antibodies as shown. [With permission of ref. 24.]

mature TGFα; in the absence of functional TACE, lower levels of a slightly larger protein were generated, apparently due to cleavage at a nonphysiologic site (FIG. 2C). In other experiments, we have observed additional larger TGFα species whose levels are enhanced in the presence of TACE; we speculate that these are due to incomplete cleavage at the membrane-distal site.

TACE Correctly Cleaves proTGFα in Vitro

We previously showed that TACE correctly cleaves peptide substrates corresponding to each of the processing sites in proTGFα.[11] To assess TACE action on a more physiologic substrate, we added partially purified, native TACE[24] to lysates of rat liver epithelial cells that were stably expressing epitope-tagged proTGFα.[25] Western blots probed with anti-HA revealed a time-dependent decrease in the levels of intact 25- and 36-kDa precursor proteins and the corresponding appearance of a

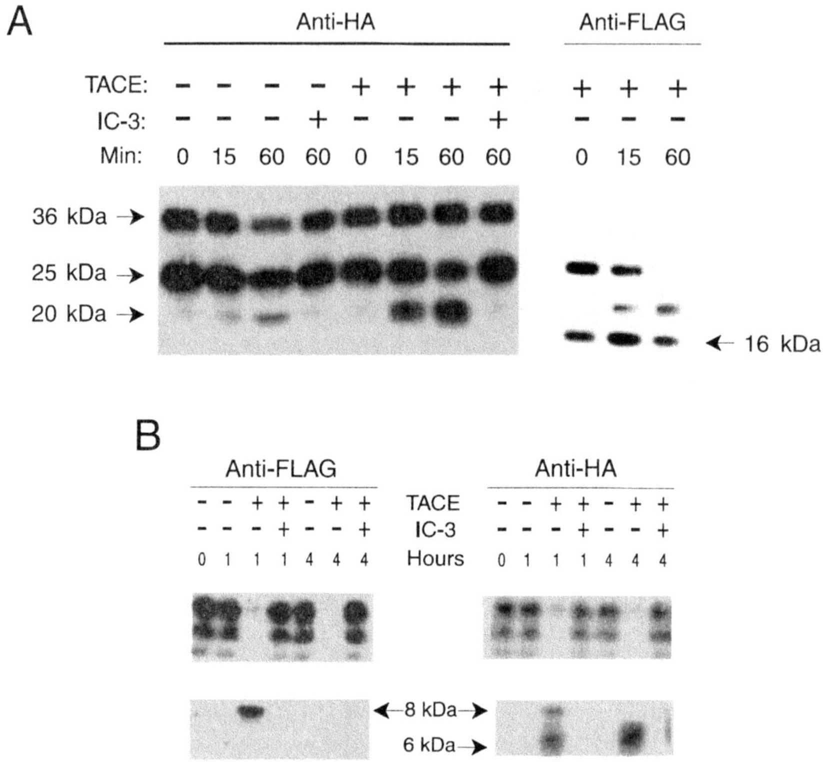

FIGURE 3. *See following page for legend.*

novel 20-kDa species (FIG. 3A). Only minor amounts of this product were observed in the absence of exogenous TACE, and its appearance was prevented when inhibitor IC-3 was included in the reaction. This 20-kDa product was also recognized by anti-FLAG, indicating that TACE cleaved intact proTGFα only at the membrane-distal site upstream of the EGF-like sequence and not at the juxtamembrane site. Consistent with this conclusion, which contrasts with the impaired release of soluble growth factor from cells lacking functional TACE, we did not detect anti-HA-reactive, mature TGFα in these reactions, nor did we observe significant or reproducible increases in the levels of the anti-FLAG-reactive 16-kDa tail. Accurate cleavage at the N-terminus of mature TGFα was subsequently established by Edman degradation, which revealed the N-terminal sequence of the 20-kDa product to be VVSHY.

Because the *in vitro* reaction involved detergent-solubilized integral membrane proTGFα, we speculated that interference by the hydrophobic transmembrane domain impaired the ability of TACE to cleave the juxtamembrane site. Accordingly, we prepared a construct (proTGFecto; FIG. 2A) encoding a truncated, epitope-tagged protein identical to the ectodomain of proTGFα and transfected this construct into COS-1 cells. Addition of recombinant human TACE ectodomain to the resulting

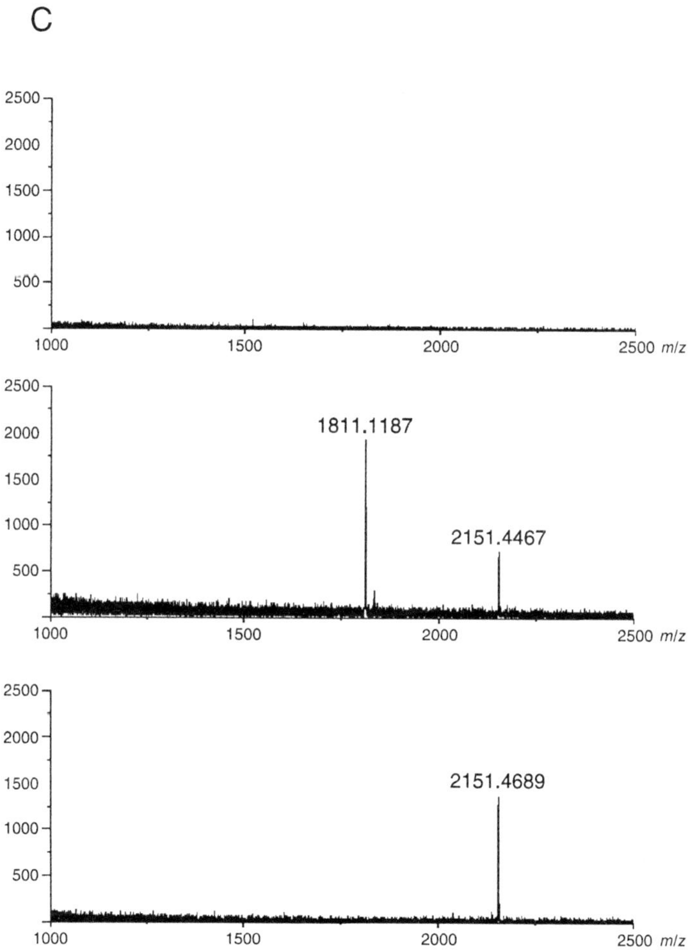

FIGURE 3. ProTGFα is cleaved by TACE *in vitro*. **(A)** Lysates from cells stably expressing epitope-tagged proTGFα were incubated with buffer alone or buffer containing ~200 U of native TACE for the indicated times at 37°C, and products were detected by Western blot using anti-HA or anti-FLAG. Where indicated, the inhibitor IC-3 was included at 50 μM. **(B)** Concentrated media from proTGFecto-transfected cells were incubated with recombinant human TACE (~1 μg) for the indicated times at 37°C in the absence or presence of IC-3. Products were separated by SDS-PAGE and detected by Western blot using anti-HA or anti-FLAG. **(C)** Anti-FLAG-reactive products from proTGFecto cleavage by TACE were analyzed by MALDI/TOF-MS: (*top*) anti-FLAG M2 affinity gel control; (*middle*) TACE reaction products; (*lower*) a synthetic peptide representing the predicted product of cleavage at the C-terminal Ala-Val site. [With permission of ref. 24.]

TABLE 1. TACE cleavage of EGF family processing site peptides

Ligand	N-terminal peptide	Cleaved?	[TACE]	C-terminal peptide	Cleaved?	[TACE]
	⇓			⇓		
TGFα	PVAAAVVSHF	Yes A/V	1.4 μM	ADLLAVVAAS	Yes A/V	1.4 μM
	⇓			⇓		
AR	SVRVEQVVKPPQ	Yes Q/V	28 μM	ERCGEKSMKTHS	No	28 μM
	⇓			⇓		
BTC	RSPETNLLCGDP	No	28 μM	RVDLFYLRGDRG	Yes Y/L	28 μM
	⇓			⇓		
EPR	NPRVAQVSITKC	Yes Q/V	28 μM	CEHFFLTVHQPL	Yes N.D.	28 μM
				⇓		
HB-EGF	N.D.	—	—	GLSLPVENRLYTYD	Yes P/V R/L	28 μM
	⇓			⇓		
EGF	HHYSVRNSDSEC	No	28 μM	KWWELRHAGHGQ	No	28 μM

NOTE: Peptides representing the cleavage sites of human EGF family members (arrows denote predicted cleavage sites) were incubated with recombinant TACE[36] at 37°C for 4 h and products analyzed by LC/MS as previously described.[23] Results of cleavage of TGFα peptides[23] are included for comparison. Observations of cleavage and the site are indicated. Approximate concentration of TACE required for reaction is indicated. Note that cleavage of TGFα peptides required approximately 10 times more TACE than cleavage of a TNF peptide.[23] [With permission of ref. 24.]

proTGFecto-containing conditioned media resulted in a time-dependent, IC-3-sensitive loss of anti-HA- and anti-FLAG-reactive 18- to 25-kDa proteins, and the initial appearance of a novel 8-kDa protein. Both antibodies recognized the latter product; hence, it was generated by cleavage at the N-terminus of mature TGFα only (FIG. 3B). With further incubation, the 8-kDa protein was replaced by a 6-kDa product that was recognized by anti-HA and not anti-FLAG (FIG. 3B). We concluded that this smaller product, indistinguishable in size from mature TGFα, was generated by TACE cleavage of both proTGFα processing sites.

Accurate cleavage of the juxtamembrane site was confirmed by mass spectrometric analysis of the predicted 2.1-kDa, anti-FLAG-reactive product (this species, corresponding to the C-terminus of proTGFecto, was not detected by Western blot [see FIG. 3B], presumably due to its small size). MALDI/TOF-MS analysis of products bound to anti-FLAG beads revealed two species with mass-to-charge ratios (*m*/*z*) of 2151 and 1811 (FIG. 3C). Since the former matched a synthetic peptide corresponding to the expected C-terminal product, we conclude that TACE accurately cleaved the stalk domain of proTGFecto at the physiologic AV site (TABLE 1). MALDI/MS-MS sequencing revealed that the second product (*m*/*z* 1811) resulted from cleavage of an AS bond located 4 residues downstream of the AV site. However, since the N-terminal sequence of the 16-kDa integral membrane tail recovered from proTGFα-expressing cells was VVAAS, we suspect that cleavage at this alternative AS site is an *in vitro* artifact.

TACE Cleaves Peptide Substrates from Other EGF Family Precursors

It was previously noted that there is an overall similarity in the phenotype of $Egfr^{-/-}$ and $Tace^{\Delta Zn/\Delta Zn}$ mice, raising the possibility that TACE functions broadly as a physiologic convertase for EGF family ligands. To address this possibility, we examined the ability of TACE to cleave peptides encompassing the known or predicted N- and C-termini of the soluble growth factors. In addition to accurately cleaving peptides representing the N- and C-terminal processing sites of proTGFα,[11] recombinant human TACE cleaved peptides representing the C-terminal sites of BTC, EPR, and HB-EGF, as well as the N-terminal sites of AR and EPR (TABLE 1). Complete cleavage of any of these peptides required 20-fold higher concentrations of TACE than was required for either the N- or C-terminal TGFα peptides, which in turn required 10-fold higher enzyme than a TNFα peptide substrate. Cleavage occurred at the expected site, except that the C-terminal peptide substrate for HB-EGF was cleaved at two positions, both with nonpolar residues in the P_1' position. One of these, a PV dipeptide, corresponds to the known C-terminus of the mature growth factor released from PMA-stimulated cells,[26] whereas evidence of physiologic cleavage at the RL dipeptide has not been reported to our knowledge. In contrast to HB-EGF, TACE did not cleave peptides representing the C-terminal site of AR, the N-terminal site of BTC, or either site of EGF (TABLE 1).

TACE Promotes Shedding of AR and HB-EGF

To further examine if TACE promotes shedding of additional EGFR ligands besides TGFα, we cotransfected $Tace^{\Delta Zn/\Delta Zn}$ cells with AR or HB-EGF expression vectors in combination with either pcDNA3 or pcDNA3-TACE. To facilitate their detection, proAR and HB-EGF were both epitope-tagged as described for proTGFα, except that a Myc epitope was inserted in place of FLAG at the C-terminus (see FIG. 2). Following transfection, we allowed the cells to condition media for 24 h prior to harvesting of media and lysates for Western blot analysis with epitope-specific antibodies. Consistent with previous reports of complex patterns of AR and HB-EGF glycosylation and processing, anti-Myc detected multiple lysate bands in each case that were not observed in cells transfected with empty pcDNA3 only (FIG. 4). In the case of AR, cotransfection of pcDNA3-TACE dramatically reduced the intensity of one or more lysate proteins of ~45 kDa, which presumably corresponded to intact, mature glycoprotein precursor. This change was accompanied by variable increases in the levels of smaller proteins of 18–22 kDa that were not detected by anti-HA, and hence presumably corresponded to the residual C-terminal tail of proAR. Analysis of media revealed that cells cotransfected with TACE released soluble, anti-HA-reactive proteins of 38–42 kDa whose sizes correspond to previously characterized soluble forms of AR. The levels of these proteins, which were not observed in the absence of TACE, were further increased upon treatment of cells with 1 μM PMA at 20 min prior to harvest (not shown).

Cotransfection of TACE reduced the levels of cell-associated, higher-molecular-weight HB-EGF proteins, with variable increases in a ~25-kDa protein (FIG. 4). Media analysis showed increased levels of several prominent anti-HA-reactive species, ranging in size from 22 to 27 kDa. These species, which were either present at much lower levels or undetectable in the absence of TACE, are similar in size to previously

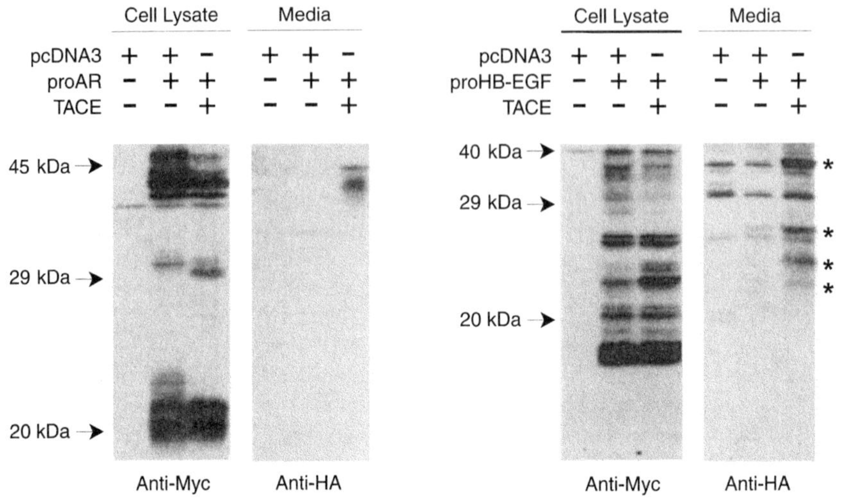

FIGURE 4. Reintroduction of TACE increases AR and HB-EGF shedding. Tace$^{\Delta Zn/\Delta Zn}$ fibroblasts were transiently transfected with epitope-tagged proAR or proHB-EGF and pcDNA3 or TACE as indicated. Media were conditioned for 24 h, and lysates and media were analyzed by Western blot as in FIGURE 2, except that HB-EGF conditioned media were directly immunoprecipitated with anti-HA and not concentrated by Sep-Pak chromatography. The migration of molecular mass standards is indicated. Asterisks indicate media HB-EGF forms that increased in the presence of TACE. [With permission of ref. 24.]

characterized soluble forms of HB-EGF. We also observed a TACE-dependent increase in the level of a 41-kDa protein that may correspond to the entire extensively glycosylated ectodomain of proHB-EGF. PMA treatment did not increase the levels of HB-EGF media proteins observed in the presence of TACE (not shown).

TACE Deficiency Exacerbates the Open-Eye Phenotype of Egfr$^{wa\text{-}2/wa\text{-}2}$ Mice

In a genetic approach to assessing its importance in EGFR signaling, we examined the impact of limiting TACE levels on the frequency with which Egfr$^{wa\text{-}2/wa\text{-}2}$ mice are born with open eye. This phenotype, which stems from delayed eyelid closure in late gestation,[27] is linked to impairment in EGFR signaling. The phenotype is only partially penetrant in TGF$\alpha^{-/-}$ mice, but is observed with higher frequency in triple-null-mice that lack additional EGF family ligands.[28] It is only occasionally observed with Egfr$^{wa\text{-}2/wa\text{-}2}$ mice, which bear a modestly impaired EGFR kinase.[29,30] Importantly, while Tace$^{\Delta Zn/\Delta Zn}$ mice are also born with open eyes, Tace$^{\Delta Zn/+}$ mice are not. Hence, we reasoned that, if TACE is a physiological convertase for multiple EGF family members, reducing its levels would exacerbate the weakly penetrant open-eye phenotype of the Egfr$^{wa\text{-}2/wa\text{-}2}$ mice by reducing the levels of soluble ligand. To test this, we bred the Tace$^{\Delta Zn}$ mutation to heterozygosity on the Egfr$^{wa\text{-}2/wa\text{-}2}$ background (FIG. 5). The data in FIGURE 5 reveal that none of 38 Tace$^{\Delta Zn/+}$ pups were born with eyes open, and only 3 of 25 Tace$^{+/+}$Egfr$^{wa\text{-}2/wa\text{-}2}$ pups

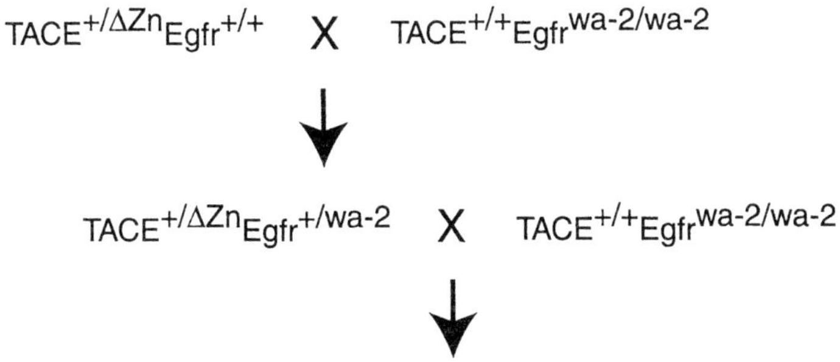

TACE:	+/+	+/ΔZn	+/+	+/ΔZn
Egfr:	+/wa-2	+/wa-2	wa-2/wa-2	wa-2/wa-2
Pups born:	38	38	25	24
Open-eyed pups:	0	0	3	11

FIGURE 5. The open-eye phenotype of Egfr$^{wa\text{-}2/wa\text{-}2}$ mice is exacerbated by TACE deficiency. Tace$^{+/\Delta Zn}$ females and Efgr$^{wa\text{-}2/wa\text{-}2}$ males were mated to produce pups of the indicated genotypes, and these were scored for open eye at birth. This phenotype was restricted to Egfr$^{wa\text{-}2/wa\text{-}2}$ animals and occurred with much greater frequency in TACE heterozygotes ($p = 0.012$). [With permission of ref. 24.]

had this phenotype. In marked contrast, 11 of 24 Tace$^{\Delta Zn/+}$Egfr$^{wa\text{-}2/wa\text{-}2}$ pups had open eyes at birth. This dramatic and statistically significant ($p = 0.012$) increase is consistent with the hypothesis that TACE functions upstream of EGFR, regulating EGFR signaling by availability of multiple ligands.

DISCUSSION

TACE Is a Physiological TGFα Convertase

Compelling evidence that TACE is a physiologically relevant TGFα convertase includes (1) the overlap in phenotypes displayed by TGFα knockout and Tace$^{\Delta Zn/\Delta Zn}$ mice, namely open eyes at birth, wavy coats, and abnormal positioning of hair follicles in neonatal skin; (2) dramatically (10- to 20-fold) impaired release of soluble TGFα from two different cell types (fibroblasts and keratinocytes) derived from Tace$^{\Delta Zn/\Delta Zn}$ mice, without alterations to TGFα expression levels; (3) enhanced shedding of TGFα upon reintroduction of TACE into Tace$^{\Delta Zn/\Delta Zn}$ cells; and (4) accurate TACE cleavage *in vitro* of TGFα peptide substrates, proTGFecto, and (in the case of intact proTGFα) the N-terminal site. We suspect that the failure of TACE to cleave the C-terminal site of intact proTGFα is due to steric hindrance caused by improper folding of the detergent-solubilized integral membrane precursor or by improper presentation of the integral membrane ligand to soluble enzyme. However, even in the case of proTGFecto, TACE cleaved the N-terminal site more efficiently than the C-terminal site, a finding consistent with more rapid processing of the N-terminal site of proTGFα in cell culture.[17] Finally, a role for TACE as TGFα convertase is supported by our recent unpublished work showing that G-protein coupled receptor (GPCR)–induced shedding of TGFα, which presumably contributes to GPCR-mediated transactivation of EGFR, is dependent on TACE.

Our findings do not exclude the possible existence of one or more additional TGFα convertases that could have important roles in other contexts. Indeed, we observed residual shedding of TGFα in $Tace^{\Delta Zn/\Delta Zn}$ cells that were inhibited by hydroxamate-based compounds and hence may be attributable to the activity of one or more related metalloproteinases. Other studies have also pointed to the existence of additional TGFα convertases.[31,32]

TACE May Be a General Convertase for EGF Family Ligands

A variety of evidence suggests a role for TACE in the conversion of other EGF family ligands to soluble forms as well. The frequency of the open-eye phenotype of $Egfr^{wa\text{-}2/wa\text{-}2}$ neonates was dramatically increased when TACE gene dosage was halved, which is consistent with our previous finding that the $Egfr^{wa\text{-}2}$ defect is masked at relatively high ligand concentrations.[29] The relatively high frequency with which $Tace^{\Delta Zn/+}Egfr^{wa\text{-}2/wa\text{-}2}$ pups were born with open eye supports the view that TACE is a physiological convertase for multiple EGF family ligands that have redundant roles in the migration of the fetal eyelids, and that cells with homozygous wild-type TACE have EGFR ligand levels high enough to overcome the $Egfr^{wa\text{-}2}$ defect. This gene dosage effect also suggests that TACE expression can be regulated at the transcriptional level.

Other evidence for a broad role for TACE in EGF family ligand processing includes *in vitro* cleavage of peptides corresponding to both processing sites of EPR, as well as the C-terminal juxtamembrane sites of BTC and HB-EGF and the N-terminal processing site of AR. Although TACE failed to cleave a peptide spanning the C-terminus of mature AR, cotransfection of the protease resulted in increased shedding of AR and HB-EGF proteins into the media. Perhaps because AR lacks an obvious cleavage site with a P_1' nonpolar residue, other structural determinants not present in the peptide substrate are important for TACE-mediated AR cleavage. Others have also implicated TACE in HB-EGF shedding.[31]

Recently, in unpublished studies, we have found further evidence of a role for TACE as a physiological HB-EGF convertase: a phenotypic link in mouse heart and lung development. $Tace^{\Delta Zn/\Delta Zn}$ and $HB\text{-}EGF^{-/-}$ embryos or newborns display similarly enlarged and stenotic heart valves, apparently as a result of hyperproliferation and impaired remodeling. Both atrioventricular and outflow valves are similarly affected. Additionally, newborn $HB\text{-}EGF^{-/-}$ lung contained significantly fewer alveoli, and the mesenchymal tissue between alveoli was abnormally thickened. Moreover, the number of surfactant-positive cells was reduced and there was a corresponding increase in the fraction of immature, glycogen-containing cells. A similar phenotype has been described for $Tace^{\Delta Zn/\Delta Zn}$ lung,[33] consistent with the hypothesis that TACE is a physiological HB-EGF convertase. ADAM9 (MDC9/meltrin-γ), ADAM10, and ADAM12, as well as MMP-7, have also been implicated in HB-EGF shedding,[34–38] although in these cases supporting evidence from knockout studies of the protease gene has not been provided.

More Questions

Although our studies point to a broad role for TACE as an EGF family convertase, they also raise a number of intriguing issues. First, since TACE is now implicated in the processing of diverse substrates including L-selectin,[11] TNFRII,[11] β-amyloid precursor,[39,40] ErbB4,[41] and collagen XVII,[42] it is unclear how regulated processing of specific substrates, and in particular the different EGF family members, is achieved. Conceivably, this could be accomplished by regulating interactions between TACE and its various substrates on the cell surface, but there is as yet no data to distinguish between this and other possible mechanisms. Second, the structural features recognized by TACE are not yet known. In the EGF family, cleavage sites are frequently characterized by a small nonpolar amino acid (often Val) in the P_1' position, but the membrane stalk regions are not otherwise conserved. Third, TACE cleaves TGFα peptide substrates with considerably (10-fold) reduced efficiency compared to TNFα, and it cleaves other EGF family substrates with even lower efficiency (see TABLE 1). Cleavage of these substrates on intact cell surfaces also occurs relatively slowly, but (as is true generally of ectodomain shedding) can be rapidly induced by activation of PKC and GPCRs. Thus, a major issue in the field currently is how TACE activity is altered in response to signaling events. The TACE cytoplasmic domain contains conserved SH3 motifs, one of which contains a Thr residue that is phosphorylated by Erk1/2.[43] In addition, the soluble phosphatase PTPH1 interacts with the cytoplasmic domain and negatively regulates TNF shedding.[44] In contrast, at least one report suggests that the TACE cytoplasmic domain is not essential for shedding of TNFα in a transfection model.[45] Understanding the mechanisms that regulate TACE activity and substrate selection are critical future goals.

TACE as a Therapeutic Target

As noted above, integral membrane precursors to EGF family growth factors are biologically active in model studies and hence may participate in juxtacrine signaling. They may also couple cell-cell adhesion and mitogenesis,[46] or induce different biological responses than their soluble counterparts.[47] On the other hand, the phenotypic similarities between $Tace^{\Delta Zn/\Delta Zn}$, $Egfr^{-/-}$, and EGF family knockout mice underscore the importance of soluble EGFR ligands, and thus proteolytic processing, for normal development. This in turn raises the possibility that TACE might be an important therapeutic target for hyperproliferative diseases, including cancer, that are frequently accompanied by increased expression of EGF family growth factors and their receptor. However, complete evaluation of this possibility will require a more complete understanding of TACE biology, including its substrate specificity and mode of regulation, as well as whether it is also a target of deregulation in cancer.

ACKNOWLEDGMENTS

We thank Christoph Borchers, Carol Parker, Christina Raska, and the UNC Proteomics Core Facility for the mass spectrometry analyses. This work was supported by NIH Grant Nos. CA43793 and CA85410 (to D. C. Lee), Department of Defense Grant No. DAMD17-00-1-0378 (to L. F. Jackson), and NIH Training Grant No. CA71341 (to C. L. Hinkle and T. J. Meyers).

REFERENCES

1. RIESE, D.J., II & D.F. STERN. 1998. Specificity within the EGF family/ErbB receptor family signaling network. Bioessays **20:** 41–48.
2. OLAYIOYE, M.A. *et al.* 2000. The ErbB signaling network: receptor heterodimerization in development and cancer. EMBO J. **19:** 3159–3167.
3. GROENEN, L.C., E.C. NICE & A.W. BURGESS. 1994. Structure-function relationships for the EGF/TGF-alpha family of mitogens. Growth Factors **11:** 235–257.
4. STEIN, R.A. & J.V. STAROS. 2000. Evolutionary analysis of the ErbB receptor and ligand families. J. Mol. Evol. **50:** 397–412.
5. LEE, D.C. *et al.* 2003. EGF family ligands. *In* The Cytokine Handbook. Fourth Edition. Elsevier Science. Amsterdam/London/New York. In press.
6. JONES, J.T., R.W. AKITA & M.X. SLIWKOWSKI. 1999. Binding specificities and affinities of egf domains for ErbB receptors. FEBS Lett. **447:** 227–231.
7. BEERLI, R.R. *et al.* 1995. Neu differentiation factor activation of ErbB-3 and ErbB-4 is cell specific and displays a differential requirement for ErbB-2. Mol. Cell. Biol. **15:** 6496–6505.
8. RIESE, D.J., II *et al.* 1995. The cellular response to neuregulins is governed by complex interactions of the erbB receptor family. Mol. Cell. Biol. **15:** 5770–5776.
9. PINKAS-KRAMARSKI, R. *et al.* 1996. Diversification of Neu differentiation factor and epidermal growth factor signaling by combinatorial receptor interactions. EMBO J. **15:** 2452–2467.
10. MASSAGUE, J. & A. PANDIELLA. 1993. Membrane-anchored growth factors. Annu. Rev. Biochem. **62:** 515–541.
11. PESCHON, J.J. *et al.* 1998. An essential role for ectodomain shedding in mammalian development. Science **282:** 1281–1284.
12. DERYNCK, R. *et al.* 1984. Human transforming growth factor-alpha: precursor structure and expression in *E. coli.* Cell **38:** 287–297.
13. LEE, D.C. *et al.* 1985. Cloning and sequence analysis of a cDNA for rat transforming growth factor-alpha. Nature **313:** 489–491.
14. IGNOTZ, R.A. *et al.* 1986. Biologically active precursor for transforming growth factor type alpha, released by retrovirally transformed cells. Proc. Natl. Acad. Sci. USA **83:** 6307–6311.
15. LUETTEKE, N.C. *et al.* 1988. Characterization of high molecular weight transforming growth factor alpha produced by rat hepatocellular carcinoma cells. Biochemistry **27:** 6487–6494.
16. TEIXIDO, J. & J. MASSAGUE. 1988. Structural properties of a soluble bioactive precursor for transforming growth factor-alpha. J. Biol. Chem. **263:** 3924–3929.
17. TEIXIDO, J. *et al.* 1990. Generation of transforming growth factor-alpha from the cell surface by an *O*-glycosylation-independent multistep process. J. Biol. Chem. **265:** 6410–6415.
18. ARRIBAS, J. *et al.* 1996. Diverse cell surface protein ectodomains are shed by a system sensitive to metalloprotease inhibitors. J. Biol. Chem. **271:** 11376–11382.
19. PANDIELLA, A. & J. MASSAGUE. 1991. Multiple signals activate cleavage of the membrane transforming growth factor-alpha precursor. J. Biol. Chem. **266:** 5769–5773.
20. BASELGA, J. *et al.* 1996. Autocrine regulation of membrane transforming growth factor-alpha cleavage. J. Biol. Chem. **271:** 3279–3284.
21. BLACK, R.A. *et al.* 1997. A metalloproteinase disintegrin that releases tumour-necrosis factor-alpha from cells. Nature **385:** 729–733.
22. MOSS, M.L. *et al.* 1997. Cloning of a disintegrin metalloproteinase that processes precursor tumour-necrosis factor-alpha. Nature **385:** 733–736.

23. SEALS, D.F. & S.A. COURTNEIDGE. 2003. The ADAMs family of metalloproteases: multidomain proteins with multiple functions. Genes Dev. **17:** 7–30.
24. SUNNARBORG, S.W. *et al.* 2002. Tumor necrosis factor-alpha converting enzyme (TACE) regulates epidermal growth factor receptor ligand availability. J. Biol. Chem. **277:** 12838–12845.
25. BRILEY, G.P. *et al.* 1997. The carboxyl-terminal valine residues of proTGF alpha are required for its efficient maturation and intracellular routing. Mol. Biol. Cell **8:** 1619–1631.
26. GOISHI, K. *et al.* 1995. Phorbol ester induces the rapid processing of cell surface heparin-binding EGF-like growth factor: conversion from juxtacrine to paracrine growth factor activity. Mol. Biol. Cell **6:** 967–980.
27. BERKOWITZ, E.A. *et al.* 1996. Characterization of the mouse transforming growth factor alpha gene: its expression during eyelid development and in waved 1 tissues. Cell Growth Differ. **7:** 1271–1282.
28. LUETTEKE, N.C. *et al.* 1999. Targeted inactivation of the EGF and amphiregulin genes reveals distinct roles for EGF receptor ligands in mouse mammary gland development. Development **126:** 2739–2750.
29. LUETTEKE, N.C. *et al.* 1994. The mouse waved-2 phenotype results from a point mutation in the EGF receptor tyrosine kinase. Genes Dev. **8:** 399–413.
30. FOWLER, K.J. *et al.* 1995. A mutation in the epidermal growth factor receptor in waved-2 mice has a profound effect on receptor biochemistry that results in impaired lactation. Proc. Natl. Acad. Sci. USA **92:** 1465–1469.
31. MERLOS-SUAREZ, A. *et al.* 2001. Metalloprotease-dependent protransforming growth factor-alpha ectodomain shedding in the absence of tumor necrosis factor-alpha–converting enzyme. J. Biol. Chem. **276:** 48510–48517.
32. HINKLE, C.L. *et al.* 2003. Multiple metalloproteases process protransforming growth factor-α (proTGF-α). Biochemistry **42:** 2127–2136.
33. ZHAO, J. *et al.* 2001. Pulmonary hypoplasia in mice lacking tumor necrosis factor-alpha converting enzyme indicates an indispensable role for cell surface protein shedding during embryonic lung branching morphogenesis. Dev. Biol. **232:** 204–218.
34. IZUMI, Y. *et al.* 1998. A metalloprotease-disintegrin, MDC9/meltrin-gamma/ADAM9, and PKCdelta are involved in TPA-induced ectodomain shedding of membrane-anchored heparin-binding EGF-like growth factor. EMBO J. **17:** 7260–7272.
35. LEMJABBAR, H. & C. BASBAUM. 2002. Platelet-activating factor receptor and ADAM10 mediate responses to *Staphylococcus aureus* in epithelial cells. Nat. Med. **8:** 41–46.
36. YAN, Y., K. SHIRAKABE & Z. WERB. 2002. The metalloprotease Kuzbanian (ADAM10) mediates the transactivation of EGF receptor by G protein–coupled receptors. J. Cell Biol. **158:** 221–226.
37. ASAKURA, M. *et al.* 2002. Cardiac hypertrophy is inhibited by antagonism of ADAM12 processing of HB-EGF: metalloproteinase inhibitors as a new therapy. Nat. Med. **8:** 35–40.
38. YU, W.H. *et al.* 2002. CD44 anchors the assembly of matrilysin/MMP-7 with heparin-binding epidermal growth factor precursor and ErbB4 and regulates female reproductive organ remodeling. Genes Dev. **16:** 307–323.
39. BUXBAUM, J.D. *et al.* 1998. Evidence that tumor necrosis factor alpha converting enzyme is involved in regulated alpha-secretase cleavage of the Alzheimer amyloid protein precursor. J. Biol. Chem. **273:** 27765–27767.
40. SLACK, B.E., L.K. MA & C.C. SEAH. 2001. Constitutive shedding of the amyloid precursor protein ectodomain is up-regulated by tumour necrosis factor-alpha converting enzyme. Biochem. J. **357:** 787–794.
41. RIO, C. *et al.* 2000. Tumor necrosis factor-alpha–converting enzyme is required for cleavage of erbB4/HER4. J. Biol. Chem. **275:** 10379–10387.
42. FRANZKE, C.W. *et al.* 2002. Transmembrane collagen XVII, an epithelial adhesion protein, is shed from the cell surface by ADAMs. EMBO J. **21:** 5026–5035.
43. DIAZ-RODRIGUEZ, E. *et al.* 2002. Extracellular signal-regulated kinase phosphorylates tumor necrosis factor alpha–converting enzyme at threonine 735: a potential role in regulated shedding. Mol. Biol. Cell **13:** 2031–2044.

44. ZHENG, Y., J. SCHLONDORFF & C.P. BLOBEL. 2002. Evidence for regulation of the tumor necrosis factor alpha-convertase (TACE) by protein-tyrosine phosphatase PTPH1. J. Biol. Chem. **277:** 42463–42470.
45. REDDY, P. *et al.* 2000. Functional analysis of the domain structure of tumor necrosis factor-alpha converting enzyme. J. Biol. Chem. **275:** 14608–14614.
46. ANKLESARIA, P. *et al.* 1990. Cell-cell adhesion mediated by binding of membrane-anchored transforming growth factor alpha to epidermal growth factor receptors promotes cell proliferation. Proc. Natl. Acad. Sci. USA **87:** 3289–3293.
47. IWAMOTO, R., K. HANDA & E. MEKADA. 1999. Contact-dependent growth inhibition and apoptosis of epidermal growth factor (EGF) receptor–expressing cells by the membrane-anchored form of heparin-binding EGF-like growth factor. J. Biol. Chem. **274:** 25906–25912.

Characterization and Expression of Novel 60-kDa and 110-kDa EGFR Isoforms in Human Placenta

JILL L. REITER AND NITA J. MAIHLE

Tumor Biology Program, Mayo Clinic, Rochester, Minnesota 55905, USA

ABSTRACT: The epidermal growth factor receptor (EGFR) and related family members (ERBB2, ERBB3, and ERBB4) previously have been shown to play pivotal roles in the development of female reproductive tissues, in blastocyst implantation, and in placental differentiation. We have cloned and sequenced several naturally occurring alternative transcripts of the human and mouse *EGFR* genes, which encode novel receptor isoforms containing varying portions of the extracellular ligand-binding domain, but lacking the transmembrane and cytoplasmic domain sequences. The human 1.8-kb and 3-kb alternative *EGFR* transcripts encode secreted 60-kDa and cell surface–associated 110-kDa EGFR isoforms, respectively. We have developed quantitative ribonuclease protection assays to study the expression of these alternative transcripts in human tissues. Similar to the full-length *EGFR* mRNAs, the highest expression level of these alternative transcripts occurs in placenta. We speculate that both of these EGFR isoforms may be important regulators of EGF-mediated cell growth and differentiation in human placenta.

KEYWORDS: alternative splicing; cell fractionation; epidermal growth factor receptor; glycoprotein; isoforms; placenta; posttranslational protein modification; receptor tyrosine kinase

INTRODUCTION

Epidermal growth factor receptor (EGFR) is the prototypic member of the ERBB family of receptor tyrosine kinases that consists of four members: EGFR (ERBB1), ERBB2, ERBB3, and ERBB4. All four ERBB family members are structurally similar to each other and consist of a large highly glycosylated extracellular ligand-binding domain, a single transmembrane-spanning domain, an intracellular tyrosine kinase domain, and a carboxy-terminal tail that contains multiple tyrosine phosphoacceptor sites. Each of these receptors is able to form dimers with other ERBB family members, and these dimeric interactions are mediated by the binding of EGF- or neuregulin-related ligands through autocrine, paracrine, and juxtacrine mechanisms. Ligand-dependent activation of ERBB family members initiates a variety of complex signaling cascades that culminate in diverse biological endpoints such as

Address for correspondence: Nita J. Maihle, Ph.D., 200 First Street SW, Rochester, MN 55905. Voice: 507-284-0279; fax: 507-284-1767.
maihle@mayo.edu

Ann. N.Y. Acad. Sci. 995: 39–47 (2003). © 2003 New York Academy of Sciences.

cell survival, proliferation, differentiation, adhesion, and migration. Indeed, the ERBB signaling network is crucial during mammalian development, and deregulation of this network is frequently involved in the initiation and progression of cancer in diverse human tissues.

In addition to these full-length receptors, the ERBB family also contains naturally occurring receptor isoforms that consist of varying portions of the extracellular domain (ECD), but which lack the transmembrane and intracellular domain sequences. These receptor isoforms can be generated either by alternative RNA splicing or by proteolytic cleavage of the full-length receptor. EGFR isoforms containing only the ECD have now been isolated from tissues of diverse species including human, mouse, rat, and chicken, and all of these receptor isoforms are encoded by alternatively spliced *EGFR* mRNAs.[1–4] In addition to these naturally occurring gene products, the A431 human carcinoma-derived cell line expresses a secreted EGFR ECD protein that arises from a complex chromosomal rearrangement and translocation.[5–7] Specifically, the chimeric 2.8-kb A431 mRNA results from RNA splicing from *EGFR* exon 16 to a recombinant 3′-terminal exon, which includes sequences from *EGFR* intron 16, two separate sequences on chromosome 7 (~67 and 83 kb downstream of the *EGFR* 3′-end), and a sequence from chromosome 4 (J. L. Reiter, unpublished observation). Hence, the carboxy-terminus of this A431-specific protein is distinct from the naturally occurring alternative *EGFR* transcripts, and this aberrant transcript has not been described in any other tissue or cell line.

EGFR GENOMIC STRUCTURE

To better understand the synthesis of the naturally occurring alternative *EGFR* transcripts, we have analyzed the structure of both human and murine *EGFR* genes,[4] and these observations are briefly summarized below. The human gene spans ~193 kb and resides on chromosome 7p11.2. The mouse *EGFR* gene maps to chromosome 11A2 and, although the complete genomic sequence is not yet available, it is estimated to be of comparable size to the human gene. In general, both the exon and intron sizes of these genes are very similar, and the positioning of the exon-intron boundaries is highly conserved. Both the human and mouse full-length EGFR are encoded by mRNAs that contain 28 exons. In addition to these 28 exons, both genes also contain alternatively spliced 3′-terminal exons, which are only included in the mRNAs encoding the EGFR isoforms that are described in greater detail below.

CLONING AND EXPRESSION OF ALTERNATIVE *EGFR* TRANSCRIPTS

To date, we have cloned several human and mouse alternative *EGFR* transcripts (FIG. 1).[3,4] All of these alternative transcripts contain sequences that are identical to the full-length *EGFR* mRNA at the 5′-end. Each of these transcripts also contains a unique 3′-sequence, which includes a stop codon and at least one polyadenylation signal. One interesting observation is that, although human, mouse, and rat tissues all express transcripts that diverge from the full-length *EGFR* mRNA after exon 15, the sequences of these alternative 3′-exons are not conserved. In addition to the

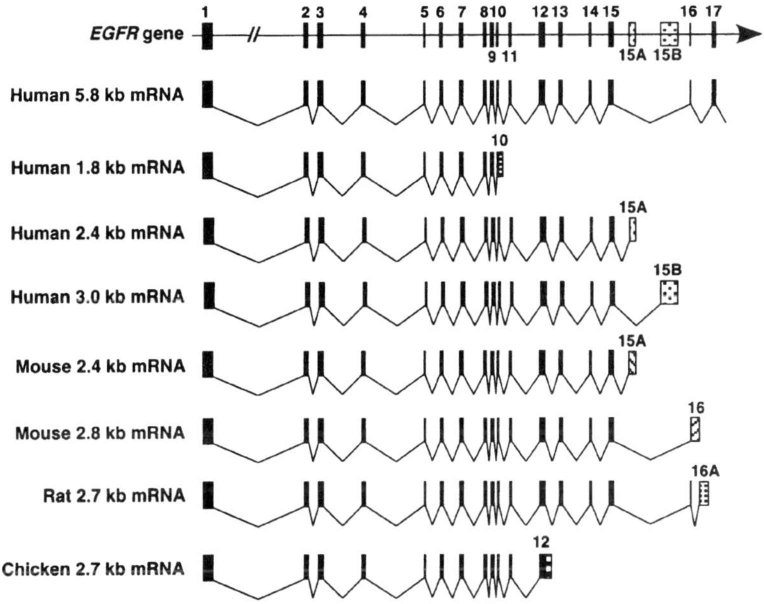

FIGURE 1. Alternative RNA splicing of human, mouse, rat, and chicken *EGFR* genes. The top line shows a portion of the human *EGFR* gene. The stippled boxes represent 2 alternative 3′-terminal exons, which have been designated 15A and 15B. The full-length EGFR is encoded by either a 5.8-kb or 10-kb mRNA, both of which include 28 exons and only differ from each other in the 3′-untranslated region; the splicing pattern of these transcripts is shown on the second line. The human 1.8-kb, mouse 2.8-kb, and chicken 2.7-kb mRNAs read through the splice donor sites in introns 10, 16, and 12, respectively; human 2.4-kb and 3-kb, mouse 2.4-kb, and rat 2.7-kb mRNAs splice to novel 3′ terminal exons that are not included in the full-length *EGFR* transcripts. None of the alternative *EGFR* mRNAs shown here contain exon 17, which encodes the transmembrane-spanning domain in the full-length EGFR.

alternative *EGFR* transcripts described here, mouse 1.8-kb and rat 2-kb mRNAs have been reported;[8,9] however, the splicing patterns for these transcripts have not been determined yet.

IN VITRO CHARACTERIZATION OF HUMAN 60-KDA AND 110-KDA EGFR ISOFORMS

The human 1.8-kb and 3-kb *EGFR* transcripts encode proteins that include the amino-terminal signal peptide and portions of the extracellular ligand-binding domain of the receptor. In addition to the amino acid sequence it shares in common with the full-length EGFR, the 1.8-kb mRNA encodes a product that contains 2 unique carboxy-terminal amino acids. Similarly, the 3-kb transcript encodes 78 unique carboxy-terminal amino acids. After proteolytic removal of the signal

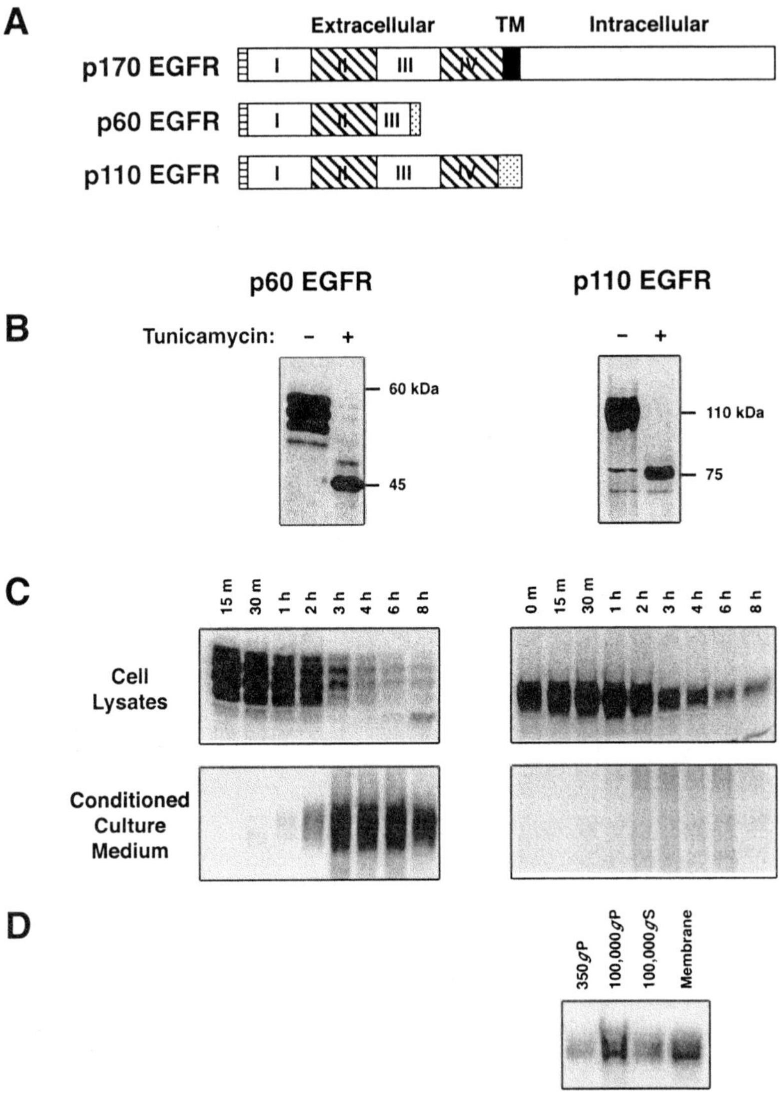

FIGURE 2. Characterization of p60 and p110 EGFR isoforms. **(A)** Schematic representation of the products encoded by human full-length (170 kDa), 1.8-kb (60 kDa), and 3-kb (110 kDa) *EGFR* transcripts. The extracellular domain is composed of four subdomains: two are cysteine-rich (II & IV) and two are involved in binding ligand (I & III). p60 and p110 EGFR contain 2 and 78 unique amino acids at their carboxy-termini (stippled boxes), respectively. **(B)** Quail fibroblasts (QT6) transfected with expression vectors containing 1.8-kb (*left panel*) or 3-kb (*right panel*) *EGFR* cDNAs were metabolically labeled with [^{35}S]-methionine in the absence (–) or presence (+) of tunicamycin. Cell lysates were immunoprecipitated with the EGFR.1 antibody, and immunoprecipitated proteins were analyzed by SDS-PAGE and autoradiography. **(C)** Pulse-chase analysis of QT6 cells transfected with p60

peptide, the protein products of these two transcripts are predicted to be 44 kDa and 75 kDa, respectively.[3,4]

Ectopic expression of the 1.8-kb and 3-kb *EGFR* cDNAs in cultured cells produces proteins with apparent molecular weights of 60 kDa and 110 kDa, respectively (FIG. 2A).[3,4] We have confirmed that, similar to the full-length 170-kDa EGFR, the addition of N-linked carbohydrates to both of these receptor isoforms accounts for the difference between their predicted and apparent molecular weights. Specifically, in cells transfected with either the 1.8-kb or 3-kb *EGFR* cDNAs and cultured in the presence of tunicamycin, an inhibitor of N-linked glycosylation, protein products of 44 kDa and 75 kDa, respectively, are observed in denaturing acrylamide gels (FIG. 2B).

Because both p60 and p110 EGFR contain signal peptides and neither protein contains a transmembrane-spanning domain, we predicted that both of these proteins would be secreted. Pulse-chase analysis of cells transfected with the 1.8-kb cDNA revealed an increasing amount of p60 EGFR in the culture medium and a compensatory decrease of this protein in cell lysates with time, confirming that this protein is secreted. In contrast, only trace amounts of p110 EGFR could be detected in the conditioned culture medium of cells transfected with the 3-kb cDNA, suggesting that this protein is not secreted efficiently (FIG. 2C). In fact, we have demonstrated recently that p110 EGFR is associated with cell membranes of transfected cells (FIG. 2D, and data not shown).

POTENTIAL MECHANISM OF p110 EGFR CELL SURFACE LOCALIZATION

The unique carboxy-terminal sequence of p110 EGFR does not contain a sufficiently long contiguous hydrophobic sequence that would be predicted to function as a transmembrane domain.[4] How then can p110 EGFR associate with the plasma membrane? One possibility is that p110 EGFR is a peripheral membrane protein and perhaps binds to transmembrane forms of EGF-family ligands and/or other cell surface proteins. Another possible mechanism of membrane anchoring involves the covalent attachment of lipid moieties to a sequence within the unique carboxy-terminus of p110 EGFR. Inspection of this carboxy-terminal sequence revealed that

(*left panel*) or p110 (*right panel*). Transfected cells were metabolically labeled with [^{35}S]methionine for 10 min and chased in complete medium in the absence of radioactivity for the indicated times. Cell lysates and conditioned media were immunoprecipitated with the EGFR.1 antibody. Immunoprecipitation of cells transfected with the 1.8-kb cDNA revealed a ladder of peptides (48–60 kDa) in cell lysates and a heterogeneous 55- to 65-kDa species in the conditioned media, which appeared after ~2–3 h. Immunoprecipitation of cells transfected with the 3-kb cDNA revealed a 110-kDa protein in the cell lysates, but not in the conditioned media. **(D)** Cell fractionation of Chinese hamster ovary cells transfected with p110 EGFR. Transfected cells were lysed in hypotonic buffer (0.25 M sucrose, 50 mM Tris, pH 7.4, 1 mM EDTA, 1 mM EGTA), and cell fractions (P, pellet; S, supernatant) were obtained by differential centrifugation. Plasma membranes were collected from the 100,000*g* pellet by flotation on 35% (w/v) sucrose. Equal protein concentrations were separated by SDS-PAGE, electroblotted to PVDF membrane, and blotted with anti-EGFR specific monoclonal antibody, 15E11.[20] p110 EGFR was enriched in the membrane-containing fractions, that is, the 100,000*g* pellet and the membrane fraction.

```
PGNESLKAML FCLFKLSSCN QSNDGSVSHQ SGSPAAQESC
LGWIPSLLPS EFQLGWGGCS HLHAWPSASV IITASSCH
```

FIGURE 3. Unique carboxy-terminal sequence of p110 EGFR. Three potential GPI-anchor cleavage and attachment sites are underscored. The tryptic peptide from purified serum p110 EGFR that was identified by mass spectrometry is boxed.[1]

there are three potential sites for the addition of a glycosylphosphatidylinositol (GPI) anchor (FIG. 3). The carboxy-terminal GPI-anchor modification signal consists of three small amino acids, followed by a relatively hydrophobic sequence. GPI anchor addition occurs by a transamidation reaction within the endoplasmic reticulum; the carboxy-terminal hydrophobic peptide is cleaved and the GPI moiety is attached to the new carboxy-terminal end of the protein. Our current working hypothesis is that p110 EGFR is lipid-modified and may become membrane-associated via a GPI anchor. However, the precise mechanism of p110 EGFR cell surface association will require further investigation.

EXPRESSION OF EGFR ISOFORMS IN HUMAN TISSUES

We have analyzed the expression of both the 1.8-kb and 3-kb *EGFR* mRNAs in several human tissues, as well as in numerous carcinoma-derived cell lines using quantitative ribonuclease protection assays[4] (and data not shown). Separate riboprobes were synthesized that simultaneously detected the full-length *EGFR* mRNA and either the 1.8-kb or the 3-kb alternatively spliced transcripts. The full-length *EGFR* mRNA was detected with both probes in all samples examined[4] (and data not shown). Similarly, the 1.8-kb mRNA also was detected in all samples examined, although it is consistently detected at lower levels compared to the full-length *EGFR* transcripts (data not shown). In contrast, the 3-kb mRNA is expressed at much lower levels compared to the full-length *EGFR* mRNA and is detectable only in samples known to express high levels of EGFR,[4] that is, in placenta and in two carcinoma-derived cell lines, A431 and MDA-MB-468 (data not shown).

Despite the apparent low level of 3-kb *EGFR* mRNA expression observed in these tissues, human serum contains relatively abundant levels of circulating 110-kDa EGFR, which is recognized by EGFR ECD-specific antibodies.[10] Moreover, mass spectrometry analysis of tryptic peptides generated from purified serum p110 EGFR has revealed that this serum protein is the product of the 3-kb *EGFR* mRNA (FIG. 3).[11] The mechanism of p110 EGFR release from cells as well as the identification of the source of circulating p110 EGFR isoform in human blood are active areas of investigation. In addition, studies also are under way to assess the potential utility of serum p110 EGFR levels as a marker of disease activity in cancer patients.[11–14]

EGFR has been implicated in many placental functions, and decreased EGFR activity has been reported in placentas from pathological pregnancies, specifically in preeclampsia (PE) and intrauterine growth retardation (IUGR).[15,16] Interestingly, when the expression of full-length and alternatively spliced *EGFR* mRNAs was analyzed in placentas from pregnancies associated with PE and IUGR, expression of the ECD-encoding mRNAs has been shown to be significantly higher in placental tissues from subjects with PE and IUGR, while expression of the full-length *EGFR* mRNA was significantly lower in the IUGR group.[17] Because this study by Faxen *et al.* was completed before additional *EGFR* transcripts were reported (i.e., 2.4-kb and 3-kb *EGFR* mRNAs[4]), the authors assumed that they were detecting full-length and 1.8-kb *EGFR* transcripts. However, their ECD-specific probe consisted of sequence spanning exons 3–18 and, therefore, would be expected to detect all *EGFR* mRNAs. This is an important point and suggests that caution may be needed in interpreting EGFR expression data resulting from studies using ECD-specific probes and/or antibodies.

While we do not yet know the distribution of the p110 EGFR isoform in human endometrial tissues, differential expression of the full-length vs. murine p110 EGFR has been reported in the preimplantation mouse uterus.[18] Specifically, murine EGFR is expressed by the blastocyst, but not by the luminal epithelium of the uterus, whereas the opposite expression pattern has been observed for p110 EGFR; that is, it is expressed in the luminal epithelium of the uterus, but not in the blastocyst. We predict that p110 EGFR will be a cell surface protein in both human and mouse tissues, and we further speculate that, if these proteins are GPI-anchored, they will be routed to the apical membrane surface of these epithelial cells. Interestingly, this polarized localization of p110 EGFR would position this protein to potentially interact with cell surface proteins of the blastocyst, including potential interaction with the full-length EGFR. Several EGF-family transmembrane ligands, as well as other ERBB family members, are known to be expressed on the surface of these cells at the time of blastocyst implantation and have been proposed to play an important role in this process.[19] Therefore, we speculate that this differential expression of p110 EGFR may result in the modulation of receptor-ligand interactions important in the process of blastocyst implantation.

SUMMARY

In addition to the full-length EGFR, other EGFR isoforms are produced by alternative RNA splicing. The structural and functional characterization of the 60-kDa and 110-kDa receptor isoforms is ongoing; however, we have demonstrated that p60 is a secreted protein, while p110 is associated with the plasma membrane. Given the functional importance of both EGF- and EGFR/ERBB-family members in placenta and the expression of both p60 and p110 EGFR isoforms in this tissue, we propose that these proteins may be involved in mediating important EGFR/ERBB-related functions. Given the growing complexity of EGFR family interactions and signaling mechanisms, these novel receptor isoforms should be considered in future studies examining either the expression or function of EGFR in placenta, as well as in other tissues.

ACKNOWLEDGMENTS

This research was supported in part by National Institutes of Health Grant No. CA85133 and by a scholarship to J. L. Reiter from the Regis Foundation for Breast Cancer Research.

REFERENCES

1. FLICKINGER, T.W., N.J. MAIHLE & H.J. KUNG. 1992. An alternatively processed mRNA from the avian c-erbB gene encodes a soluble, truncated form of the receptor that can block ligand-dependent transformation. Mol. Cell. Biol. **12:** 883–893.
2. PETCH, L.A., J. HARRIS, V.W. RAYMOND *et al.* 1990. A truncated, secreted form of the epidermal growth factor receptor is encoded by an alternatively spliced transcript in normal rat tissue. Mol. Cell. Biol. **10:** 2973–2982.
3. REITER, J.L. & N.J. MAIHLE. 1996. A 1.8 kb alternative transcript from the human epidermal growth factor receptor gene encodes a truncated form of the receptor. Nucleic Acids Res. **24:** 4050–4056.
4. REITER, J.L., D.W. THREADGILL, G.D. ELEY *et al.* 2001. Comparative genomic sequence analysis and isolation of human and mouse alternative EGFR transcripts encoding truncated receptor isoforms. Genomics **71:** 1–20.
5. HUNTS, J.H., N. SHIMIZU, T. YAMAMOTO *et al.* 1985. Translocation chromosome 7 of A431 cells contains amplification and rearrangement of EGF receptor gene responsible for production of variant mRNA. Somatic Cell Mol. Genet. **11:** 477–484.
6. MERLINO, G.T., S. ISHII, J. WHANG-PENG *et al.* 1985. Structure and localization of genes encoding aberrant and normal epidermal growth factor receptor RNAs from A431 human carcinoma cells. Mol. Cell. Biol. **5:** 1722–1734.
7. ULLRICH, A., L. COUSSENS, J.S. HAYFLICK *et al.* 1984. Human epidermal growth factor receptor cDNA sequence and aberrant expression of the amplified gene in A431 epidermoid carcinoma cells. Nature **309:** 418–425.
8. RHO, O., L.M. BELTRAN, I.B. GIMENEZ-CONTI *et al.* 1994. Altered expression of the epidermal growth factor receptor and transforming growth factor-alpha during multistage skin carcinogenesis in SENCAR mice. Mol. Carcinog. **11:** 19–28.
9. YU, Y., A.K. RISHI, J.R. TURNER *et al.* 2001. Cloning of a novel EGFR-related peptide: a putative negative regulator of EGFR. Am. J. Physiol. Cell Physiol. **280:** C1083–C1089.
10. BARON, A.T., J.M. LAFKY, D.C. CONNOLLY *et al.* 1998. A sandwich type acridinium-linked immunosorbent assay (ALISA) detects soluble ErbB1 (sErbB1) in normal human sera. J. Immunol. Methods **219:** 23–43.
11. BARON, A.T., E.M. CORA, J.M. LAFKY *et al.* 2003. Soluble epidermal growth factor receptor (sEGFR/sErbB1) as a potential risk, screening, and diagnostic serum biomarker of epithelial ovarian cancer. Cancer Epidemiol. Biomarkers Prev. **12:** 103–113.
12. BARON, A.T., J.M. LAFKY, C.H. BOARDMAN *et al.* 1999. Serum sErbB1 and epidermal growth factor levels as tumor biomarkers in women with stage III or IV epithelial ovarian cancer. Cancer Epidemiol. Biomarkers Prev. **8:** 129–137.
13. BARON, A.T., J.M. LAFKY, V.J. SUMAN *et al.* 2001. A preliminary study of serum concentrations of soluble epidermal growth factor receptor (sErbB1), gonadotropins, and steroid hormones in healthy men and women. Cancer Epidemiol. Biomarkers Prev. **10:** 1175–1185.
14. MAIHLE, N.J., A.T. BARON, B.A. BARRETTE *et al.* 2002. EGF/ErbB receptor family in ovarian cancer. Cancer Treat. Res. **107:** 247–258.
15. FONDACCI, C., E. ALSAT, R. GABRIEL *et al.* 1994. Alterations of human placental epidermal growth factor receptor in intrauterine growth retardation. J. Clin. Invest. **93:** 1149–1155.
16. FUJITA, Y., H. KURACHI, K. MORISHIGE *et al.* 1991. Decrease in epidermal growth factor receptor and its messenger ribonucleic acid levels in intrauterine growth-retarded and diabetes mellitus–complicated pregnancies. J. Clin. Endocrinol. Metab. **72:** 1340–1345.

17. FAXEN, M., J. NASIELL, A. BLANCK *et al.* 1998. Altered mRNA expression pattern of placental epidermal growth factor receptor (EGFR) in pregnancies complicated by preeclampsia and/or intrauterine growth retardation. Am. J. Perinatol. **15:** 9–13.
18. TONG, B.J., S.K. DAS, D. THREADGILL *et al.* 1996. Differential expression of the full-length and truncated forms of the epidermal growth factor receptor in the pre-implantation mouse uterus and blastocyst. Endocrinology **137:** 1492–1496.
19. LIM, H., H. SONG, B.C. PARIA *et al.* 2002. Molecules in blastocyst implantation: uterine and embryonic perspectives. Vitam. Horm. **64:** 43–76.
20. BARON, A.T., B.K. HUNTLEY, J.M. LAFKY *et al.* 1997. Monoclonal antibodies specific for peptide epitopes of the epidermal growth factor receptor's extracellular domain. Hybridoma **16:** 259–271.

The Roles of Activin in Cytoprotection and Tissue Repair

MIRIAM WANKELL,[a] SILKE WERNER,[a] CHRISTIAN ALZHEIMER,[b] AND SABINE WERNER[a]

[a]*Institute of Cell Biology, ETH Zürich, Hönggerberg, CH-8093 Zürich, Switzerland*

[b]*Institute of Physiology, Ludwig-Maximilians-Universität München, München, Germany*

ABSTRACT: Activin is a member of the transforming growth factor beta family of growth and differentiation factors. Initially discovered as a protein that stimulates release of follicle-stimulating hormone, it is now well accepted as an important regulator of cell growth and differentiation. Most interestingly, a series of previous studies have revealed novel roles of activin in inflammation and repair. Our own results have provided evidence for an important function of activin in cutaneous wound repair as well as in neuroprotection, and these data will be summarized and discussed in this chapter.

KEYWORDS: activin; follistatin; injury; Mad1; neuroprotection; wound healing

INTRODUCTION

Polypeptides of the transforming growth factor β (TGF-β) family are known to be key players in embryonic development and tissue repair. In mammals, this family includes, for example, three types of TGF-β (TGF-β1, β2, and β3), various bone morphogenetic proteins (BMPs), mullerian inhibiting substance, inhibins, and activins.[1] A series of studies have provided insight into the important function of TGF-βs in skin morphogenesis and wound repair.[2,3] However, an *in vivo* function of activin in these processes has only recently been demonstrated (see below).

Like the majority of other members of the TGF-β family, activins are dimeric proteins, the monomeric polypeptides of which are connected by disulfide linkage. Three different forms of activin—the homodimeric activin A (βAβA) and activin B (βBβB), as well as the heterodimeric activin AB (βAβB)—have been described.[1] Recently, βC, βD, and βE chains have been discovered, although little is as yet known about the corresponding proteins. The biological activities of activins are mediated by heteromeric receptor complexes consisting of two different types of receptor, the type I (ActRIA and ActRIB) and type II (ActRIIA and ActRIIB) receptors, which are all characterized by the presence of an intracellular serine/threonine

Address for correspondence: Dr. Sabine Werner, Institute of Cell Biology, ETH Zürich, Hönggerberg, CH-8093 Zürich, Switzerland. Voice: +41 1 633 3941; fax: +41 1 633 1174.
sabine.werner@cell.biol.ethz.ch

Ann. N.Y. Acad. Sci. 995: 48–58 (2003). © 2003 New York Academy of Sciences.

kinase domain.[4] Besides these transmembrane receptors, soluble activin-binding proteins called follistatins have been discovered that bind to activin and thereby inhibit its biological effects.[5,6] Follistatins bind activin with high affinity and a few BMPs (BMP-2, 4, 7, and 11) with significantly lower affinity.[7,8]

A NOVEL ROLE OF ACTIVIN IN SKIN MORPHOGENESIS AND WOUND REPAIR

Recent studies from our laboratory have provided evidence for a novel role of activin in cutaneous wound repair. The first indication of a possible role of activin in the skin, however, resulted from the phenotypes of knockout mice. Besides various defects that caused death within the first 24 hours after birth, mice lacking the activin βA chain lacked whiskers, and whisker follicles were abnormal,[9] and mice deficient in follistatin showed disturbed whisker development and hyperkeratotic skin.[10] Furthermore, *in vitro* studies demonstrated a weak inhibitory effect of activin on keratinocyte proliferation, and keratinocyte differentiation was induced by this factor.[11,12] These findings indicated an important and unique role of activin in the skin, which differs from the known effects of TGF-βs in this tissue. However, the mechanisms that underlie the effects of activin on keratinocytes and other cell types in the skin have yet to be defined. Most importantly, the *in vivo* functions of activin in cutaneous wound repair and in other types of tissue repair processes remain to be determined.

To gain insight into a possible role of activin in normal and wounded skin, we first analyzed its expression during the wound repair process. Interestingly, expression of both the activin βA and βB chains was hardly detectable in normal skin. However, we found a strong induction of activin βA and βB expression in the granulation tissue and in suprabasal keratinocytes of the hyperproliferative epithelium after skin injury.[13] This increase in activin expression is likely to be the result of increased levels of serum growth factors and proinflammatory cytokines in the wound, which are strong inducers of activin expression in fibroblasts and keratinocytes.[14] In addition to the ligands, all known activin receptors were expressed in normal and wounded skin, although their expression was not induced after injury.[13]

To determine the activities of activin in the skin, we generated transgenic mice overexpressing the activin βA chain in the epidermis under the control of a keratin 14 promoter.[15] The transgenic mice were significantly smaller than control littermates, and they had smaller ears and shorter tails with a necrotic tip. As expected from the activity of the keratin 14 promoter, the most remarkable abnormalities were observed in the skin. The thickness of the epidermis was significantly increased, resembling the hyperplastic neonatal mouse epidermis. Interestingly, we found enhanced levels of endogenous activin βA mRNA in back skin of newborn control animals compared to adult back skin, indicating a role of activin in the hyperplasia seen in neonatal epidermis. Consistent with the morphology of the epidermis in our transgenic animals, the number of proliferating keratinocytes was 2- to 3-fold higher compared to control mice.[15] The observed stimulation of keratinocyte proliferation by activin is probably indirect since activin inhibited proliferation of human keratinocytes *in vitro*.[11,12] In addition to the enhanced mitotic activity of the keratinocytes, the increased levels of activin also affected the program of keratinocyte differentiation.

The number of spinous layers was significantly increased, and these cells expressed keratin 16, a marker for abnormal keratinocyte differentiation. Although the transgene is only expressed in keratinocytes, we also observed fibrosis in the dermis, most likely due to diffusion of activin from the basal keratinocytes to the underlying connective tissue.[15]

The most striking result of our study was the strongly enhanced healing rate of full-thickness skin wounds in the activin-overexpressing mice. The process of granulation tissue formation was particularly enhanced, and increased reepithelialization was also observed in most of the animals.[15] Although the enhanced wound-healing process appears to be beneficial, first analyses of late wounds revealed a significantly extended area of scar tissue, indicating that the increased levels of activin stimulate fibrotic processes in not only nonwounded, but also wounded skin (our unpublished data). Taken together, the results obtained with the activin-overexpressing mice revealed novel activities of activin in the regulation of keratinocyte proliferation and differentiation as well as in dermal fibrosis and cutaneous wound repair. However, the roles of the endogenous activin in these processes remained to be determined.

To further address this question, we attempted to block the endogenous activin in normal and wounded skin. Since activin βA knockout mice die within the first 24 hours after birth from multiple abnormalities,[9] it was not possible to analyze the wound-healing process in these mice. Furthermore, a tissue-specific deletion of the activin βA gene in the skin was not suitable for the analysis of activin function in the wound because activin is expressed in the dermis and in the epidermis, and dermis-specific gene inactivation is as yet not possible due to the lack of a suitable dermis-specific promoter. For this purpose, we generated transgenic mice that overexpress the activin antagonist, follistatin, in the epidermis under the control of the keratin 14 promoter.[16] Since follistatin is a soluble protein, it should diffuse from the basal keratinocytes and hair follicles to the interfollicular dermis and to the granulation tissue of wounded skin.

Transgenic mice that express high levels of the follistatin transgene in the skin were characterized by reduced body size and weight, possibly as a result of the expression of the transgene in the tongue, oral epithelium, and esophagus. In addition, we observed significantly deformed and longer incisors, which might impede food uptake and therefore result in a nutrition defect.

In addition to the reduced body weight, we observed several other macroscopic abnormalities. The animals were born with open eyes, a feature that was also observed in activin βB knockout mice.[17] This finding indicates that our recombinant follistatin does indeed block activin action *in vivo*. Furthermore, the mice had particularly large ears and longer tails. This is opposite to the phenotype observed in our activin βA-transgenic mice.[15]

In the skin, a minor thinning of the epidermis and the dermis was observed, and the dermal matrix was less dense. In contrast to the activin-overexpressing mice, however, no obvious abnormalities in keratinocyte proliferation and differentiation were found, indicating that the low levels of activin present in normal adult skin are obviously of minor importance for skin morphogenesis.[16]

Since activin is highly expressed in wounded skin, we expected a significant effect of the overexpressed follistatin on the wound-healing process. Indeed, a thinner hyperproliferative epithelium and particularly a reduced granulation tissue were observed in these animals at day 5 after injury (FIG. 1). These histological abnormalities had

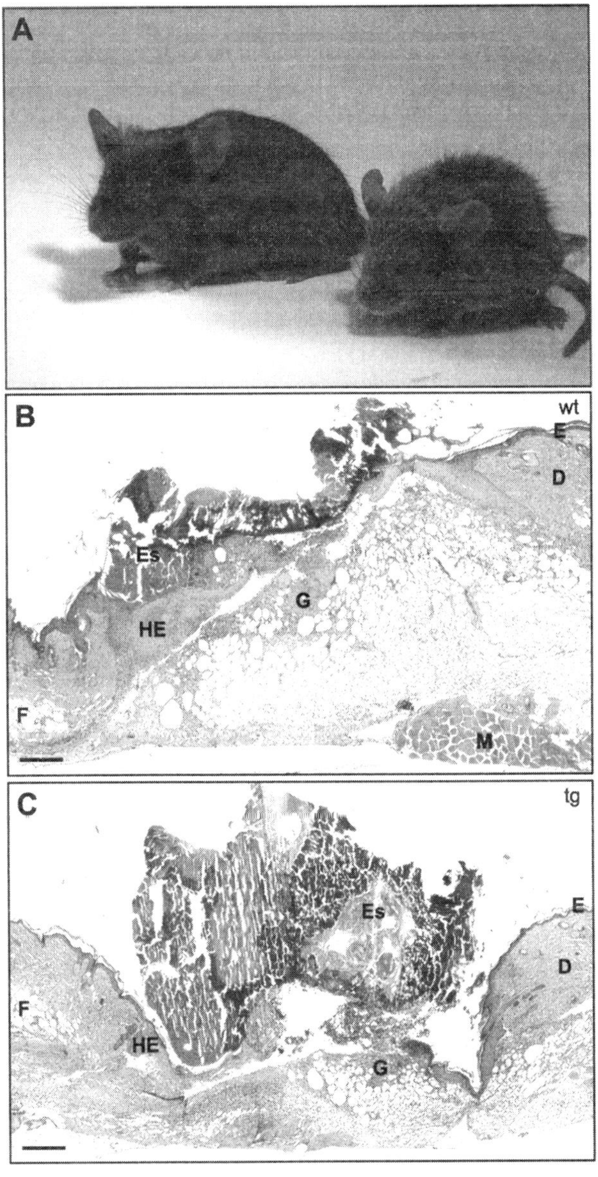

FIGURE 1. Phenotypical abnormalities in transgenic mice overexpressing follistatin in the epidermis. The macroscopic phenotype of the animals is shown in **A**. Note the rough fur and the eye abnormalities in the follistatin-transgenic animal (right-hand side) compared to the control littermate (left-hand side). **(B, C)** Paraffin sections from 5-day-old full-thickness excisional wounds from control mice (B; wt) and follistatin-transgenic mice (C; tg) were stained with hematoxylin/eosin. Note the thinner hyperproliferative epithelium (HE) and the reduced granulation tissue (G) in the transgenic animals. Terms—D: dermis; E: epidermis; Es: eschar; F: fatty tissue. Bars indicate 250 μm.

severe functional consequences as demonstrated by a significant reduction in wound breaking strength at this time point after injury.[16]

In spite of the observed wound-healing abnormalities, the density of blood vessels in the granulation tissue appeared normal in the transgenic mice, but the individual vessels were enlarged. Vascular endothelial growth factor, angiopoietin-1, and angiopoietin-2 are unlikely to be responsible for this effect on the vessels since they were normally expressed in the wounds of the follistatin-overexpressing mice. By contrast, the enlarged vessels could be explained by a direct effect of follistatin on blood vessel morphogenesis since the latter has previously been shown to be angiogenic.[18]

Interestingly, the area of scar tissue was reduced and the epidermis had returned to the normal thickness of unwounded skin. These results suggest that the levels of bioactive activin in a wound determine not only the speed of the repair process, but also the quality of healing.

As described above, the wound-healing phenotype seen in the follistatin-overexpressing mice was complementary to the one observed in the activin-transgenic mice, suggesting that the observed abnormalities in the follistatin-overexpressing mice are indeed a result of the inhibition of endogenous activin action. To further test this hypothesis, we tried to rescue the phenotype by crossing the follistatin-overexpressing mice with the animals that express activin under the control of the same promoter. First results obtained with these animals revealed that the reduced granulation tissue formation observed in the follistatin-transgenic mice was indeed rescued by the elevated levels of activin. This result further supports our hypothesis that the observed wound-healing phenotype is due to the inhibition of activin action by follistatin. By contrast, BMPs seem to play a minor role in the repair phenotype because (i) their expression was slightly downregulated after skin injury and (ii) follistatin binds to activins with much higher affinity than to BMPs.

Interestingly, the phenotype seen in the follistatin-transgenic animals was strikingly different from those lacking Smad3, a major signaling molecule that is activated in response to TGF-β and activin. In the Smad3 knockout mice, wound closure and reepithelialization were strongly enhanced.[19] This discrepancy suggests that Smad3 is not the predominant signal transducer used by activin in keratinocytes and possibly in other cells of the wound. Rather, this molecule seems to be activated in response to TGF-β, which in contrast to activin is a potent inhibitor of keratinocyte proliferation. Activin, on the other hand, appears to use Smad2 for signal transduction in keratinocytes since overexpression of this signaling molecule in the epidermis of transgenic mice caused abnormalities that strongly resembled those seen in our activin-overexpressing mice.[20] Taken together, these recent *in vivo* findings suggest that the different TGF-β family members exert distinct functions in wound repair, which are mediated by different signaling pathways.

In summary, our data suggest that increased levels of activin in the wound enhance the healing rate, but cause increased scarring. On the other hand, inhibition of activin action delays the repair process, but the quality of the healed wound seems to be improved. These results indicate that modification of the levels of bioactive activin could be therapeutically explored for the treatment of wound-healing disorders. Thus, overexpression of activin or inhibition of follistatin might be useful to enhance the speed of repair, whereas inhibition of activin action could be used for

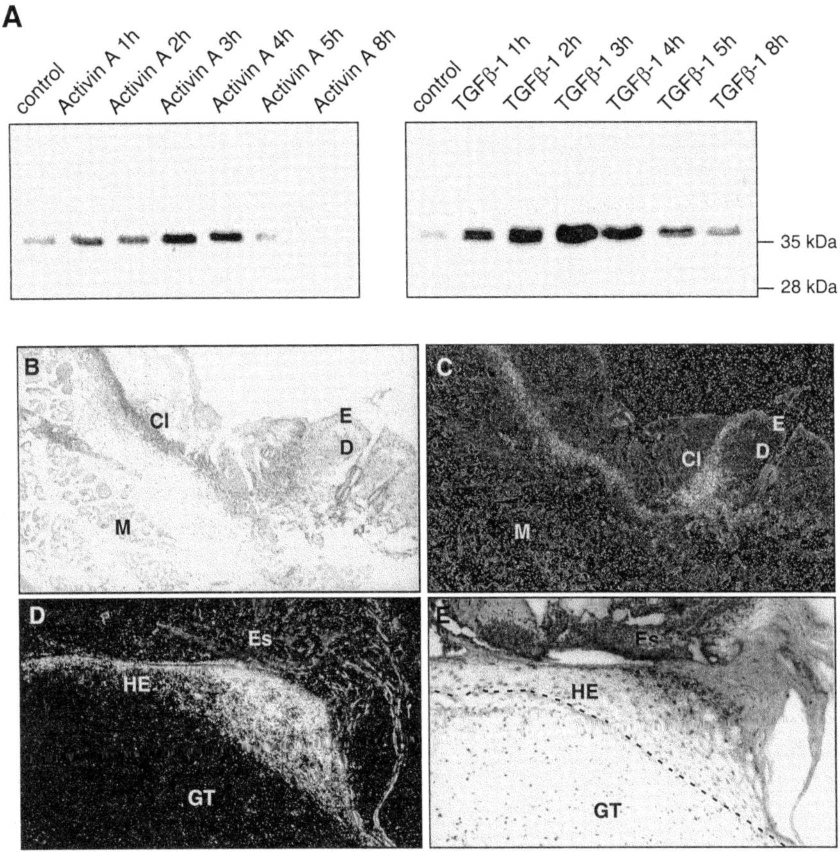

FIGURE 2. (A) Activin A and TGF-β1 induce Mad1 expression in cultured keratinocytes. HaCaT keratinocytes were rendered quiescent by serum starvation and subsequently treated with 18 ng/mL activin A or 1 ng/mL TGF-β1. Total cell lysates were prepared at different time points after addition of activin A or TGF-β1. Then, 120 μg protein was analyzed by Western blotting for the presence of Mad1. Bands were visualized with the ECL detection system. **(B–E)** Expression of *mad1* mRNA in healing wounds. Paraformaldehyde-fixed frozen sections were taken from the middle of 1-day (B, C) and 5-day (D, E) wounds and analyzed by *in situ* hybridization for the presence of *mad1* mRNA using an [α-^{35}S]-labeled mouse *mad1* antisense RNA. Hybridization with the corresponding sense RNA was used as a negative control (data not shown). Signals appear as white dots in the dark field surveys (C, D) and as black dots in the bright field surveys (B, E). Note the presence of *mad1* mRNA in polymorphonuclear leukocytes of 1-day wounds and in suprabasal keratinocytes of the hyperproliferative epithelium of 5-day wounds. Magnifications: ×100 (B, C), ×200 (D, E). Terms—E: epidermis; Cl: clot; HE: hyperproliferative epithelium; GT: granulation tissue; Es: eschar; M: muscle (*Panniculus carnosus*). [Reprinted with permission from ref. 23, © Nature Pub. Group (www.nature.com).]

the prevention of scarring in patients suffering from hypertrophic scars and keloids or other types of fibrotic disease.

MECHANISMS OF ACTIVIN ACTION IN KERATINOCYTES

Having identified an important role of activin in skin morphogenesis and wound repair, we are now interested in its mechanisms of action. As a first step, we are studying the effects of activin on keratinocytes, and we compare these effects with those exerted by TGF-β. For this purpose, we attempted to identify genes that are regulated by these TGF-β family members in cultured keratinocytes. Using the differential display RT-PCR technology, we identified several genes that are only regulated by TGF-β1, including calpactin I[21] and keratin 15.[22] Most importantly, the first activin target gene was recently identified; it encodes the Mad1 transcriptional regulator. We demonstrated a direct regulation of the *mad1* gene by both activin A and TGF-β1 in keratinocytes. The levels of Mad1 mRNA and protein were induced by these growth factors (FIG. 2A). Most importantly, the increased Mad1 expression correlated with increased DNA-binding activity of this protein.[23]

Mad1 is one of four members of the Mad family that act as antagonists of Myc proteins. Their expression is usually associated with growth inhibition and terminal differentiation of various cell types, including epidermal keratinocytes.[24] In the skin, Mad family members are predominantly expressed by suprabasal, differentiated keratinocytes,[25–27] where they seem to play a role in the inhibition of cell proliferation as determined by transfection of a mouse keratinocyte cell line with *mad1*. The colony formation of these transfected cells was strongly reduced.[26] Thus, the increased expression of *mad1* in TGF-β1- or activin A–treated keratinocytes is likely to underlie, at least in part, the growth inhibitory effect of these factors for keratinocytes. Thereby, the stronger induction of *mad1* expression by TGF-β1 correlates with the more potent growth inhibitory effect of the latter in comparison to activin. This inhibitory effect of Mad1 on keratinocyte proliferation is likely to be a prerequisite for subsequent differentiation of these cells.

To determine a possible role of activin and TGF-β1 in the regulation of *mad1* expression *in vivo*, we analyzed its expression during cutaneous wound repair where high levels of these factors are present.[13,28] Indeed, mRNA and protein levels of Mad1 increased significantly after skin injury in a biphasic manner, with maximal expression levels being observed at day 1 and day 5 after injury. *In situ* hybridization revealed that neutrophils are responsible for the first peak and suprabasal keratinocytes of the hyperthickened wound epidermis for the second peak (FIGS. 2B–2E).[23]

The temporal and spatial expression of *mad1* correlates with the presence of activin and TGF-β1 in wounded skin. Thus, TGF-β1, which is released in large amounts from degranulating platelets after injury, could be responsible for the early increase in *mad1* expression. Furthermore, both activin and TGF-β (in particular, TGF-β3) are highly expressed in the *mad1*-positive suprabasal keratinocytes of the hyperproliferative wound epithelium at day 5 after injury.[13,28] These data suggest that activin and TGF-β are indeed involved in the regulation of *mad1* expression *in vivo*. Due to the antimitotic effect of Mad1 for keratinocytes, it seems possible that the presence of this transcription factor in the hyperthickened epidermis of wounds limits the proliferation of keratinocytes under these conditions, a prerequisite for their subsequent differentiation.

ACTIVIN: A GENERAL REGULATOR OF INFLAMMATORY PROCESSES

Besides its role in cutaneous wound repair, recent studies from us and from others have provided evidence for a general role of activin in inflammation and repair. Thus, activin has been shown to promote fracture healing in the rat fibula fracture model.[29] By contrast, it has been identified as an inhibitor of tubular regeneration after renal ischemia in rats,[30] suggesting that activin affects healing processes in a tissue-specific manner. In addition to tissue repair, recent studies suggest a role of activin in various types of inflammatory disease since it is overexpressed in lung fibrosis, liver cirrhosis, arteriosclerosis, and inflammatory arthropathies.[31,32] In addition, we have recently demonstrated a novel role of activin in neuroprotection after lesion to the central nervous system.

ACTIVIN: A NOVEL NEUROPROTECTIVE GROWTH FACTOR

Previous findings from our and other laboratories already pointed to a possible involvement of activin A in the early neuronal response to injury.[33–35] Since activin A promotes survival of midbrain and hippocampal neurons *in vitro*,[36,37] reduces ischemic brain injury in infant rats,[38] and protects striatal and midbrain neurons against neurotoxic damage,[36,39] we speculated that the lesion-induced upregulation of activin A might serve an excitoprotective function. Using an animal model of acute CNS injury, in which the intracerebroventricular injection of the excitotoxin kainic acid (KA) in mouse brain produces selective neuronal loss in the CA3 region of the ipsilateral hippocampus, we found that induction of activin A is essential for the neuroprotective action of FGF-2 *in vivo*.[40] A first indication for a possible interaction between FGF-2 and activin A came from a comparison between the activin β_A mRNA expression pattern in lesioned and in FGF-2-protected hippocampi. In our lesion model, exogenous FGF-2 abrogated the neuronal damage in the CA3 region of the ipsilateral hippocampus. Unexpectedly, FGF-2 also strongly potentiated the KA-induced upregulation of β_A mRNA. Two lines of experimental evidence corroborated the hypothesis that activin A is crucially involved in the neuroprotective effects ascribed to FGF-2. First, recombinant activin A was as effective as exogenous FGF-2 in preventing excitotoxic neuronal loss. Second, the activin-binding protein follistatin, which neutralizes activin *in vitro* and *in vivo*,[7,8] consistently abrogated the beneficial action of FGF-2 in the KA lesion model. Control experiments ruled out that follistatin caused neuronal damage by a mechanism other than its activin-neutralizing action. A different type of interaction between FGF-2 and the TGF-β family was reported from cultured midbrain dopaminergic neurons, in which the neurotrophic effect of FGF-2 was mediated by TGF-β1–3.[41] However, in contrast to activin A, which is of neuronal origin in lesioned hippocampus, the effect of the TGF-βs was mediated by cocultured glial cells.

As evidence accumulates that activin A promotes neuronal survival after acute brain injury, development of substances that influence activin expression or receptor binding should offer new strategies to fight neuronal loss in stroke and brain trauma. Also, the identification of the molecular targets of activin, few of which are known at present, should provide new insights into basic mechanisms of neuroprotection.

ACKNOWLEDGMENTS

Miriam Wankell and Silke Werner equally contributed to this work. The work described in this review was supported by the ETH Zürich (to S. Werner), the Swiss National Science Foundation (to S. Werner), the German Ministry for Education and Research (BMBF) (to S. Werner and C. Alzheimer), the Deutsche Forschungsgemeinschaft (to S. Werner and C. Alzheimer), and a Boehringer Ingelheim predoctoral fellowship (to M. Wankell).

REFERENCES

1. MASSAGUE, J. 1990. The transforming growth factor-β family. Annu. Rev. Cell. Biochem. **6:** 597–641.
2. ROBERTS, A.B. & M.B. SPORN. 1996. Transforming growth factor-β. *In* The Molecular and Cellular Biology of Wound Repair. Second ed., pp. 275–308. Plenum. New York.
3. O'KANE, S. & M.W. FERGUSON. 1997. Transforming growth factor betas and wound healing. Int. J. Biochem. Cell Biol. **29:** 63–78.
4. MATHEWS, L.S. & W.W. VALE. 1993. Molecular and functional characterization of activin receptors. Receptor **3:** 173–181.
5. NAKAMURA, T., K. TAKIO, Y. ETO *et al.* 1990. Activin-binding protein from rat ovary is follistatin. Science **247:** 836–838.
6. DE WINTER, J.P., P. TEN DIJKE, C.J.M. DE VRIES *et al.* 1996. Follistatins neutralize activin bioactivity by inhibition of activin binding to its type II receptors. Mol. Cell. Endocrinol. **116:** 105–114.
7. SUGINO, H., K. SUGINO, O. HASHIMOTO *et al.* 1997. Follistatin and its role as an activin-binding protein. J. Med. Invest. **44:** 1–14.
8. IEMURA, S., T.S. YAMAMOTO, C. TAKAGI *et al.* 1998. Direct binding of follistatin to a complex of bone-morphogenetic protein and its receptor inhibits ventral and epidermal cell fates in early *Xenopus* embryo. Proc. Natl. Acad. Sci. USA **95:** 9337–9342.
9. MATZUK, M.M., T.R. KUMAR, A. VASSALLI *et al.* 1995. Functional analysis of activins during mammalian development. Nature **374:** 354–356.
10. MATZUK, M.M., N. LU, H. VOGEL *et al.* 1995. Multiple defects and perinatal death in mice deficient in follistatin. Nature **374:** 360–363.
11. SHIMIZU, A., M. KATO, A. NAKAO *et al.* 1998. Identification of receptors and Smad proteins involved in activin signalling in a human epidermal keratinocyte cell line. Genes Cells **3:** 125–134.
12. SEISHIMA, M., M. NOJIRI, C. ESAKI *et al.* 1999. Activin A induces terminal differentiation of cultured human keratinocytes. J. Invest. Dermatol. **112:** 432–436.
13. HÜBNER, G., Q. HU, H. SMOLA *et al.* 1996. Strong induction of activin expression after injury suggests an important role of activin in wound repair. Dev. Biol. **173:** 490–498.
14. HÜBNER, G. & S. WERNER. 1996. Serum growth factors and pro-inflammatory cytokines are potent inducers of activin expression in cultured fibroblasts and keratinocytes. Exp. Cell Res. **228:** 103–113.
15. MUNZ, B., H. SMOLA, F. ENGELHARDT *et al.* 1999. Overexpression of activin in the epidermis of transgenic mice reveals new activities of activin in keratinocyte differentiation, cutaneous fibrosis, and wound repair. EMBO J. **18:** 5205–5215.

16. WANKELL, M., B. MUNZ, G. HÜBNER *et al.* 2001. Impaired wound healing in transgenic mice overexpressing the activin antagonist follistatin in the epidermis. EMBO J. **20:** in press.
17. SCHREWE, H., M. GENDRON-MAGUIRE, M.L. HARBISON *et al.* 1994. Mice homozygous for a null mutation of activin beta B are viable and fertile. Mech. Dev. **47:** 43–51.
18. KOZIAN, D.H., M. ZICHE & H.G. AUGUSTIN. 1997. The activin-binding protein follistatin regulates autocrine endothelial cell activity and induces angiogenesis. Lab. Invest. **76:** 267–276.
19. ASHCROFT, G.S., X. YANG, A.B. GLICK *et al.* 1999. Mice lacking Smad3 show accelerated wound healing and an impaired local inflammatory response. Nat. Cell Biol. **1:** 260–266.
20. ITO, Y., P. SARKAR, M. QUINGLI *et al.* 2001. Overexpression of Smad2 reveals its concerted action with Smad4 in regulating TGF-β-mediated epidermal homeostasis. Dev. Biol. **236:** 181–194.
21. MUNZ, B., V. GERKE, R. GILLITZER *et al.* 1997. Differential regulation of annexin II and p11 in cultured keratinocytes and during wound healing. J. Invest. Dermatol. **108:** 307–312.
22. WERNER, S., S. WERNER & B. MUNZ. 2000. Suppression of keratin 15 expression by transforming growth factor-β *in vitro* and by cutaneous injury *in vivo*. Exp. Cell Res. **254:** 80–90.
23. WERNER, S., H-D. BEER, C. MAUCH *et al.* 2001. The Mad1 transcription factor is a novel target of activin and TGF-β action in keratinocytes: possible role of Mad1 in wound repair and psoriasis. Oncogene **20:** 7494–7504.
24. GRANDORI, C., S.M. COWLEY, L.P. JAMES *et al.* 2000. The Myc/Max/Mad network and the transcriptional control of cell behaviour. Annu. Rev. Cell Dev. Biol. **16:** 653–699.
25. HURLIN, P.J., K.P. FOLEY, D.E. AYER *et al.* 1995. Regulation of Myc and Mad during epidermal differentiation and HPV-associated tumorigenesis. Oncogene **11:** 2487–2501.
26. VÄSTRIK, I., A. KAIPAINEN, T.L. PENTTILÄ *et al.* 1995. Expression of the mad gene during cell differentiation *in vivo* and its inhibition of cell growth *in vitro*. J. Cell Biol. **128:** 1197–1208.
27. LYMBOUSSAKI, A., A. KAIPAINEN, E. HATVA *et al.* 1996. Expression of Mad, an antagonist of Myc oncoprotein function, in differentiating keratinocytes during tumorigenesis of the skin. Br. J. Cancer **73:** 1347–1355.
28. FRANK, S., M. MADLENER & S. WERNER. 1996. Transforming growth factors β1, β2, and β3 are differentially regulated by glucocorticoids *in vitro* and during wound healing. J. Biol. Chem. **271:** 10188–10193.
29. SAKAI, R., M. MIWA & Y. ETO. 1999. Local administration of activin promotes fracture healing in the rat fibula fracture model. Bone **25:** 191–196.
30. MAESHIMA, A., Y.Q. ZHANG, Y. NOJIMA *et al.* 2001. Involvement of the activin-follistatin system in tubular regeneration after renal ischemia in rats. J. Am. Soc. Nephrol. **12:** 1685–1695.
31. HÜBNER, G., C. ALZHEIMER & S. WERNER. 1999. Activin: a novel player in tissue repair processes. Histol. Histopathol. **14:** 295–304.
32. MUNZ, B., G. HÜBNER, Y. TRETTER *et al.* 1999. A novel role of activin in inflammation and repair. J. Endocrinol. **161:** 187–193.
33. TRETTER, Y.P., B. MUNZ, G. HÜBNER *et al.* 1996. Strong induction of activin expression after hippocampal lesion. Neuroreport **7:** 1819–1823.
34. LAI, M., E. SIRIMANNE, C.E. WILLIAMS *et al.* 1996. Sequential patterns of inhibin subunit gene expression following hypoxic-ischemic injury in the rat brain. Neuroscience **70:** 1013–1024.
35. LAI, M., P. GLUCKMAN, M. DRAGUNOW *et al.* 1997. Focal brain injury increases activin βA mRNA expression in hippocampal neurons. Neuroreport **8:** 2691–2694.
36. KRIEGLSTEIN, K., C. SUTER-CRAZZOLARA, W.H. FISCHER *et al.* 1995. TGF-beta superfamily members promote survival of midbrain dopaminergic neurons and protect them against MPP+ toxicity. EMBO J. **14:** 736–742.
37. IWAHORE, Y., H. SAITO, K. TORIJ *et al.* 1997. Activin exerts a neurotrophic effect on cultured hippocampal neurons. Brain Res. **760:** 52–58.
38. WU, D.D., M. LAI, P.E. HUGHES *et al.* 1999. Expression of the activin axis and neuronal rescue effects of recombinant activin A following hypoxic-ischemic brain injury in the infant rat. Brain Res. **835:** 369–378.

39. Hughes, P.E., T. Alexi, C.E. Williams *et al.* 1999. Administration of recombinant human activin-A has powerful neurotrophic effects on select striatal phenotypes in the quinolinic acid lesion model of Huntington's disease. Neuroscience **92:** 197–209.
40. Tretter, Y.P., M. Hertel, B. Munz *et al.* 2000. Induction of activin A is essential for the neuroprotective action of basic fibroblast growth factor *in vivo*. Nat. Med. **6:** 812–815.
41. Krieglstein, K., B. Reuss, D. Maysinger *et al.* 1998. Transforming growth factor-beta mediates the neurotrophic effect of fibroblast growth factor-2 on midbrain dopaminergic neurons. Eur. J. Neurosci. **10:** 2746–2750.

Tissue Growth and Remodeling of the Embryonic and Adult Adrenal Gland

MICHELLE L. BLAND,[a] MARION DESCLOZEAUX,[b]
AND HOLLY A. INGRAHAM[b]

[a]Biomedical Sciences Graduate Program, [b]Department of Physiology, University of California, San Francisco, San Francisco, California 94143-0444, USA

ABSTRACT: The adrenal gland provides a model system for the study of tissue remodeling in endocrine physiology. For example, proper adrenal development requires proliferation of the adrenogonadal primordia, separation of adrenal and gonadal precursors, and cell migration that unites the adrenal cortex and adrenal medulla. In the adult, normal adrenal function is assured by the adrenal gland's unique capacity for growth in response to both tissue injury and physiological demand. Identification of the molecular and genetic programs underlying tissue remodeling in the adrenal is important for understanding basic aspects of development and regeneration, as well as adrenal disease. Here, we will highlight the roles that nuclear receptors and pituitary hormones play in regulating fetal adrenal development and adult adrenal growth. In addition, we will review the most current data on how extracellular signaling pathways are coupled to the function of these important regulators of adrenal development and function.

KEYWORDS: adrenal; tissue remodeling; SF-1; Dax1; orphan nuclear receptor; MAP kinase; ACTH

TISSUE REMODELING IN ADRENAL DEVELOPMENT

Hormone secretion from the adrenal glands mediates the body's response to stress. The adrenals are composed of two distinct tissues, the cortex and medulla, that respectively synthesize glucocorticoids and catecholamines. The adrenal cortex and gonads arise from a common population of precursor cells derived from the coelomic epithelium of the intermediate mesoderm at approximately embryonic day (E) 9.0 in the mouse. Adrenal cortical precursors separate from gonadal precursors at E10.5. From E12.5 to E13.5, neural crest cells migrate to the growing adrenal cortex and infiltrate it to form the medulla. The adrenal cortex is encapsulated by E14.5 and, by E16.5, distinct cortical and medullary zones have formed, with the cortex surrounding

Address for correspondence: Holly A. Ingraham, Department of Physiology, University of California, San Francisco, Box 0444, San Francisco, CA 94143-0444. Voice: 415-476-2731; fax: 415-476-4929.
hollyi@itsa.ucsf.edu

Ann. N.Y. Acad. Sci. 995: 59–72 (2003).

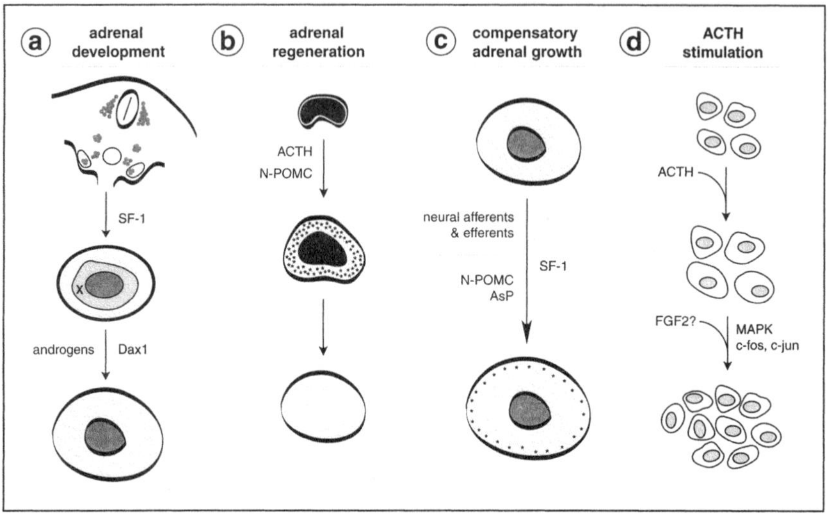

FIGURE 1. Types of tissue remodeling in fetal and adult adrenal growth. **(a)** During development, neural crest cells (gray) migrate to the developing cortex (light gray). SF-1 promotes survival and growth of the adrenal cortex as well as normal cortical-medullary interactions in a dose-dependent manner. After birth, Dax1 and androgens are required for degeneration of the fetal zone (X) to produce the structure of the mature adrenal gland. **(b)** Enucleation of the adrenal gland leaves a capsule (black) lined by the remaining cortical cells (light gray) as well as a large blood clot (dark gray). ACTH and N-terminal POMC peptides are thought to induce cellular proliferation (asterisks) during adrenal regeneration. Three weeks after enucleation in rats, the regenerated adrenal cortex has full functional capacity. **(c)** Removal of one adrenal results in compensatory growth of the remaining adrenal via cellular proliferation (asterisks) in the adrenal cortex. Afferent and efferent neural signals regulate this response. N-terminal POMC peptides are processed in the adrenal by adrenal secretory protease (AsP) to produce the adrenal growth factor, N-POMC 1–52. SF-1 is required for compensatory adrenal growth in a dose-dependent manner. **(d)** Elevated ACTH levels produce a hypertrophic response in Y1 cells *in vitro* and in adrenocortical cells *in vivo*. Prolonged ACTH administration results in cellular proliferation, which may be a direct or indirect effect of ACTH. For example, ACTH may induce expression of growth factors such as FGF2 that promote proliferation via MAPK, c-fos, and c-jun.

the medulla.[1] In mice and humans, the fetal adrenal cortex is composed of a definitive (adult) zone that surrounds a fetal zone. After birth, cell proliferation in the definitive zone and degeneration of cells in the fetal zone lead to the structure of the adult cortex (FIG. 1a). In mice, degeneration of the fetal zone occurs at puberty in males and after the first pregnancy in females.[2] The definitive adrenal cortex is composed of four distinct, concentric zones that synthesize different hormones. The zona glomerulosa is situated directly under the capsule and synthesizes the mineralocorticoid aldosterone. The zona intermedia lies between the zona glomerulosa and the zona fasciculata and does not appear to synthesize hormones, based on lack of expression of enzymes that carry out late steps of steroidogenesis. The zona fasciculata and the zona reticularis lie between the zona intermedia and the adrenal medulla. These zones synthesize the glucocorticoid corticosterone (in rats and mice; cortisol in humans). In rats, the adult pattern of adrenocortical zonation occurs during late stages of fetal development (E18 – postnatal day 1; gestation is 21 days).[3]

The nuclear receptor superfamily of transcription factors directs development of multiple organ systems by regulating expression of genes essential for proliferation and differentiation. Nuclear receptors are activated by small molecule ligands and bind to response elements in promoters of target genes to activate transcription. A class of nuclear receptors designated "orphans" function in the same manner; however, their activating ligands have not yet been identified. The orphan nuclear receptor steroidogenic factor-1 (SF-1) is an essential regulator of adrenal development. *SF-1* knockout mice lack adrenals, gonads, pituitary gonadotropes, and the ventromedial hypothalamus and die soon after birth due to adrenal insufficiency.[4–7] Apoptosis in the developing adrenals and gonads at E12.0 results in complete organ agenesis in *SF-1* knockout mice.[7] The essential target genes regulated by SF-1 in early adrenal development are currently unknown.

To date, three human patients are known to have loss-of-function mutations in *SF-1* associated with adrenal insufficiency; two of these patients are XY and also exhibit XY sex reversal.[8–10] Analysis of these mutations illustrates that SF-1 functions in a dose-dependent manner in human endocrine development. In two patients, the mutations are heterozygous: the R255L mutant is transcriptionally inactive, while the G35E mutation impairs SF-1's ability to bind DNA.[8,10] The third mutation discovered, R92Q, also decreases SF-1's ability to bind DNA, but less so than the G35E mutation. Consistent with this milder phenotype, persons heterozygous for the R92Q mutation do not present with adrenal phenotypes, while a homozygous patient exhibits adrenal insufficiency and XY sex reversal.[9]

We found that SF-1 also functions in a dose-dependent manner in mice. *SF-1* heterozygous mice, although viable and fertile, exhibit profound defects in adrenal growth and function.[11] *SF-1* haploinsufficiency results in an early embryonic defect in the size of the adrenal primordium, such that at E12.5 the cortex is eight times smaller in *SF-1* heterozygotes compared with wild-type mice. The small size of the *SF-1* heterozygous adrenal cortex does not result from an increase in apoptosis. Moreover, cell proliferation is increased in *SF-1* heterozygous adrenals during development (Bland and Ingraham, unpublished results). However, increased proliferation can only partially compensate for the early growth disadvantage faced by *SF-1* heterozygous adrenals, given that adult *SF-1* heterozygous adrenals are two to three times smaller than wild-type adrenals. Additionally, depending on the background strain and sex of the mice, *SF-1* heterozygotes have an anatomically eccentric adrenal medulla that is not surrounded by the cortex (FIG. 2).[11] These results demonstrate that SF-1 functions both non-cell autonomously and in a dose-dependent manner during adrenal development.

Impaired adrenal development in *SF-1* heterozygotes leads to compromised adrenal function in these mice. The adrenal glands are required for physiological adaptation to stressors such as starvation, injury, and infection via secretion of the cortical hormones, aldosterone and corticosterone, and the medullary hormones, epinephrine and norepinephrine. Adrenal hormones regulate electrolyte balance, gluconeogenesis, lipolysis, thymocyte development, inflammation, and blood pressure. The adrenocortical stress response is initiated by the hypothalamic neuropeptide corticotropin-releasing hormone (CRH), which stimulates pituitary secretion of the

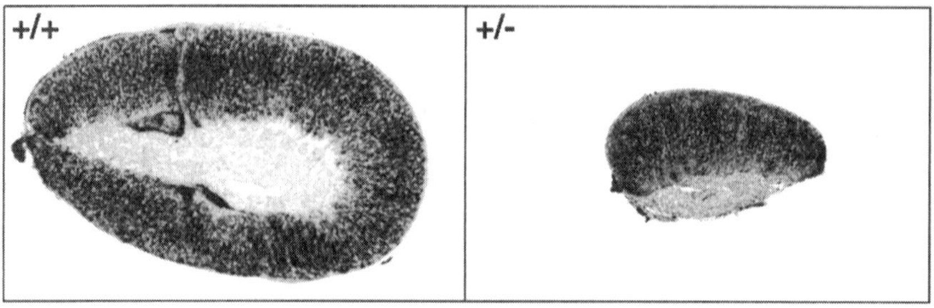

FIGURE 2. Oil red O staining of *SF-1* wild-type and heterozygous adult adrenal glands. In wild-type mice, the adrenal cortex (dark staining) surrounds the adrenal medulla (unstained). In male *SF-1* heterozygous mice on a C57BL/6 × FVB background, the adrenal medulla is located eccentrically with respect to the cortex.

pro-opiomelanocortin (POMC) derivative adrenocorticotropic hormone (ACTH). ACTH binds to its receptor in the adrenal cortex and promotes glucocorticoid synthesis and secretion. Glucocorticoids, in addition to mediating physiological responses to stress, inhibit further pituitary ACTH release via negative feedback, thereby terminating the stress response. In response to both acute and chronic stressors, *SF-1* heterozygous mice exhibit elevated ACTH secretion but impaired corticosterone secretion compared with wild-type mice.[11,12] We showed that the physiological consequences of decreased corticosterone secretion in *SF-1* heterozygous mice include reduced glucocorticoid-induced CD4(+), CD8(+) thymocyte apoptosis in the thymus, impaired lipolysis of fat stores, and decreased plasma glucose levels in response to food deprivation. Adrenal medullary function is also impaired in *SF-1* heterozygotes, as revealed by an 80% decrease in adrenal content of both norepinephrine and epinephrine compared with wild-type mice.[11]

Work by several groups has established that SF-1 regulates a number of genes essential for steroidogenesis.[13] Since transcript and protein levels of SF-1 are decreased in *SF-1* heterozygous adrenals, it was predicted that SF-1 target gene expression would also be reduced, thereby decreasing the ability of *SF-1* heterozygous adrenals to synthesize corticosterone.[11] Unexpectedly, *SF-1* heterozygous adrenals were found to have increased expression of SF-1 target genes such as the steroidogenic acute regulatory (StAR) protein and the ACTH receptor, despite decreased levels of SF-1.[11,12] Upregulation of these SF-1 target genes presumably allows *SF-1* heterozygous adrenals to compensate for their small size by increasing steroidogenic capacity on a per cell basis, although this hypothesis remains to be tested directly. It seems likely that the high basal levels of ACTH observed in SF-1 heterozygotes (presumably due to decreased negative feedback by corticosterone) are required to elevate adrenal expression of SF-1 target genes.[11,12] Indeed, dexamethasone treatment to block pituitary ACTH secretion results in equivalent levels of SF-1 target gene expression in *SF-1* wild-type and heterozygous adrenals (Bland and Ingraham, unpublished results). Increased expression of SF-1 target genes in *SF-1* heterozygous adrenals could be achieved through altered levels of transcription factors other than SF-1, increased availability of a putative SF-1 ligand, or increased SF-1 activity due to posttranslational modifications (discussed below).

One transcription factor that could modulate SF-1 target gene expression in *SF-1* heterozygotes is the orphan nuclear receptor Dax1. Dax1 and SF-1 overlap in their expression patterns, SF-1 regulates expression of *Dax1*, and Dax1 represses SF-1 activity at a number of promoters *in vitro*.[14–18] Genetic evidence in humans and mice has established that *Dax1* is an important regulator of adrenal development. In humans, loss-of-function mutations in *Dax1* lead to hypogonadotropic hypogonadism and adrenal hypoplasia congenita, in which the adult adrenal lacks a definitive zone and contains a large fetal zone that does not degenerate. Patients with *Dax1* mutations suffer from severe adrenocortical insufficiency and must be treated with exogenous glucocorticoids throughout their lives.[19–21] In mice, deletion of *Dax1* results in adrenals that contain both a definitive zone and a permanent fetal zone.[22] In contrast to the severe adrenal insufficiency associated with loss-of-function *Dax1* mutations in humans, the presence of a definitive zone preserves adrenal function in *Dax1* knockout mice. In fact, these mice exhibit increased corticosterone responses to ACTH stimulation compared with wild-type mice.[12] Although Dax1 has been shown *in vitro* to repress SF-1 activity, loss of Dax1 function does not rescue the hypoplasia and adrenal insufficiency observed in *SF-1* heterozygous mice. Furthermore, many SF-1 target genes are not upregulated in *Dax1* null mice.[12] These findings underscore the complex nature of the interaction between SF-1 and Dax1 *in vivo*.

The developmental phenotype of *Dax1* knockout adrenals is especially interesting when the Dax1 expression pattern is considered. In the mouse adrenal, Dax1 is expressed throughout the cortex in females and is restricted to the zona glomerulosa in males. This expression pattern is regulated by androgens: testosterone treatment recapitulates the male Dax1 expression pattern in female adrenals, and gonadectomy plus estradiol result in a female expression pattern of Dax1 in male adrenals.[23] *Dax1* knockout mice have increased testes aromatase activity that results in high levels of testicular estradiol and increased plasma estradiol. Plasma testosterone is not significantly decreased in these mice.[24] Since testosterone promotes and gonadectomy plus estradiol prevent fetal zone degeneration, increased circulating estradiol might repress fetal zone degeneration in *Dax1* knockouts; this hypothesis remains to be tested.

In addition to Dax1, another orphan nuclear receptor, nerve growth factor I-B (NGFI-B, Nur77), is expressed in the adrenal and may regulate the expression of steroidogenic enzymes such as P450c21.[25–27] ACTH treatment results in rapid induction of NGFI-B expression followed by steroidogenic enzyme synthesis, suggesting that NGFI-B might mediate ACTH-responsive gene expression.[28–30] However, NGFI-B is not essential for adrenal function since *NGFI-B* knockout mice exhibit normal adrenals, although the lack of an overt adrenal phenotype may be due to compensation by SF-1 and/or other nuclear receptors such as Nor1.[25,30]

TISSUE REMODELING IN THE ADULT ADRENAL

In the adult adrenal, three distinct types of growth have been studied: the ability of the adrenal cortex to regenerate, the compensatory adrenal growth response to removal of one adrenal, and cellular growth responses to ACTH. Like the liver, the adrenal cortex is capable of complete functional regeneration when damaged.[31] After bilateral adrenal enucleation, cortical cells lining the interior of the capsule remain and a large blood clot forms in the center of the gland (FIG. 1b). Dedifferentiation of these remaining subcapsular cortical cells is thought to play an important role in adrenal regeneration. Two to ten days after enucleation in rats, subcapsular cells lose expression of the zona glomerulosa marker, aldosterone synthase, and assume a zona intermedia phenotype. These dedifferentiated cells proliferate extensively so that three weeks after bilateral adrenal enucleation in rats, the adrenal cortex has regenerated, subcapsular cells again express aldosterone synthase, and adrenocortical responses to stress are relatively normal.[32,33] The medulla does not regenerate, although some chromaffin cells remain and survive in enucleated adrenals.[34]

Derivatives of the anterior- and intermediate-lobe pituitary hormone precursor POMC, such as ACTH and pro-γ-MSH (N-POMC 1–77), are thought to be important mediators of adrenal regeneration. Bilateral adrenal enucleation results in a sharp decrease in plasma corticosterone levels and a concomitant increase in plasma ACTH levels due to decreased negative feedback control by corticosterone. Treatment of bilaterally enucleated rats with corticosterone suppresses ACTH secretion and prevents dedifferentiation of subcapsular cortical cells and adrenal regeneration, suggesting that ACTH is required for adrenal regeneration.[32,35] Pro-γ-MSH may also play an important role in adrenal regeneration. Plasma levels of pro-γ-MSH are increased following bilateral adrenal enucleation, and immunoneutralization of pro-γ-MSH results in decreased adrenal mitogenesis at 14 and 21 days of adrenal regeneration.[36,37]

Removal of one adrenal results in proliferation of cells in the zona glomerulosa and outer zona fasciculata of the remaining gland, a phenomenon known as compensatory adrenal growth (CAG) (FIG. 1c). CAG is neurally mediated as lesions of the ventromedial hypothalamus (VMH) ipsilateral to the removed adrenal or hemisection of the spinal cord contralateral to the removed adrenal block the response.[38] A full dose of *SF-1* is required for a normal compensatory adrenal growth response. *SF-1* heterozygous adrenocortical cells fail to proliferate following removal of one adrenal, while wild-type adrenals exhibit normal CAG.[39] The failure of *SF-1* heterozygous adrenals to grow in response to unilateral adrenalectomy is intriguing, and the underlying defect is unknown. For example, *SF-1* heterozygous adrenals may lack cortical stem cells, or *SF-1* heterozygous cortical cells may be incapable of reentering the cell cycle. The VMH is involved in compensatory adrenal growth, and this hypothalamic structure fails to develop normally in *SF-1* knockout mice.[4,40–42] Although the structure of the VMH is thought to be relatively normal in *SF-1* heterozygotes, it is possible that subtle defects in VMH function might prevent a normal CAG response.[39,40]

Alternatively, SF-1 may be an essential component of mitogenic signaling by N-terminal POMC peptides that promote CAG. Immunoneutralization studies demonstrated that pro-γ-MSH from the pituitary positively regulates CAG, but only after local proteolysis produces a shorter form that has mitogenic activity in the adrenal.[43] Recently, Bicknell *et al.* identified an extracellular serine protease expressed in the adrenal cortex that cleaves circulating pro-γ-MSH to produce the adrenal mitogen N-POMC 1–52. Following unilateral adrenalectomy, this enzyme, adrenal secretory protease (AsP), is upregulated in the zona glomerulosa and outer zona fasciculata of the remaining adrenal.[44] This increase in AsP expression is not observed in *SF-1* heterozygous adrenals during CAG.[39] It will be of interest to determine whether AsP or the unidentified N-POMC 1–52 receptor is a target gene of SF-1 or whether SF-1 is required for N-POMC 1–52 signaling.

PEPTIDE HORMONE AND GROWTH FACTOR SIGNALING IN ADRENOCORTICAL CELLS

It is well established that pituitary hormones are needed for maintenance of adrenal weight. High levels of the pituitary hormone ACTH produce adrenocortical cellular hypertrophy *in vivo* (FIG. 1d). This effect is observed in rats and mice treated with ACTH and in genetic models that have elevated ACTH, such as *glucocorticoid receptor* and *dopamine D2 receptor* knockout mice and *SF-1* heterozygous mice.[11,12,35,45,46] Prolonged treatment with ACTH also leads to proliferation of adrenocortical cells, although it is not clear whether this is a direct effect of ACTH or secondary to hypertrophy and/or activation of some unknown growth factor.[35] Conversely, low ACTH levels result in adrenal atrophy as seen in *CRH*, *CRH receptor 1*, and *POMC* knockout mice.[47–50] Accordingly, administration of dexamethasone to block ACTH secretion in *SF-1* wild-type and heterozygous mice results in adrenal cell hypotrophy (Bland and Ingraham, unpublished results).

The signaling pathways that regulate cell size and cell number in response to ACTH stimulation are complex and a focus of much investigation. Over the last two decades, biochemical and cellular studies have established that some of the same signals that regulate steroidogenesis also markedly affect growth of adrenocortical cells. However, while stimulation of steroidogenesis by ACTH is well established, ACTH-induced growth in model adrenocortical cell lines remains controversial. The two most common cellular models used to study adrenal cortex growth are the mouse Y1 and human NCI-H295 adrenocortical cell lines.[51,52] Other studies have been carried out in isolated primary adrenal cortical cells. ACTH inhibits proliferation in both Y1 and primary adrenocortical cells.[35,53,54] Indeed, this observation is consistent with the fact that ACTH blocks progression of adrenocortical cells through the G_1 phase of the cell cycle and inhibits DNA synthesis in S phase, even in the continued presence of serum or growth factors.[55] Despite these inhibitory effects on growth, ACTH induces expression of genes associated with cell cycle progression such as *c-fos* and *c-jun* in Y1 cells.[56,57] This paradox was partially explained by studies showing that short ACTH pulses prior to the G_1 checkpoint induce cell cycle progression by promoting the transition from G_1 to S phase; this results in increased cell division.[58,59] In contrast, prolonged administration of ACTH during G_1 phase arrests cell growth by blocking the transition from G_1 to S phase.[59] These data illustrate that the underlying mechanisms of ACTH-induced adrenocortical proliferation are still incompletely understood and are likely to involve cross-talk among multiple signaling pathways.

Another factor important for adrenocortical growth is fibroblast growth factor 2 (FGF2). This polypeptide growth factor belongs to the large FGF family that regulates tissue remodeling in wound healing, angiogenesis, and bone repair.[60–62] FGF2 triggers progression of G_0 arrested fibroblasts to the G_1 and S phases.[63,64] Furthermore, FGF2 may be a target of ACTH signaling as evidenced by ACTH-induced upregulation of FGF2 transcripts in human fetal adrenal cortex.[65] Interestingly, as with ACTH, FGF2 also induces c-fos and c-jun.[66] Induction of these two intermediate regulators appears to be an essential aspect of both ACTH- and FGF2-induced adrenocortical cell proliferation since blocking c-fos or c-jun expression using antisense RNA inhibition attenuates Y1 cell proliferation.[67] Thus, ACTH and FGF2 employ common intracellular signals that promote cell cycle progression in adrenocortical cells.

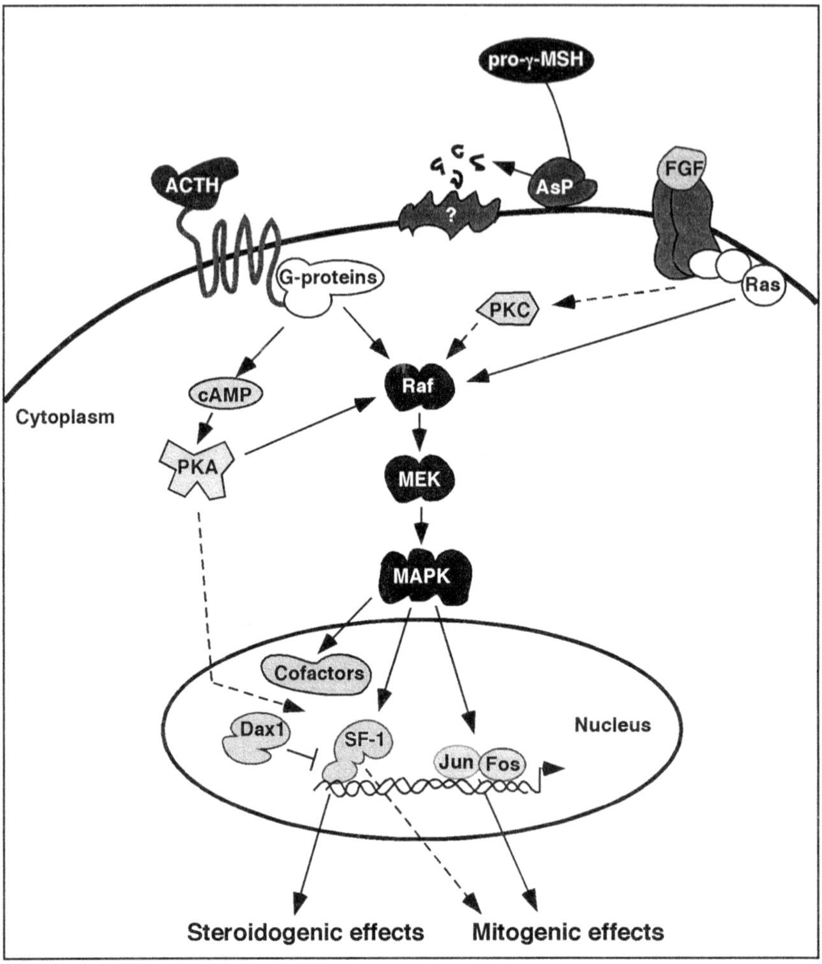

FIGURE 3. Model for signaling cross-talk in adrenal growth and function. A schematic representation is shown for both the known and postulated intracellular signaling pathways that are involved in regulating adrenocortical cell growth. ACTH and FGF2 bind to their cognate receptors and induce growth by activating the MAPK cascade as well as expression of c-fos and c-jun. ACTH also induces steroidogenesis by activating the cAMP-PKA pathway. *In vivo* and *in vitro* data show that SF-1 is phosphorylated by MAPK and positively regulates adrenal growth and steroidogenesis. SF-1 cofactors are also phosphorylated as a result of PKA and MAPK activation and can result in enhanced transcription of SF-1 target genes. The N-terminal POMC peptide, pro-γ-MSH, undergoes processing by the serine protease AsP to produce the adrenal growth factor, N-POMC 1–52; a receptor for this peptide has yet to be identified. Direct links between different signaling pathways that have been experimentally demonstrated are indicated with solid arrows, while those that remain to be firmly established are shown with dashed arrows.

INTEGRATION OF PEPTIDE HORMONE AND GROWTH FACTOR SIGNALING VIA THE MAPK CASCADE

Growth factors such as FGF2 bind their cognate tyrosine kinase receptors and transduce signals via the mitogen-activated protein kinase (MAPK) pathway. In addition, it is established that G-protein coupled receptors are capable of activating the MAPK pathway. For instance, binding of ACTH to its cognate G-protein coupled receptor (MC2R) activates not only PKA, but also Ras-Raf-MEK, which leads to phosphorylation of p42/44 MAPK (Erk1/2).[68] Inhibitors of MEK block both FGF2 and ACTH signaling pathways in Y1 cells.[68,69] The exact molecular cross-talk between G-protein coupled receptors and tyrosine kinase receptor signaling pathways is not fully understood and is likely to involve other protein kinase cascades.[70,71] For instance, both cAMP and PKC signaling pathways have been shown to activate MAPK.[72,73] In the adrenal, stimulation of steroidogenesis by ACTH requires the cAMP-PKA pathway.[74] However, studies using a PKA-mutant Y1 cell line show that ACTH-induced cell proliferation is independent of PKA activation and possibly dependent on MAPK.[59] Taken together, these experiments suggest that the MAPK signaling pathway may play a central role in modulating adrenal growth in response to extracellular stimulation by peptide hormones and growth factors (FIG. 3).

ORPHAN NUCLEAR RECEPTORS AND THE MAPK SIGNALING PATHWAY IN ADRENAL GROWTH

The essential role of SF-1 in adrenal organogenesis and growth in mice and humans suggests that SF-1 expression levels and/or activity may be the target of growth factor and peptide hormone signaling. Intracellular signaling pathways have been shown to regulate SF-1 activity, directly or indirectly. SF-1 mediates cAMP-induced upregulation of several genes involved in steroidogenesis, but does not appear to be directly phosphorylated by the cAMP-dependent PKA.[75–79] On the other hand, increases in cAMP influence the dynamics of SF-1 cofactor recruitment by direct phosphorylation of these cofactors.[75,78] Previous experiments from our laboratory showed that the MAPK Erk2 (p42 MAPK) phosphorylates SF-1 *in vitro* on a single serine residue (Ser 203) located in the hinge region that connects the DNA binding domain (DBD) to the ligand binding domain (LBD).[80] More recently, using a phospho-specific antibody directed against phosphorylated Ser 203, we have also shown that SF-1 is phosphorylated on Ser 203 in Y1 cells and in adult and embryonic adrenals[81] (Bland and Ingraham, unpublished results). This posttranslational modification is needed for full transcriptional activity of SF-1, most likely because phosphorylation enhances cofactor recruitment. Replacement of Ser 203 with a non-phosphorylatable alanine residue (S203A) results in a reduced basal level of SF-1 transcriptional activation in some cell lines and eliminates the increased activity stimulated by the MAPK pathway. Additionally, transfection of Y1 cells with MKP, a specific inhibitor of MAPK, decreases SF-1 phosphorylation at Ser 203.[81] These data suggest that extracellular signaling events that promote phosphorylation of SF-1 enhance its activity, similar to activation of nuclear receptors by ligand binding.

Structural data have revealed the mechanisms underlying ligand-dependent activation of nuclear receptors. The LBD contains multiple α-helical structural motifs and one β-sheet that together form a ligand binding cavity with restricted molecular volume. Two major structural changes in the LBD are apparent upon ligand binding: first, the hydrophobic ligand binding cavity is greatly compacted when occupied by ligand; second, the position of helix 12 becomes ordered and helps to form a "charged clamp" after ligand binding. These conformational changes result in formation of a hydrophobic cleft or interaction surface that binds to LxxLL motifs present in all nuclear receptor coactivators. Recruitment of coactivators is the first step in forming a multiprotein complex that promotes gene expression (reviewed in refs. 82 and 83).

While structural studies have clarified how ligand-dependent receptors are activated after ligand binding, it is still not clear how orphan receptors, such as SF-1, function in the apparent absence of ligand. Indeed, biochemical and biophysical studies suggest that SF-1 adopts an active conformation in the apparent absence of ligand and thus exhibits many features of ligand-bound receptors.[81] Posttranslational modification in response to extracellular signaling may provide a mechanism for ligand-independent activation of receptors. Pissios *et al.*[84] established that, for ligand-dependent receptors, helix 1 of the LBD assembles in *trans* with the remainder of the LBD (helices 2–12) in a ligand-dependent manner. Using this assay, we have shown that Ser 203 phosphorylation of SF-1 increases helix assembly in the absence of ligand.[81] These data suggest that Ser 203 phosphorylation stabilizes the LBD similar to ligand binding to a ligand-dependent receptor. Whether modification of the hinge region influences the conformation of the LBD helical bundle or alters the small ligand binding pocket predicted by homology modeling of SF-1 is still not understood.[81] Further insights into the underlying function of Ser 203 phosphorylation both *in vitro* and *in vivo* are likely to shed new insights on how extracellular signaling pathways are coupled to SF-1 function in regulating adrenal growth and development.

SUMMARY

In this review, we have described how similar signaling pathways are used in the embryonic and adult adrenal gland when undergoing tissue remodeling. Genetic studies in mice and humans have established that the orphan nuclear receptors SF-1 and Dax1 regulate survival of adrenal precursors, growth of the embryonic adrenal, steroidogenesis, and adult adrenal growth processes such as compensatory adrenal growth. Endocrine signals such as the pituitary hormones ACTH and pro-γ-MSH are important mediators of adrenal growth and function. There is growing evidence that peptide hormone signaling activates orphan nuclear receptors in the adrenal gland. Our recent studies suggest that activation of the MAPK signaling cascade by ACTH is coupled directly to SF-1 activity via a specific phosphorylation event at Ser 203. Defining the nature and mechanism of intracellular signals that mediate growth in the embryonic and adult adrenal should provide new insights into adrenal disease and adrenal cancer.

ACKNOWLEDGMENTS

M. L. Bland and M. Desclozeaux contributed equally to this work.

REFERENCES

1. IKEDA, Y., W.H. SHEN, H.A. INGRAHAM *et al.* 1994. Developmental expression of mouse steroidogenic factor-1, an essential regulator of the steroid hydroxylases. Mol. Endocrinol. **8:** 654–662.
2. DEACON, C.F., W. MOSLEY & I.C. JONES. 1986. The X zone of the mouse adrenal cortex of the Swiss albino strain. Gen. Comp. Endocrinol. **61:** 87–99.
3. WOTUS, C., B.K. LEVAY-YOUNG, L.M. ROGERS *et al.* 1998. Development of adrenal zonation in fetal rats defined by expression of aldosterone synthase and 11beta-hydroxylase. Endocrinology **139:** 4397–4403.
4. SHINODA, K., H. LEI, H. YOSHII *et al.* 1995. Developmental defects of the ventromedial hypothalamic nucleus and pituitary gonadotroph in the Ftz-F1 disrupted mice. Dev. Dyn. **204:** 22–29.
5. SADOVSKY, Y., P.A. CRAWFORD, K.G. WOODSON *et al.* 1995. Mice deficient in the orphan receptor steroidogenic factor 1 lack adrenal glands and gonads, but express P450 side-chain-cleavage enzyme in the placenta and have normal embryonic serum levels of corticosteroids. Proc. Natl. Acad. Sci. USA **92:** 10939–10943.
6. LUO, X., Y. IKEDA, D.A. SCHLOSSER *et al.* 1995. Steroidogenic factor 1 is the essential transcript of the mouse Ftz-F1 gene. Mol. Endocrinol. **9:** 1233–1239.
7. LUO, X., Y. IKEDA & K.L. PARKER. 1994. A cell-specific nuclear receptor is essential for adrenal and gonadal development and sexual differentiation. Cell **77:** 481–490.
8. ACHERMANN, J.C., M. ITO, P.C. HINDMARSH *et al.* 1999. A mutation in the gene encoding steroidogenic factor-1 causes XY sex reversal and adrenal failure in humans. Nat. Genet. **22:** 125–126.
9. ACHERMANN, J.C., G. OZISIK, M. ITO *et al.* 2002. Gonadal determination and adrenal development are regulated by the orphan nuclear receptor steroidogenic factor-1, in a dose-dependent manner. J. Clin. Endocrinol. Metab. **87:** 1829–1833.
10. BIASON-LAUBER, A. & E.J. SCHOENLE. 2000. Apparently normal ovarian differentiation in a prepubertal girl with transcriptionally inactive steroidogenic factor 1 (NR5A1/SF-1) and adrenocortical insufficiency. Am. J. Hum. Genet. **67:** 1563–1568.
11. BLAND, M.L., C.A. JAMIESON, S.F. AKANA *et al.* 2000. Haploinsufficiency of steroidogenic factor-1 in mice disrupts adrenal development leading to an impaired stress response. Proc. Natl. Acad. Sci. USA **97:** 14488–14493.
12. BABU, P.S., D.L. BAVERS, F. BEUSCHLEIN *et al.* 2002. Interaction between Dax-1 and steroidogenic factor-1 *in vivo*: increased adrenal responsiveness to ACTH in the absence of Dax-1. Endocrinology **143:** 665–673.
13. PARKER, K.L. & B.P. SCHIMMER. 1997. Steroidogenic factor 1: a key determinant of endocrine development and function. Endocr. Rev. **18:** 361–377.
14. CRAWFORD, P.A., C. DORN, Y. SADOVSKY *et al.* 1998. Nuclear receptor DAX-1 recruits nuclear receptor corepressor N-CoR to steroidogenic factor 1. Mol. Cell. Biol. **18:** 2949–2956.
15. IKEDA, Y., A. SWAIN, T.J. WEBER *et al.* 1996. Steroidogenic factor 1 and Dax-1 colocalize in multiple cell lineages: potential links in endocrine development. Mol. Endocrinol. **10:** 1261–1272.
16. ITO, M., R. YU & J.L. JAMESON. 1997. DAX-1 inhibits SF-1-mediated transactivation via a carboxy-terminal domain that is deleted in adrenal hypoplasia congenita. Mol. Cell. Biol. **17:** 1476–1483.
17. KAWABE, K., T. SHIKAYAMA, H. TSUBOI *et al.* 1999. Dax-1 as one of the target genes of Ad4BP/SF-1. Mol. Endocrinol. **13:** 1267–1284.
18. NACHTIGAL, M.W., Y. HIROKAWA, D.L. ENYEART-VANHOUTEN *et al.* 1998. Wilms' tumor 1 and Dax-1 modulate the orphan nuclear receptor SF-1 in sex-specific gene expression. Cell **93:** 445–454.
19. MERKE, D.P., T. TAJIMA, J. BARON *et al.* 1999. Hypogonadotropic hypogonadism in a female caused by an X-linked recessive mutation in the DAX1 gene. N. Engl. J. Med. **340:** 1248–1252.
20. ZANARIA, E., F. MUSCATELLI, B. BARDONI *et al.* 1994. An unusual member of the nuclear hormone receptor superfamily responsible for X-linked adrenal hypoplasia congenita. Nature **372:** 635–641.

21. ZHANG, Y.H., W. GUO, R.L. WAGNER *et al.* 1998. DAX1 mutations map to putative structural domains in a deduced three-dimensional model. Am. J. Hum. Genet. **62:** 855–864.
22. YU, R.N., M. ITO, T.L. SAUNDERS *et al.* 1998. Role of Ahch in gonadal development and gametogenesis. Nat. Genet. **20:** 353–357.
23. MUKAI, T., M. KUSAKA, K. KAWABE *et al.* 2002. Sexually dimorphic expression of Dax-1 in the adrenal cortex. Genes Cells **7:** 717–729.
24. WANG, Z.J., B. JEFFS *et al.* 2001. Aromatase (Cyp19) expression is up-regulated by targeted disruption of Dax1. Proc. Natl. Acad. Sci. USA **98:** 7988–7993.
25. CRAWFORD, P.A., Y. SADOVSKY, K. WOODSON *et al.* 1995. Adrenocortical function and regulation of the steroid 21-hydroxylase gene in NGFI-B-deficient mice. Mol. Cell. Biol. **15:** 4331–4316.
26. WILSON, T.E., A.R. MOUW, C.A. WEAVER *et al.* 1993. The orphan nuclear receptor NGFI-B regulates expression of the gene encoding steroid 21-hydroxylase. Mol. Cell. Biol. **13:** 861–868.
27. ZHANG, P. & S.H. MELLON. 1997. Multiple orphan nuclear receptors converge to regulate rat P450c17 gene transcription: novel mechanisms for orphan nuclear receptor action. Mol. Endocrinol. **11:** 891–904.
28. DAVIS, I.J. & L.F. LAU. 1994. Endocrine and neurogenic regulation of the orphan nuclear receptors Nur77 and Nurr-1 in the adrenal glands. Mol. Cell. Biol. **14:** 3469–3483.
29. ENYEART, J.J., R.T. BOYD & J.A. ENYEART. 1996. ACTH and AII differentially stimulate steroid hormone orphan receptor mRNAs in adrenal cortical cells. Mol. Cell. Endocrinol. **124:** 97–110.
30. FERNANDEZ, P.M., F. BRUNEL, M.A. JIMENEZ *et al.* 2000. Nuclear receptors Nor1 and NGFI-B/Nur77 play similar, albeit distinct, roles in the hypothalamo-pituitary-adrenal axis. Endocrinology **141:** 2392–2400.
31. MICHALOPOULO, G.K. & M.C. DEFRANCES. 1997. Liver regeneration. Science **276:** 60–66.
32. ENGELAND, W.C. & B.K. LEVAY-YOUNG. 1999. Changes in the glomerulosa cell phenotype during adrenal regeneration in rats. Am. J. Physiol. **276:** R1374–R1382.
33. WILKINSON, C.W., J. SHINSAKO & M.F. DALLMAN. 1981. Return of pituitary-adrenal function after adrenal enucleation or transplantation: diurnal rhythms and responses to ether. Endocrinology **109:** 162–169.
34. ULRICH-LAI, Y.M. & W.C. ENGELAND. 2000. Hyperinnervation during adrenal regeneration influences the rate of functional recovery. Neuroendocrinology **71:** 107–123.
35. DALLMAN, M.F. 1984. Control of adrenocortical growth *in vivo*. Endocr. Res. **10:** 213–242.
36. ALFANO, J., L.G. DRESSLER & A.C. BROWNIE. 1985. Immunoreactive ACTH and gamma-MSH in rat plasma during early stages of adrenal regeneration. Am. J. Physiol. **248:** E471–E474.
37. ESTIVARIZ, F.E., M.I. MORANO, M. CARINO *et al.* 1988. Adrenal regeneration in the rat is mediated by mitogenic N-terminal pro-opiomelanocortin peptides generated by changes in precursor processing in the anterior pituitary. J. Endocrinol. **116:** 207–216.
38. DALLMAN, M.F., W.C. ENGELAND & M.H. MCBRIDE. 1977. The neural regulation of compensatory adrenal growth. Ann. N.Y. Acad. Sci. **297:** 373–392.
39. BEUSCHLEIN, F., C. MUTCH, D.L. BAVERS *et al.* 2002. Steroidogenic factor-1 is essential for compensatory adrenal growth following unilateral adrenalectomy. Endocrinology **143:** 3122–3135.
40. DELLOVADE, T.L., M. YOUNG, E.P. ROSS *et al.* 2000. Disruption of the gene encoding SF-1 alters the distribution of hypothalamic neuronal phenotypes. J. Comp. Neurol. **423:** 579–589.
41. IKEDA, Y., X. LUO, R. ABBUD *et al.* 1995. The nuclear receptor steroidogenic factor 1 is essential for the formation of the ventromedial hypothalamic nucleus. Mol. Endocrinol. **9:** 478–486.
42. TRAN, P., M. LEE, O. MARIN *et al.* 2003. Requirement of the orphan nuclear receptor SF-1 in terminal differentiation of hypothalamic neurons. MCN. In press.
43. LOWRY, P.J., L. SILAS, C. MCLEAN *et al.* 1983. Pro-gamma-melanocyte-stimulating hormone cleavage in adrenal gland undergoing compensatory growth. Nature **306:** 70–73.
44. BICKNELL, A.B., K. LOMTHAISONG, R.J. WOODS *et al.* 2001. Characterization of a serine protease that cleaves pro-gamma-melanotropin at the adrenal to stimulate growth. Cell **105:** 903–912.

45. COLE, T.J., J.A. BLENDY, A.P. MONAGHAN *et al.* 1995. Targeted disruption of the glucocorticoid receptor gene blocks adrenergic chromaffin cell development and severely retards lung maturation. Genes Dev. **9:** 1608–1621.
46. SAIARDI, A. & E. BORRELLI. 1998. Absence of dopaminergic control on melanotrophs leads to Cushing's-like syndrome in mice. Mol. Endocrinol. **12:** 1133–1139.
47. MUGLIA, L., L. JACOBSON, P. DIKKES *et al.* 1995. Corticotropin-releasing hormone deficiency reveals major fetal, but not adult glucocorticoid need. Nature **373:** 427–432.
48. SMITH, G.W., J.M. AUBRY, F. DELLU *et al.* 1998. Corticotropin releasing factor receptor 1–deficient mice display decreased anxiety, impaired stress response, and aberrant neuroendocrine development. Neuron **20:** 1093–1102.
49. TIMPL, P., R. SPANAGEL, I. SILLABER *et al.* 1998. Impaired stress response and reduced anxiety in mice lacking a functional corticotropin-releasing hormone receptor. Nat. Genet. **19:** 162 166.
50. YASWEN, L., N. DIEHL, M.B. BRENNAN *et al.* 1999. Obesity in the mouse model of pro-opiomelanocortin deficiency responds to peripheral melanocortin. Nat. Med. **5:** 1066–1070.
51. SCHIMMER, B.P. 1979. Adrenocortical Y1 cells. Methods Enzymol. **58:** 570–574.
52. GAZDAR, A.F., H.K. OIE, C.H. SHACKLETON *et al.* 1990. Establishment and characterization of a human adrenocortical carcinoma cell line that expresses multiple pathways of steroid biosynthesis. Cancer Res. **50:** 5488–5496.
53. HORNSBY, P.J., S.E. HARRIS & K.A. ALDERN. 1985. The role of ascorbic acid in the function of the adrenal cortex: studies in adrenocortical cells in culture. Endocrinology **117:** 1264–1271.
54. MASUI, H. & L.D. GARREN. 1971. Inhibition of replication in functional mouse adrenal tumor cells by adrenocorticotropic hormone mediated by adenosine 3′:5′-cyclic monophosphate. Proc. Natl. Acad. Sci. USA **68:** 3206–3210.
55. GILL, G.N. & E.R. WEIDMAN. 1977. Hormonal regulation of initiation of DNA synthesis and of differentiated function in Y-1 adrenal cortical cells. J. Cell. Physiol. **92:** 65–75.
56. KIMURA, E., M.H. SONOBE, M.C. ARMELIN *et al.* 1993. Induction of FOS and JUN proteins by adrenocorticotropin and phorbol ester, but not by 3′,5′-cyclic adenosine monophosphate derivatives. Mol. Endocrinol. **7:** 1463–1471.
57. VIARD, I., A. PENHOAT, R. OUALI *et al.* 1994. Peptide hormone and growth factor regulation of nuclear proto-oncogenes and specific functions in adrenal cells. J. Steroid Biochem. Mol. Biol. **50:** 219–224.
58. ARMELIN, H.A., C. LOTFI & A.P. LEPIQUE. 1996. Regulation of growth by ACTH in the Y-1 line of mouse adrenocortical cells. Endocr. Res. **22:** 373–383.
59. LOTFI, C., Z. TODOROVIC, H.A. ARMELIN *et al.* 1997. Unmasking a growth-promoting effect of the adrenocorticotropic hormone in Y1 mouse adrenocortical tumor cells. J. Biol. Chem. **272:** 29886–29891.
60. DIGNASS, A.U. & A. STURM. 2001. Peptide growth factors in the intestine. Eur. J. Gastroenterol. Hepatol. **13:** 763–770.
61. LIEBERMAN, J.R., A. DALUISKI & T.A. EINHORN. 2002. The role of growth factors in the repair of bone: biology and clinical applications. J. Bone Jt. Surg. Am. Vol. **84A:** 1032–1044.
62. ORNITZ, D.M. & P.J. MARIE. 2002. FGF signaling pathways in endochondral and intramembranous bone development and human genetic disease. Genes Dev. **16:** 1446–1465.
63. KOVARY, K. & R. BRAVO. 1991. The jun and fos protein families are both required for cell cycle progression in fibroblasts. Mol. Cell. Biol. **11:** 4466–4472.
64. KOVARY, K. & R. BRAVO. 1991. Expression of different Jun and Fos proteins during the G0-to-G1 transition in mouse fibroblasts: *in vitro* and *in vivo* associations. Mol. Cell. Biol. **11:** 2451–2459.
65. MESIANO, S., S.H. MELLON, D. GOSPODAROWICZ *et al.* 1991. Basic fibroblast growth factor expression is regulated by corticotropin in the human fetal adrenal: a model for adrenal growth regulation. Proc. Natl. Acad. Sci. USA **88:** 5428–5432.
66. ANGEL, P. & M. KARIN. 1991. The role of Jun, Fos, and the AP-1 complex in cell-proliferation and transformation. Biochim. Biophys. Acta **1072:** 129–157.
67. LOTFI, C.F. & H.A. ARMELIN. 1998. c-Fos protein is a mediator in mitogenic response to ACTH. Endocr. Res. **24:** 421–424.

68. LE, T. & B.P. SCHIMMER. 2001. The regulation of MAPKs in Y1 mouse adrenocortical tumor cells. Endocrinology **142:** 4282–4287.
69. LOTFI, C.F., E.T. COSTA, T.T. SCHWINDT *et al.* 2000. Role of ERK/MAP kinase in mitogenic interaction between ACTH and FGF2 in mouse Y1 adrenocortical tumor cells. Endocr. Res. **26:** 873–877.
70. LIEBMANN, C. 2001. Regulation of MAP kinase activity by peptide receptor signalling pathway: paradigms of multiplicity. Cell Signal **13:** 777–785.
71. RICHARDS, J.S. 2001. New signaling pathways for hormones and cyclic adenosine 3′,5′- monophosphate action in endocrine cells. Mol. Endocrinol. **15:** 209–218.
72. LOTFI, C.F., A.P. LEPIQUE, F.L. FORTI *et al.* 2000. Proliferative signaling initiated in ACTH receptors. Braz. J. Med. Biol. Res. **33:** 1133–1140.
73. STOCCO, D.M. & B.J. CLARK. 1996. Role of the steroidogenic acute regulatory protein (StAR) in steroidogenesis. Biochem. Pharmacol. **51:** 197–205.
74. GYLES, S.L., C.J. BURNS, B.J. WHITEHOUSE *et al.* 2001. ERKs regulate cyclic AMP–induced steroid synthesis through transcription of the steroidogenic acute regulatory (StAR) gene. J. Biol. Chem. **276:** 34888–34895.
75. BORUD, B., T. HOANG, M. BAKKE *et al.* 2002. The nuclear receptor coactivators p300/CBP/cointegrator-associated protein (p/CIP) and transcription intermediary factor 2 (TIF2) differentially regulate PKA-stimulated transcriptional activity of steroidogenic factor 1. Mol. Endocrinol. **16:** 757–773.
76. HU, M.C., N.C. HSU, C.I. PAI *et al.* 2001. Functions of the upstream and proximal steroidogenic factor 1 (SF-1)–binding sites in the CYP11A1 promoter in basal transcription and hormonal response. Mol. Endocrinol. **15:** 812–818.
77. ITO, M., J.C. ACHERMANN & J.L. JAMESON. 2000. A naturally occurring steroidogenic factor-1 mutation exhibits differential binding and activation of target genes. J. Biol. Chem. **275:** 31708–31714.
78. LIU, Z. & E.R. SIMPSON. 1999. Molecular mechanism for cooperation between Sp1 and steroidogenic factor-1 (SF-1) to regulate bovine CYP11A gene expression. Mol. Cell. Endocrinol. **153:** 183–196.
79. LUND, J., M. BAKKE, G. MELLGREN *et al.* 1997. Transcriptional regulation of the bovine CYP17 gene by cAMP. Steroids **62:** 43–45.
80. HAMMER, G.D., I. KRYLOVA, Y. ZHANG *et al.* 1999. Phosphorylation of the nuclear receptor SF-1 modulates cofactor recruitment: integration of hormone signaling in reproduction and stress. Mol. Cell. **3:** 521–526.
81. DESCLOZEAUX, M., I. KRYLOVA, F. HORN *et al.* 2002. Phosphorylation and intramolecular stabilization of the ligand binding domain in the orphan nuclear receptor, SF-1. Mol. Cell. Biol. **22:** 2193–2203.
82. GLASS, C.K. & M.G. ROSENFELD. 2000. The coregulator exchange in transcriptional functions of nuclear receptors. Genes Dev. **14:** 121–141.
83. RACHEZ, C. & L.P. FREEDMAN. 2001. Mediator complexes and transcription. Curr. Opin. Cell Biol. **13:** 274–280.
84. PISSIOS, P., I. TZAMELI, P. KUSHNER *et al.* 2000. Dynamic stabilization of nuclear receptor ligand binding domains by hormone or corepressor binding. Mol. Cell. **6:** 245–253.

The Human Placenta Remodels the Uterus by Using a Combination of Molecules That Govern Vasculogenesis or Leukocyte Extravasation

YAN ZHOU,[a] OLGA GENBACEV,[a] AND SUSAN J. FISHER[a,b,c,d]

Departments of [a]Stomatology, [b]Anatomy, [c]Obstetrics, Gynecology, and Reproductive Sciences, and [d]Pharmaceutical Chemistry, University of California, San Francisco, San Francisco, California, USA

ABSTRACT: After fertilization, the next major hurdle for human reproduction is trophoblast differentiation, which is required for implantation, followed in lockstep by rapid assembly of these embryonic cells into a functional placenta. During this process, cytotrophoblast stem cells invade the uterus, anchoring the conceptus to the mother and establishing blood flow to the placenta. Cytotrophoblast invasion is actually a differentiation process that yields cells with many unusual attributes, that is, their tumor-like ability to invade the uterus and engraft maternal blood vessels. We discovered that once cytotrophoblasts commit to invasion they turn on expression of adhesion receptors characteristic of endothelium. They also begin to express and activate matrix metalloproteinase-9. Together, these phenotypic changes facilitate invasion while enabling cytotrophoblasts to present an endothelium-like surface to maternal blood. Currently, we are trying to understand the factors that play a role in cytotrophoblast differentiation/invasion. In keeping with the vascular characteristics that differentiated cytotrophoblasts assume, our recent data suggest that the key regulators include an unusual subset of vascular endothelial growth factor family members that play important roles in conventional vasculogenesis/angiogenesis. Surprisingly, we also discovered that these cells express functional L-selectin, which mediates neutrophil rolling and tethering, under shear stress, on inflamed endothelium. Trophoblast L-selectin likely interacts with carbohydrate selectin ligands that are upregulated on uterine glandular epithelium during the window of receptivity. Together, these data suggest that differentiating cytotrophoblasts have co-opted portions not only of vasculogenesis, but also of the process that facilitates leukocyte emigration from the blood into tissues, additional evidence of these cells' amazing plasticity.

KEYWORDS: cytotrophoblast; uterus; VEGF; selectin; mucin

The success or failure of human reproduction depends on formation of the placenta.[1–3] In turn, placentation entails a highly unusual differentiation process involving its specialized stem cells, ectodermal derivatives termed cytotrophoblasts.

Address for correspondence: Susan Fisher, University of California, San Francisco, 513 Parnassus Avenue, HSW-604, San Francisco, CA 94143-0512. Voice: 415-476-5297; fax: 415-502-7338. sfisher@cgl.ucsf.edu

Ann. N.Y. Acad. Sci. 995: 73–83 (2003). © 2003 New York Academy of Sciences.

Through an amazing series of interactions, these cells insert the placenta, a hemi-allogeneic organ as large as the adult liver, into the uterine circulation, which balloons to accommodate this temporary arrangement. These cellular interactions are paralleled by equally important maternal hemodynamic alterations that shunt the mother's blood flow to the placenta. As a result, this transient fetal organ is ideally positioned to perform a wide variety of functions during pregnancy, including nutrient, gas, and waste exchange.

We are studying the molecular underpinnings of normal placentation and aberrations in its component processes that can lead to infertility and pregnancy complications.[4] A detailed understanding of the complex cellular relationships at the maternal-fetal interface is necessary to understand why we are focusing on particular research questions. The microscopic anatomy of this region is illustrated in FIGURE 1. Chorionic villi are the basic building blocks of the human placenta. In a subset of these villi, termed anchoring villi, cytotrophoblasts begin to behave like tumor cells. Specifically, they leave the trophoblast basement membrane, where they are anchored in the placenta proper, and form aggregates that attach the conceptus to the uterine wall. Cytotrophoblasts that emanate from these sites invade the interstitium of the decidua, eventually reaching the inner third of the myometrium.

Although there are many circumstances in which normal and abnormal cells carry out interstitial invasion, several aspects of cytotrophoblast vascular invasion are unique. During this process, placental cells of fetal origin intravasate the superficial portions of uterine vessels that lie near the implantation site. Initially, these vessels are plugged by cytotrophoblast aggregates. As invasion proceeds, the placental cells form heterotypic interactions with—and then replace—the endothelium. They also intercalate within the muscular tunica, a process that substantially increases the vessel diameter while decreasing its resistance. In this manner, uterine arterioles, and to a lesser extent veins, are remodeled into unique hybrid vessels composed of fetal and maternal cells. Once the lumina reform, uterine blood flow is diverted to the intervillous space where nutrient, gas, and waste exchange occurs.

Although the histological features of this process have been studied for decades, the molecular aspects, which are likely to be complex, have only recently been

---

FIGURE 1. Diagram of the histological organization of the human maternal-fetal interface at midgestation. In this location, cytotrophoblasts, which are specialized (fetal) epithelial cells of the placenta, differentiate and invade the uterine wall, where they also breach maternal blood vessels. The basic structural unit of the placenta is the chorionic villus, composed of a stromal core with blood vessels, surrounded by a basement membrane and overlain by cytotrophoblast stem cells. As part of their differentiation program, these stem cells detach from the basement membrane and adopt one of two lineage fates. They either fuse to form the syncytiotrophoblasts that cover floating villi or join a column of extravillous cytotrophoblasts at the tips of anchoring villi. The syncytial covering of floating villi mediates the nutrient, gas, and waste exchange between fetal and maternal blood. The anchoring villi, through the attachment of cytotrophoblast columns, establish physical connections between the fetus and the mother. Invasive cytotrophoblasts penetrate the uterine wall up to the first third of the myometrium. A portion of the extravillous cytotrophoblasts home to uterine spiral arterioles and remodel these vessels by destroying the muscular wall and replacing the endothelial lining. To a lesser extent, they also remodel uterine veins. At term, few cytotrophoblast stem cells remain, the syncytiotrophoblast layer thins, and the stromal cores expand. (Diagram modified from Hoang *et al.*[20])

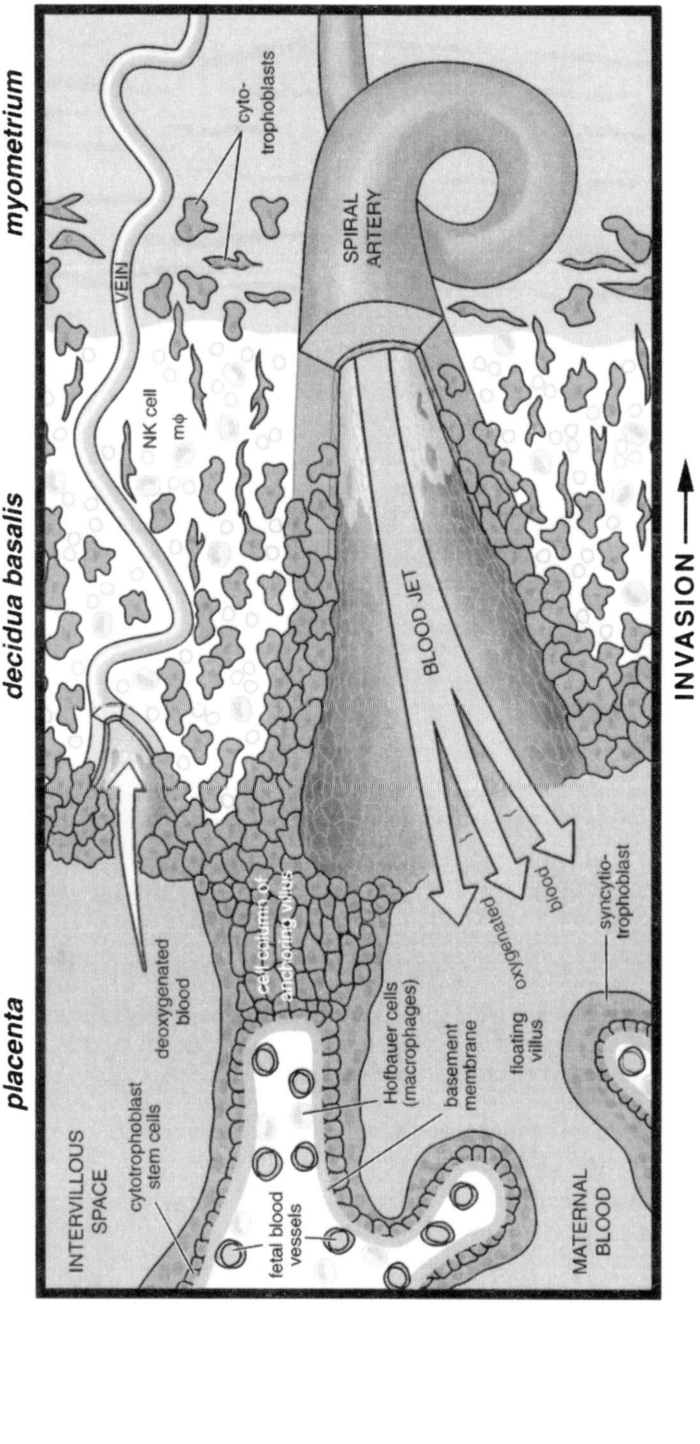

FIGURE 1. *See previous page for legend.*

addressed. Our previous work suggests that invading cytotrophoblasts execute a novel epithelial-to-endothelial phenotypic transformation that we termed pseudovasculogenesis.[5] This switch is exemplified by changes in the cells' adhesion molecule expression that take place during uterine invasion. For example, cytotrophoblast stem cells, which express adhesion molecules typical of many epithelial cells, that is, E-cadherin and $\alpha6\beta4$ integrin, dramatically alter this repertoire as they invade the uterus. In essence, they replace their epithelial-like receptors with adhesion molecules typical of endothelial cells, that is, vascular endothelial-cadherin, vascular cell adhesion molecule-1, platelet-endothelial cell adhesion molecule-1, and $\alpha V\beta3$ integrin. Other aspects of the cell surfaces of invasive cytotrophoblasts also resemble vascular cells. For example, they express urokinase plasminogen activator, the thrombin receptor, and activated matrix metalloproteinase-9.[6]

The functional importance of this transformation program is highlighted by the fact that failures in executing portions of it are associated with a subset of pregnancy complications, for example, preeclampsia, the leading cause of maternal mortality in the Western world, which also increases perinatal mortality fivefold.[5] The clinical diagnostic criteria of this syndrome include the new onset of hypertension and the appearance of proteinuria and edema during pregnancy, all of which could be explained by functional alterations in the maternal vascular endothelium.[7] Preeclampsia and about half the cases of intrauterine growth restriction are associated with particular placental pathologies. The extent of interstitial invasion by cytotrophoblasts is variable, but frequently shallow, and endovascular invasion is consistently rudimentary, making it extremely difficult to find any maternal vessels that contain cytotrophoblasts. These anatomical defects suggested to us that, in preeclampsia, cytotrophoblast differentiation along the invasive pathway is abnormal. Biopsy specimens of the uterine walls of women with this syndrome showed that invasive cytotrophoblasts retain expression of adhesion receptors characteristic of stem cells and fail to turn on receptors that promote invasion and/or assumption of an endothelial phenotype. The concept that failed cytotrophoblast invasion and pseudovasculogenesis are linked to the maternal vascular pathology appears plausible, but has yet to be proved. Thus, the chain of events that link a defect in placentation to the maternal systemic disorder is under intense investigation.

We are very interested in understanding the molecular pathways that regulate cytotrophoblast pseudovasculogenesis. A vast array of physiological factors (e.g., hypoxia and shear stress) and effector pathways (e.g., transcription factors, growth factors, chemokines, cytokines, and protein fragments) govern blood vessel development and growth, either directly or indirectly. To focus our investigation, we have been considering cytotrophoblast pseudovasculogenesis, which involves cells derived from the extraembryonic lineages, in the context of master regulatory pathways that govern *de novo* differentiation and assembly of blood vessels within the embryo. Gene deletion studies in mice have pointed to the particular importance of three families of ligands and their tyrosine kinase receptors—vascular endothelial growth factors (VEGFs) and their receptors, angiopoietins and their Tie receptors, and ephrins and their Eph receptors—in vasculogenesis and angiogenesis. The discrete phenotypes of the null animals suggest distinct roles for individual families, with VEGF family members and their receptors having important actions during the initial stages of vasculogenesis and angiogenesis, both during development and as a result of pathological processes.[8,9]

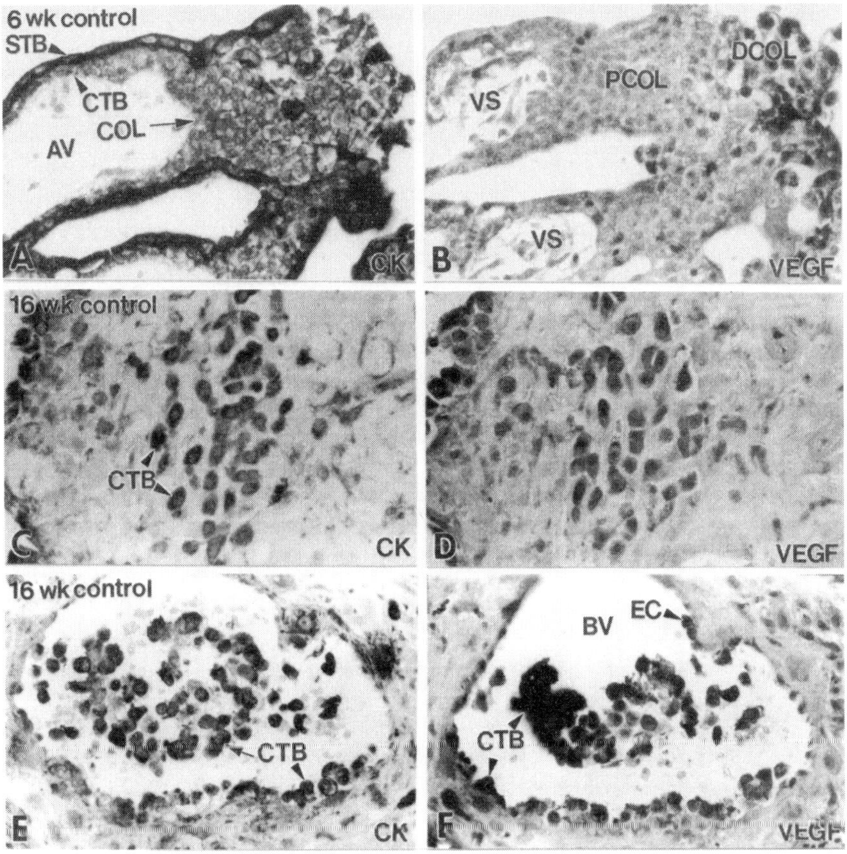

FIGURE 2. Cytotrophoblast staining for VEGF-A was upregulated as the cells differentiated and invaded the uterus *in situ*. Serial paraffin sections of the maternal-fetal interface were stained with anti-cytokeratin (CK) to identify all the trophoblast populations **(A, C, E)** and with anti-VEGF-A **(B, D, F)**. Essentially the same staining pattern was observed during the first and second trimesters (6 and 16 weeks of gestation, respectively). A few cytotrophoblast (CTB) stem cells and cells in the proximal column (PCOL) region stained with anti-VEGF-A **(B)**, but much more intense staining was observed in association with a majority of cytotrophoblasts in the distal regions of columns (DCOL) and with those that invaded the uterine wall **(D)**. Cytotrophoblasts within the lumina of uterine blood vessels (BV) also exhibited intense staining (arrowheads), as did the maternal endothelial cells (EC **[F]**). In contrast, some cytotrophoblasts in the vessel wall failed to react with anti-VEGF-A (arrows). The cells continued to stain for VEGF-A at term (not shown). AV, anchoring villus; STB, syncytiotrophoblast; VS, villous stroma; COL, cytotrophoblast column.[10]

We hypothesized that cytotrophoblast pseudovasculogenesis co-opts a subset of the ligand-receptor interactions that govern conventional vasculogenesis. To test this theory, we used a combination of *in situ* and *in vitro* analyses to characterize the cells' expression of VEGF family ligands and receptors, key regulators of conventional vasculogenesis and angiogenesis.[10] Cytotrophoblast differentiation and invasion during the first and second trimesters of pregnancy were associated with down-regulation of VEGF receptor (VEGFR) 2. Invasive cytotrophoblasts in early gestation expressed VEGF-A (FIG. 2), VEGF-C, placental growth factor (PlGF), VEGFR-1, and VEGFR-3; and, at term, VEGF-A, PlGF, and VEGFR-1. *In vitro*, the cells incorporated VEGF-A into the surrounding extracellular matrix; PlGF was secreted. We also found that cytotrophoblasts responded to the VEGF ligands they produced. Blocking ligand binding significantly decreased their expression of integrin α1, an adhesion molecule highly expressed by endovascular cytotrophoblasts, and increased apoptosis. In severe preeclampsia and HELLP syndrome (hemolysis, elevated liver enzymes, and low platelets), immunolocalization on tissue sections showed that cytotrophoblast VEGF-A and VEGFR-1 staining decreased; staining for PlGF was unaffected. Cytotrophoblast secretion of the soluble form of VEGFR-1 *in vitro* also increased. Together, the results of this study showed that VEGF family members regulate cytotrophoblast survival and that expression of a subset of family members is dysregulated in severe forms of preeclampsia. Together, the results of these experiments suggest that VEGF-VEGFR interactions play an important role in differentiation and survival of the unique cytotrophoblast subpopulation that invades the uterine wall, occupies the maternal vessels, and channels maternal blood to the placenta during pregnancy.

Interpreting these data in the context of existing information about the function of particular VEGF ligands and receptors gives some very interesting insights into the role they play in the unique differentiation pathway that forms the human maternal-fetal interface.[10] With regard to receptors, deletion of the gene encoding VEGFR-2 has the severest effects; the mice die at embryonic day (E) 8.5 because differentiation of endothelial and hematopoietic cells is blocked. VEGFR-1 appears to act downstream of VEGFR-2. Null embryos, which also die at E8.5, have an excess of endothelial cells that do not assemble into functional vessels. Unexpectedly, deletion of the receptor's kinase domain does not interfere with angiogenesis—possible evidence of cooperative interactions with VEGFR-2. Deletion of VEGFR-3, which also stimulates growth of lymphatic vessels, causes death at E9.5. In this case, vasculogenesis and angiogenesis appear to proceed normally, but large vessels are organized abnormally. Overexpression of soluble VEGFR-3 produces lymphedema.

With regard to receptor ligands, threshold levels for VEGF interactions with its receptors (VEGFR-1 and -2) are important because deletion of only one copy of the VEGF gene causes death at E11–12 due to abnormalities in both angiogenesis and blood island formation. Recently, additional family members (VEGF-B, -C, and -D; PlGF) were discovered. Mice that lack VEGF-B, which binds to VEGFR-1, have impaired cardiac function. VEGF-C, a ligand for VEGFR-2 and -3, has multiple actions. Mice engineered to overexpress this ligand in the skin have selective hyperplasia of the superficial lymphatic vasculature. However, VEGF-C also induces angiogenesis. Human VEGF-D binds and activates VEGFR-2 and -3, suggesting roles in angiogenesis and lymphangiogenesis. Finally, PlGF interactions with VEGFR-1 potentiate angiogenic responses to VEGF.

Our analyses of cytotrophoblast VEGF ligands and receptors suggested that their expression is precisely regulated because these unusual cells exit the mitotic cycle and undergo pseudovasculogenesis *in vivo*. In turn, these intricate expression patterns allowed us to generate hypotheses about the autocrine effects of VEGF on cytotrophoblast differentiation and invasion. These processes, which are key to placental development, mainly occur during the first half of pregnancy. Accordingly, we were particularly interested in ligand-receptor pairs that could function during this time period. For example, the coordinated expression of VEGF-C and VEGFR-2 by early gestation cytotrophoblast stem cells suggested that signals transmitted through the latter receptor could affect important functions, that is, proliferation. However, we were unable to test this hypothesis because early-gestation cytotrophoblasts rapidly downregulated VEGFR-2 expression as they differentiated *in vitro*. The latter observation is in accord with the results of our previous studies that show that cytotrophoblast differentiation, both *in situ* and in culture, is associated with permanent withdrawal from the cell cycle.[11]

As discussed above, our previous studies also suggest that invading cytotrophoblasts upregulate the expression of molecules that play functional roles in uterine invasion and pseudovasculogenesis. Thus, the immunolocalization data regarding VEGF family members led us to theorize that interactions of PlGF, VEGF-A, and VEGF-C with VEGFR-1 are critical for invasion and pseudovasculogenesis, as are VEGF-C signals transduced through VEGFR-3. Our *in vitro* experiments showed this to be the case because the addition of fusion proteins that specifically interfered with ligand binding to either receptor altered the cells' ability to express integrin $\alpha 1$, an adhesion molecule that is an accurate barometer of the cells' invasive capacity; this correlation was also demonstrated in the experiments described here. Finally, signals transmitted through both receptors also play important functions in cytotrophoblast survival, a finding that is in accord with the central role played by VEGF signals in assembly and maintenance of the vasculature. Thus, the reduced invasion we observed is almost certainly a secondary phenomenon related to increased apoptosis.

Cytotrophoblast-derived VEGF family ligands are also likely to have paracrine effects on other populations of both fetal and maternal cells. In this regard, it is interesting to note that PlGF and VEGF-C are more likely to play a role since cytotrophoblast VEGF-A remained tightly associated with the extracellular matrix, a property of certain isoforms of the molecule ($VEGF_{189}$, $VEGF_{165}$, $VEGF_{145}$), but not others ($VEGF_{121}$).[12] Although we did not study receptor expression by potential target cells, the intrinsic placental vasculature that forms within the stromal cores of chorionic villi could be influenced by factors produced in the trophoblast layers. This arrangement would coordinate trophoblast differentiation with that of the extraembryonic fetal vasculature, an important component of normal placental development. On the opposite side of the placenta, the maternal vasculature is another likely target. In this location, cytotrophoblast-derived angiogenesis inhibitors, such as soluble VEGFR-1, could help cytotrophoblasts gain entrance to and occupy uterine vessels without triggering the cascade of events that normally lead to angiogenesis of the same population of vessels during the normal menstrual cycle.

It is important to note that VEGF family ligands and receptors are only a portion of the molecules with angiogenic and/or vasculogenic effects that are expressed by trophoblast cells. Thus, we predict that these other factors will also play important roles in cytotrophoblast pseudovasculogenesis and occupation of the maternal vessels. Additionally, it is interesting to note that the placenta often co-opts pathways used in the development of other organs and tissues.

Our very recent data suggest that placentation may also entail elements of pathological processes. Surprisingly, trophoblast adhesion to the uterus appears to be related to the selectin system that leukocytes use to emigrate from blood into tissues.[13]

These lectin-like protein receptors bind to mucin oligosaccharide ligands that are constitutively expressed in some locations (high endothelial venules) and inducible in others (inflamed endothelia). Selectin receptor–ligand bonds have unusual kinetic and mechanical properties that permit transient tethering and shear stress–dependent rolling.[14] Upon capture, integrin activation triggers stable adhesion, bringing the cells into contact with factors that govern transmigration into tissues (e.g., chemokines and cytokines).[15]

At a morphological level, analogies can be drawn between key steps in leukocyte emigration from blood and trophoblast attachment to the uterine wall. Implantation begins with apposition: the trophectoderm of the originally free-floating embryo lies adjacent to the uterine epithelium, but the blastocyst is easily dislodged.[16] Soon thereafter, blastocyst adhesion to the uterine wall is stabilized and trophoblasts transmigrate across the uterine epithelium, a process that in humans buries the entire embryo beneath the uterine surface. Subsequent development depends on the ability of trophoblasts to adhere under conditions of shear stress created when these fetal cells breach uterine vessels, a process that diverts maternal blood flow to the placenta. At a molecular level, trophoblast adhesion from implantation onward is an integrin-dependent process that takes place in a chemokine- and cytokine-rich milieu analogous to the blood-vascular interface.[17,18] Together, these findings raised the possibility that implantation and placentation utilize other components of the leukocyte emigration system, such as selectins and their ligands.

To determine whether selectin ligands are present at appropriate times and locations to function in blastocyst adhesion to the uterus, we studied their expression at the maternal-fetal interface with antibodies that bind sulfated oligosaccharides that interact with L-selectin. We found that tissue sections of uterine biopsies obtained from patients on days 6 and 19 of the menstrual cycle reacted with the MECA-79 antibody, which recognizes carbohydrate epitopes on all L-selectin ligands (particularly $SO_3 \rightarrow 6$ GlcNAc). On day 6, when the uterus was not receptive, MECA-79 reactivity localized to the surface of glandular and luminal epithelia. Staining was patchy and weaker than the more intense immunoreactivity observed when the uterus was receptive. We also stained the same samples with the HECA-452 antibody, which reacts with sLe^x and related sulfated structures, including 6-sulfo-sLe^x. Again, antibody reactivity, limited to luminal and glandular epithelial cells, was weaker on day 6 of the cycle and stronger on day 19. Analyses of samples later in gestation showed that staining with these antibodies was lost during the second trimester, a situation that persisted until birth. Immunoblot analyses with MECA-79 confirmed upregulation of selectin oligosaccharide ligands as the window of human receptivity opens. Together, these results suggest that uterine expression of specialized carbohydrate structures that support selectin-mediated adherence is dramatically upregulated as the uterus becomes receptive. Whether these alterations in glycosylation track with changes in hormone levels in other tissues remains to be determined.

This discovery prompted us to ask whether placental trophoblast cells express L-selectin. To answer this question, we analyzed L-selectin expression in tissue sections of the human maternal-fetal interface. Between 6 and 16 weeks of gestation, cytotrophoblast stem cells, cytotrophoblasts in cell columns, and invasive cytotrophoblasts strongly reacted with an antibody that recognizes the extracellular domain of L-selectin. The pattern of reactivity was indicative of antibody binding to cell surface L-selectin. Overlying syncytiotrophoblasts also stained, as did macrophages in the

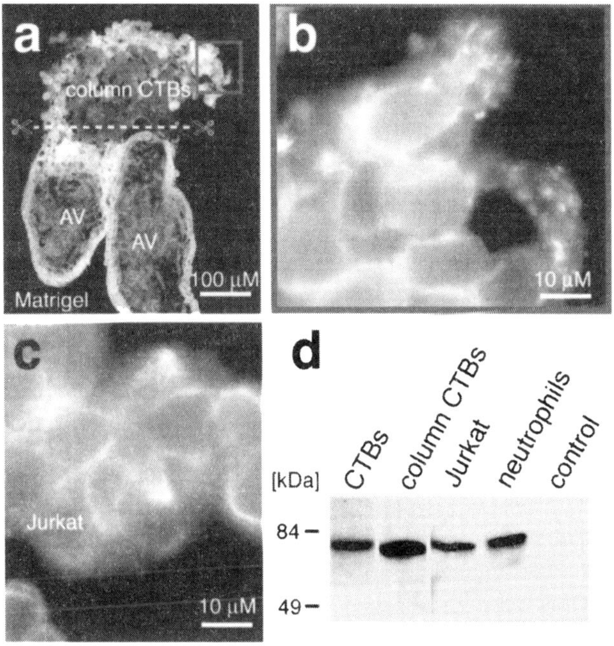

FIGURE 3. Expression of L-selectin by human trophoblasts *in vitro*. **(a)** An anchoring villus (AV) was cultured for 72 h. The trophoblast layer and cytotrophoblasts that invaded the Matrigel (box) stained with anti-L-selectin (DREG-56). **(b)** Immunoreactivity was detected at cell surfaces and cell-cell contacts. **(c)** L-Selectin–expressing Jurkat cells had a similar staining pattern. **(d)** Immunoblot analyses (anti-CD62L) of lysates from purified cytotrophoblasts, cytotrophoblasts in columns dissected from cultured anchoring villi (scissors in **a**), Jurkat cells, and neutrophils revealed a single immunoreactive band of the expected molecular mass (~75 to 90 kDa) that was absent when the primary antibody was omitted (control). CTBs, cytotrophoblasts.

villus stromal cores. To establish that trophoblast L-selectin staining was not due to adsorption of shed L-selectin from blood, antibodies that recognized the carboxyl terminus of the molecule were also used in immunolocalization experiments. The same populations of cells stained with both antibodies.

To support our *in situ* observations, we characterized human trophoblast L-selectin expression *in vitro* by using two culture systems that model cytotrophoblast differentiation along the pathway that leads to uterine attachment and invasion. In an organ culture model, explanted anchoring villi are cultured on Matrigel-coated wells; villi attach to the extracellular matrix via the remnants of severed cytotrophoblast columns that subsequently give rise to invasive cytotrophoblasts.[19] Sections of explants cultured for 72 h were stained with anti-L-selectin. The pattern of antibody reactivity was very similar to that observed *in situ*; cytotrophoblast stem cells within the villi and column cytotrophoblasts that invaded the Matrigel stained, as did the syncytial surface (FIG. 3a). At higher magnification, the pattern of cytotrophoblast staining was indistinguishable from that of Jurkat cells that express L-selectin (compare FIGS. 3b and 3c). In a second *in vitro* model, purified cytotrophoblast stem cells plated on Matrigel differentiate to invasive cells over 48–72 h. After 48 h, these cells also reacted with anti-L-selectin (data not shown). To confirm antigen identity, lysates prepared from isolated cytotrophoblasts and the columns dissected from explants (see FIG. 3a) were analyzed by immunoblotting samples that were separated on 10% polyacrylamide gels. These samples contained a broad ~75- to 90-kDa band (FIG. 3d). Jurkat and neutrophil lysates contained the same bands. Together, these data indicate that trophoblast upregulation of L-selectin expression occurs concomitantly with the opening of the window of blastocyst attachment to the uterine epithelium and cytotrophoblast invasion of the underlying parenchyma.

Next, we assessed the functional state of trophoblast L-selectin *in vitro*. Under shear stress, beads coated with 6-sulfo-sLe^X bound to explants of the human chorionic villus tree. They associated with cytotrophoblast column remnants of anchoring villi and with the syncytiotrophoblasts at the surfaces of the villi, that is, the cells that express L-selectin *in situ* and *in vitro*. Beads that displayed sLe^X also bound to villus explants, but in fewer numbers. Very few control phosphatidylcholine-conjugated beads bound to cytotrophoblast cell columns or syncytiotrophoblasts. Likewise, uncoated beads did not bind. Beads that displayed 6-sulfo-sLe^X did not adhere to explants preincubated with an antibody that blocks L-selectin function, suggesting that the binding was specific.

Our finding of L-selectin function outside the blood-vascular system has several important implications for implantation and placentation. Shear stress, required for L-selectin-mediated adhesion through its specialized oligosaccharide ligands, is likely to be an important component of both processes. During implantation, apposition may be analogous to leukocyte transient tethering and rolling. Although the magnitude of shear stress at the surface of the receptive uterus is unknown, distractive forces are likely derived from several sources, including fluid secretions and uterine contractions. During the early stages of placentation, cytotrophoblasts that invade and line maternal vessels experience the same shear stress as the resident vascular cells. Given the many and varied causes of infertility and early pregnancy loss, it seems likely that defects in the shear stress–activated selectin adhesion system at the maternal-fetal interface account for a portion of unexplained reproductive failures.

In conclusion, our data offer further insights into the highly unusual nature of trophoblasts. Previously, we showed that invasive cytotrophoblasts of ectodermal origin undergo a novel differentiation process, taking on characteristics of vascular cells. Our finding that the same cells also share leukocyte adhesion mechanisms raises the possibility that trophoblasts have characteristics of the hemangioblast stem cell population that gives rise to both blood cells and blood vessels.

ACKNOWLEDGMENTS

This work was supported by the National Institutes of Health (Grant Nos. DE 07244, HL 64597, U01 HD 42283, and HD 30367).

REFERENCES

1. Cross, J.C., Z. Werb & S.J. Fisher. 1994. Implantation and the placenta: key pieces of the development puzzle. Science **266:** 1508–1518.
2. Rossant, J. & J.C. Cross. 2001. Placental development: lessons from mouse mutants. Nat. Rev. Genet. **2:** 538–548.
3. Paria, B.C., J. Reese, S.K. Das *et al.* 2002. Deciphering the cross-talk of implantation: advances and challenges. Science **296:** 2185–2188.
4. Norwitz, E.R., D.J. Schust & S.J. Fisher. 2001. Implantation and the survival of early pregnancy. N. Engl. J. Med. **345:** 1400–1408.
5. Damsky, C.H. & S.J. Fisher. 1998. Trophoblast pseudo-vasculogenesis: faking it with endothelial adhesion receptors. Curr. Opin. Cell Biol. **10:** 660–666.
6. Librach, C.L., Z. Werb, M.L. Fitzgerald *et al.* 1991. 92-kD type IV collagenase mediates invasion of human cytotrophoblasts. J. Cell Biol. **113:** 437–449.
7. Roberts, J.M., R.N. Taylor, T.J. Musci *et al.* 1989. Preeclampsia: an endothelial cell disorder. Am. J. Obstet. Gynecol. **161:** 1200–1204.
8. Veikkola, T., M. Karkkainen, L. Claesson-Welsh *et al.* 2000. Regulation of angiogenesis via vascular endothelial growth factor receptors. Cancer Res. **60:** 203–212.
9. Carmeliet, P. & R.K. Jain. 2000. Angiogenesis in cancer and other diseases. Nature **407:** 249–257.
10. Zhou, Y., M. McMaster, K. Woo *et al.* 2002. Vascular endothelial growth factor ligands and receptors that regulate human cytotrophoblast survival are dysregulated in severe preeclampsia and hemolysis, elevated liver enzymes, and low platelets syndrome. Am. J. Pathol. **160:** 1405–1423.
11. Genbacev, O., Y. Zhou, J.W. Ludlow *et al.* 1997. Regulation of human placental development by oxygen tension. Science **277:** 1669–1672.
12. Ortega, N., F.E. L'Faqihi & J. Plouet. 1998. Control of vascular endothelial growth factor angiogenic activity by the extracellular matrix. Biol. Cell. **90:** 381–390.
13. Bertozzi, C.R. & L.L. Kiessling. 2001. Chemical glycobiology. Science **291:** 2357–2364.
14. Alon, R., S. Chen, R. Fuhlbrigge *et al.* 1998. The kinetics and shear threshold of transient and rolling interactions of L-selectin with its ligand on leukocytes. Proc. Natl. Acad. Sci. USA **95:** 11631–11636.
15. Gunn, M.D., K. Tangemann, C. Tam *et al.* 1998. A chemokine expressed in lymphoid high endothelial venules promotes the adhesion and chemotaxis of naive T lymphocytes. Proc. Natl. Acad. Sci. USA **95:** 258–263.
16. Carson, D.D., I. Bagchi, S.K. Dey *et al.* 2000. Embryo implantation. Dev. Biol. **223:** 217–237.
17. Guleria, I. & J.W. Pollard. 2000. The trophoblast is a component of the innate immune system during pregnancy. Nat. Med. **6:** 589–593.
18. Drake, P.M., M.D. Gunn, I.F. Charo *et al.* 2001. Human placental cytotrophoblasts attract monocytes and CD56(bright) natural killer cells via the actions of monocyte inflammatory protein 1 alpha. J. Exp. Med. **193:** 1199–1212.
19. Genbacev, O., S.A. Schubach & R.K. Miller. 1992. Villous culture of first trimester human placenta—model to study extravillous trophoblast (EVT) differentiation. Placenta **13:** 439–461.
20. Hoang, V.M., R. Foulk, K. Clauser *et al.* 2001. Functional proteomics: examining the effects of hypoxia on the cytotrophoblast protein repertoire. Biochemistry **40:** 4077–4086.

Chorioallantoic Morphogenesis and Formation of the Placental Villous Tree

JAMES C. CROSS, DAVID G. SIMMONS, AND ERICA D. WATSON

Genes and Development Research Group, Department of Biochemistry and Molecular Biology, University of Calgary, Calgary, Alberta T2N 4N1, Canada

Abstract: The placenta is a highly specialized organ whose primary function is to promote the exchange of nutrients and oxygen between maternal and fetal blood, essential for survival and growth of the baby. The surface area for nutrient transport is a highly convoluted villous structure that forms by branching morphogenesis. In mice, this process begins after embryonic day 8.5, following attachment of allantoic mesoderm to the chorion, and continues through the end of gestation. Gene targeting studies in mice have identified a large number of genes that are essential for chorioallantoic development to give rise to the layer of the placenta called the labyrinth. Collectively, these studies reveal that a number of signaling pathways regulate four distinct phases of labyrinth development: chorioallantoic attachment (involving VCAM1 and its receptor α4 integrin, Bmp5/7, and Wnt7b, as well as the cochaperone Mrj), initiation of branching (involving the Gcm1 transcription factor to select sites of branch initiation), extension of villous branching (involving FGF, EGF, and HGF/Met signaling, through the Grb2/Sos1/Mek1/p38α MAPK pathway), followed by vascularization of the villous tree. The restricted expression and/or action of the signaling components indicate that a series of intercellular interactions regulate chorioallantoic development.

Keywords: placenta; trophoblast; branching morphogenesis; FGF; labyrinth

INTRODUCTION

Although it also produces interesting hormones and growth factors that affect the mother and fetus, the primary function of the placenta is to facilitate the transport of nutrients and oxygen to the fetus during gestation. Some cells of the fetal placenta function to promote the angiogenesis of uterine blood vessels and their subsequent dilation to facilitate blood flow to the implantation site.[1,2] The bulk of the placenta, however, forms a highly branched villous structure for nutrient uptake. Branching morphogenesis is a general mechanism of embryonic development that allows organs to develop large surface areas for transport of nutrients and gases. For organs like the placenta, the epithelial surface undergoes branching along with the underlying stroma, and a vasculature network forms beneath it to maximize the potential

Address for correspondence: James C. Cross, Genes and Development Research Group, Department of Biochemistry and Molecular Biology, University of Calgary, HSC Room 2279, 3330 Hospital Drive, N.W., Calgary, Alberta T2N 4N1, Canada. Voice: 403-220-6876; fax: 403-270-0737.

jcross@ucalgary.ca

Ann. N.Y. Acad. Sci. 995: 84–93 (2003). © 2003 New York Academy of Sciences.

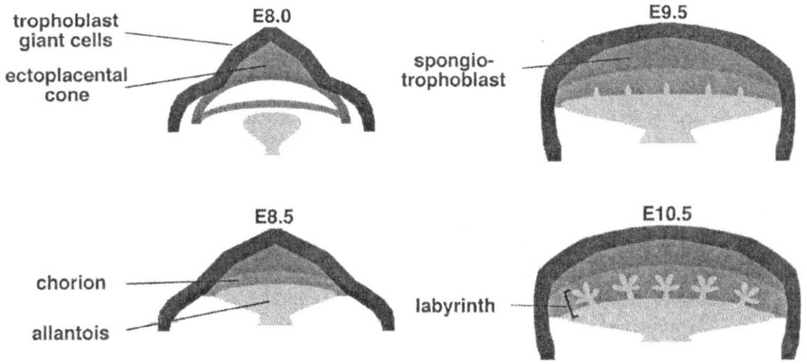

FIGURE 1. Stages of chorioallantoic attachment and morphogenesis.

for nutrient exchange between the maternal blood space and the fetal circulation.[3] Defects in branching morphogenesis in the placenta likely underlie some forms of miscarriage and also severe manifestations of preeclampsia that have an associated fetal growth restriction, as the placenta in these cases often shows greatly reduced villous structure.[4]

Studying the development of the villous part of the placenta affords the opportunity to explore several interesting types of intercellular signaling. The epithelial compartment of the placenta is derived from the trophoblast cell lineage, whereas the stromal and vascular compartments are derived from extraembryonic mesoderm (specifically the allantoic mesoderm in the mouse).[3,5] In mice, the villous part of the placenta is called the labyrinth because of the density of the villous structure and hence complex appearance in histological sections. The trophoblast progenitors of the labyrinth reside in the chorion layer and only make contact with the allantoic mesoderm at embryonic day (E) 8.5 (FIG. 1). After chorioallantoic attachment, morphogenesis is observed early on E9 with the formation of primary villi. These villi later elongate and branch, a process that continues until near term. The signaling between cell types at the chorioallantoic interface can be explored in mice using experimentally prepared chimeric embryos.[5] In chimeras, diploid blastomeres from cleavage stage embryos can contribute to both trophoblast and extraembryonic mesoderm. Embryonic stem (ES) cells, however, can contribute only to the extraembryonic mesoderm and not to trophoblast. By contrast, tetraploid blastomeres (generated by electrofusion of two-cell embryos) can contribute to trophoblast, but not to extraembryonic mesoderm. The analysis of mutant mice combined with chimera studies has given us a fairly clear picture of the major events and molecular pathways controlling chorioallantoic morphogenesis.

CHORIOALLANTOIC ATTACHMENT

The first major insights into the regulation of chorioallantoic attachment came with the analysis of mutant mouse phenotypes associated with deficiency of VCAM1 and its receptor α4 integrin. In both mutant strains, the chorion and the

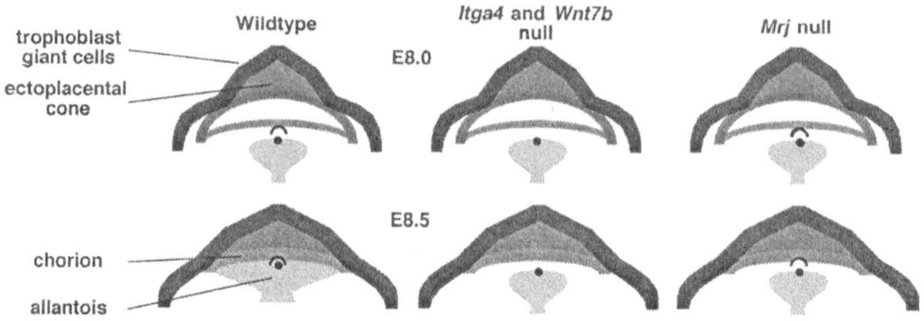

FIGURE 2. Summary of mutant phenotypes associated with defects in chorioallantoic attachment. The semicircle depicts expression of α4 integrin (encoded by the *Itga4* gene) in the chorion, whereas the dot depicts expression of VCAM1 at the tip of the allantois.

allantoic mesoderm appear to develop normally, but most do not form stable attachment at E8.5 (FIG. 2). It was subsequently discovered that VCAM1 is expressed at the tip of the outgrowing allantois,[6,7] whereas α4 integrin is expressed on the basal surface of the chorion.[8] While most of the cells in the chorion are trophoblast-derived (extraembryonic ectoderm or chorionic ectoderm), the basal surface is covered by a very thin mesothelium derived from the extraembryonic mesoderm.[9] The expression of α4 integrin protein appears in the mesothelial layer and therefore the initial interaction between the allantoic mesoderm and chorion is mediated by the chorionic mesothelium. Because the antibodies used to detect α4 integrin expression work only on frozen sections, it is difficult to determine whether expression also appears in the trophoblast compartment of the chorion. This is important to know since the chorionic mesothelium is thought to be displaced following chorioallantoic attachment, leading to direct contact between chorionic trophoblast and allantoic mesoderm cells. An ongoing requirement for VCAM1 after chorioallantoic attachment (see below and ref. 7) suggests that there might be an ongoing need for α4 integrin. Recent findings suggest that trophoblast cells do not express the mRNA for α4 integrin (H. Yamamoto and J. C. Cross, unpublished data).

Chorioallantoic attachment normally occurs precisely at E8.5 and the timing appears to be regulated by the adhesiveness of the allantois that is limited to a narrow period of time.[10] By contrast, expression of α4 integrin on the chorion surface appears during a broad window. Expression of α4 integrin on the chorion surface is, however, regulated by interactions between the trophoblast and mesothelial cells of the chorion. A member of the Wnt family of signaling molecules, Wnt7b, is expressed in the chorionic ectoderm at E7.5 and 8.5.[11] Analysis of Wnt7b mutant mice[11] showed that chorioallantoic attachment is blocked in the mutants and was associated with the absence of α4 integrin expression (FIG. 2). While the underlying signaling effect is not clear, the data support a model in which Wnt7b produced by the chorionic trophoblast cells either directly or indirectly signals to the chorionic mesothelium to turn on α4 integrin (FIG. 3). The chorionic trophoblast cells also have a slightly different shape in the Wnt7b mutants,[11] suggesting that perhaps the polarity of the cells is altered (FIG. 3). A defect in chorioallantoic fusion is also observed in double-mutant mice that are deficient for Lef1 and Tcf,[12] two Wnt-regulated transcription factors, implying that they lie downstream of Wnt7b signaling.

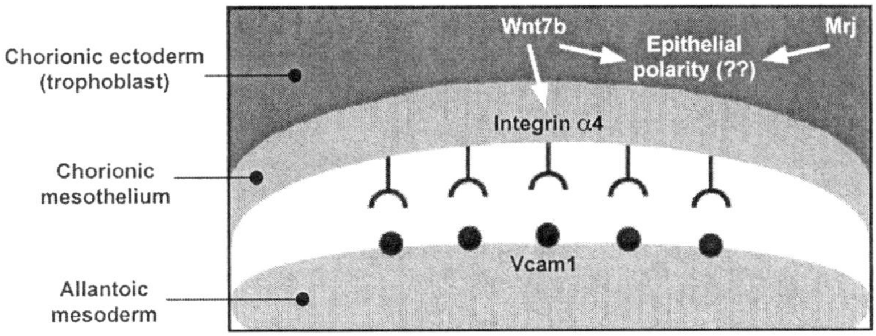

FIGURE 3. Summary of the potential interactions between cell types at the chorio-allantoic interface.

Other signaling interactions at the chorioallantoic interface around E8.5 are suggested by chorioallantoic attachment defects in mice that are deficient for Bmp5/7[13] or Fgfr2.[14] The nature of their signaling is not clear, however, as the sites of ligand and receptor actions have not been defined.

It is important to note that chorioallantoic attachment appears to occur normally in ~30% of *Vcam1*-deficient embryos, implying that the VCAM1-α4 integrin interaction is not the only pathway regulating attachment.[7] A major unanswered question is what cell adhesion molecule functions to compensate for the absence of VCAM1. The fact that the defective chorioallantoic attachment phenotype is completely penetrant in the α4 integrin mutants may suggest that the "substitute factor" is similarly a ligand for α4 integrin. However, a completely separate system is also possible as implied by the phenotype of *Mrj* mutant mice. Failure of chorioallantoic attachment is observed in mice that are deficient for the *Mrj* gene and, like the VCAM1 and α4 integrin mutants, the allantois and chorion appear to develop normally.[15] Indeed, VCAM1 and α4 integrin are still expressed in *Mrj* mutants, implying that expression of these cell adhesion molecules is not sufficient for chorioallantoic attachment.[15] The *Mrj* gene is expressed in all trophoblast cell subtypes, including those in the chorion, and encodes a cochaperone protein of the large DnaJ family.[15] DnaJ-related proteins contain a J-domain that associates with and activates the Hsp70 and Hsc70 chaperone proteins to facilitate protein folding,[16] but the precise substrates of the cochaperones are unknown. Recent studies have suggested that the Mrj protein can associate with and promote the refolding of intermediate filaments of the cytoskeleton, including keratin 8/18 in epithelial cells[17] and the Huntingtin protein in neurons.[18] It is possible that *Mrj*-deficiency is associated with a failure of the chorionic trophoblast cells to polarize and therefore to properly interact with the mesothelium in a way necessary for chorioallantoic attachment.

BRANCH-POINT SELECTION

As branching morphogenesis begins early on E9, primary villi are initiated at a few hundred sites scattered along the chorioallantoic interface. These sites are equally

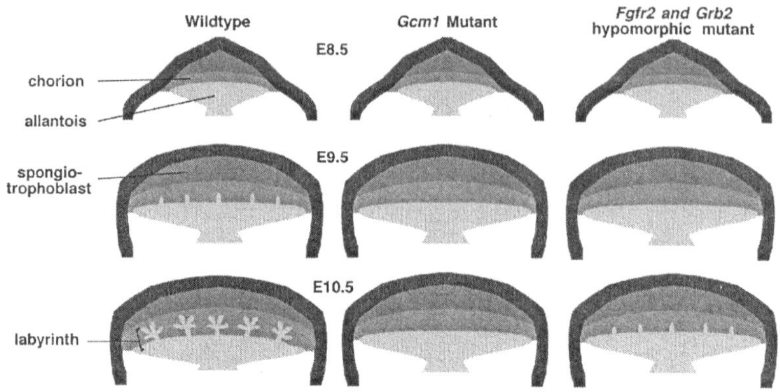

FIGURE 4. Summary of mutant phenotypes associated with defects in chorioallantoic morphogenesis.

spaced, suggesting that the sites are carefully selected, likely involving active activation and also repression of nonbranch sites. The problem of site selection is even more intriguing because the surrounding trophoblast cells of the chorion are the stem cells of the trophoblast lineage. This raises the question not only of how branch-points are selected, but also how the surrounding trophoblast stem cell population is maintained. The regulated expression and activity of the *Gcm1* transcription factor gene provides a reasonable explanation for both problems. The *Gcm1* gene was discovered as a mammalian homologue of the *glial cells missing* (*gcm*) gene in *Drosophila*.[19–21] Subsequent studies have showed that *Gcm1* expression is predominantly restricted to the placenta in both mice and humans.[21–27] Within the placenta, *Gcm1* mRNA expression is restricted to the labyrinth region in mice and to the chorionic villi in humans, and specifically to syncytiotrophoblast and its precursor cells. *Gcm1* transcripts are first detected weakly at E7.5 in small focal clusters of trophoblast cells within chorion that become more prominent on E8.[21,23,24] After E8.5, the *Gcm1*-expressing cell clusters are localized precisely at sites of primary branch-points.[24] As the branches elongate, the *Gcm1*-expressing cells are restricted to the distal tip. Not only is *Gcm1* a specific marker of branch-point selection, but it is also required. The chorioallantoic interface remains flat in *Gcm1* mutants[24] and, as a result, the labyrinth completely fails to develop[23,24] (FIG. 4). While several other genes are required for labyrinth development, the *Gcm1* mutation blocks morphogenesis at a much earlier stage.

The expression of *Gcm1* is regulated at a number of levels. The patterned early expression in the chorion limited to clusters of equally spaced cells is extremely interesting. One obvious candidate mechanism to regulate such a pattern is Notch signaling, but the evidence does not support such a hypothesis. First, the onset of *Gcm1* expression appears as a clustered pattern[21,23,28] rather than first being uniform and then being restricted into clusters, the expected pattern for a lateral inhibition type of mechanism. Second, while Notch signaling components and Notch-activated transcription factors of the HES and Hey/HesR/Hrt family are expressed in the developing placenta,[29] none appear to show an expression pattern that is obviously similar or complementary to *Gcm1*. Third, none of the mutations in Notch signaling components have been reported to be associated with early placental labyrinth defects, although the latter could also be a matter of omission rather than fact. After initiation of *Gcm1* mRNA expression, the maintenance of its expression after E8.5 appears to be dependent on signal(s) from the allantoic mesoderm. This was first demonstrated by downregulation of *Gcm1* mRNA in *Mrj* mutant mice associated with failure of chorioallantoic attachment,[15] and has been more recently supported by coculture experiments of chorion and allantois[28] (J.C.P. Kingdom and J.C. Cross, in preparation). In addition to regulating mRNA expression, contact with the allantois appears also to be associated with regulating the onset of Gcm1 protein expression.[28] The signals from the allantois that regulate *Gcm1* expression in chorionic trophoblast are unknown.

PROGRESSION OF CHORIOALLANTOIC MORPHOGENESIS

After initiation of chorioallantoic morphogenesis, a number of different molecular pathways are implicated in the progression of chorioallantoic branching (for review, see ref. 5). Mutations in components of the pathway result in a very similar phenotype in which the labyrinth begins to develop, but is either too slow or blocks at a subsequent step, resulting in inadequate placental transport function. Some publications describing these phenotypes summarize it as a failure of "vascular invasion" of the labyrinth. This interpretation is inaccurate because it overlooks the fact that the entire labyrinth layer is smaller than normal in the mutants, and it is not simply a lack of vascularity in an otherwise normally developed labyrinth. There are a few mutants in which the labyrinth layer is disproportionately undervascularized,[30,31] but this type of phenotype is still relatively rare.

The term "vascular invasion" is also extremely misleading because it creates a false impression of how the development of the labyrinth is actually driven. In the mouse, the morphogenesis of the chorioallantoic interface and the development of the placental vascular network occur simultaneously.[3] However, the chorionic villi in the developing human placenta first develop as trophoblast outgrowths that then become associated with stromal cells (called primary and secondary villi).[32] Only later do the villi become vascularized to become tertiary villi. This temporal separation of villous morphogenesis and vascularization makes the mechanistic distinction between the two processes quite clear. The formation of other branched organs such as the kidney and lung can similarly be distinguished into epithelial-mesenchymal interactions leading to branching morphogenesis, followed by vascularization.[33,34] This information collectively suggests that vascular development is not the driving force of labyrinth development in mice. The fact that the initiating molecular event is associated with trophoblast cells in the chorion (i.e., *Gcm1* expression) provides the final evidence.

SIGNALING PATHWAYS AND MORPHOGENESIS

Three general classes of signaling pathways have been implicated in the progression of chorioallantoic morphogenesis, mediated by the cell adhesion molecule

VCAM1, Wnt's, and receptor tyrosine kinases. As described above, a fraction of VCAM1-deficient embryos undergo chorioallantoic attachment apparently on cue, but still die due to severe delay/block in labyrinth development.[7] The signaling events downstream of VCAM1 that mediate this effect are unknown. Signaling between the allantoic mesoderm and chorionic trophoblast is suggested by the expression of Wnt ligands (Wnt2 and Wnt5a) and the Wnt receptor Fzd5. Both the *Wnt2* and *Wnt5a* genes are expressed broadly in allantoic mesoderm prior to chorioallantoic attachment,[35,36] and *Wnt2* expression has been shown to persist in the labyrinth. *Fzd5* mRNA is localized to the presumptive labyrinth layer[37] and, although it has not been precisely localized, based on the broad expression it is likely to be in the trophoblast compartment if not also in the derivatives of the allantoic mesoderm. Wnt2 can interact with Fzd5, and mutations in either gene result in labyrinth defects. Notably, the *Wnt2* mutants show a much less severe phenotype, perhaps reflecting overlapping function with a related Wnt ligand. The precise function of Wnt signaling at the chorioallantoic interface is unclear, but Wnt functions in other systems would be consistent with potential roles in regulating polarity or differentiation of trophoblast cells in the chorion.

Several receptor tyrosine kinases have been implicated in labyrinth development, including receptors for HGF (c-Met), EGF (EGFR), and FGF (FGFR2).[5,38] HGF signaling has been implicated in tissue morphogenesis in other systems in mice,[39–41] and FGF signaling is a highly conserved pathway associated with branching morphogenesis.[42] There are three general conclusions that emerge from the work. First, one of the most satisfying results is that loss-of-function mutations in components of receptor tyrosine kinase signaling cascades,[5,38] which had been identified using biochemical approaches, support the same pathway hierarchies. Second, it is notable that even null mutations affecting widely expressed signaling components such as Gab1, Sos1, Mek1, and p38α MAP kinase do not affect early development and only manifest as midgestation embryonic lethal phenotypes.[5,38] The one exception is the signaling adaptor Grb2, in which the null mutants fail to progress beyond the blastocyst stage due to endoderm differentiation defects.[43] Grb2 has a later essential function, similar to that of its known downstream effector Sos1,[44] during the progression of labyrinth development.[45] The unique early requirement for Grb2 implies that its effects at this stage are independent of Sos1, Mek1, and p38α MAP kinase. Third, the effects of the receptor tyrosine kinase signaling events appear to be largely essential within the trophoblast compartment of the labyrinth. This conclusion is based on the results of expression studies that show restricted expression (e.g., Fgfr2) or chimera studies with mutant cells that prove that the function of a particular signaling component is required in the trophoblast, but not extraembryonic mesoderm, derived cells (e.g., p38α) (for review, see ref. 5).

The labyrinth phenotypes associated with mutations in the signaling pathways described above are severe and usually associated with embryonic lethality. As a group, however, the block in labyrinth development occurs after the initiation of chorioallantoic morphogenesis, in contrast to *Gcm1* mutants. This implies that the signaling pathways act later than *Gcm1*. It is possible that they may act to maintain *Gcm1* expression (see above), although there is no direct evidence to support this. Alternatively, the signaling pathways may regulate independent events. Detailed analysis of *Grb2* hypomorphic mutants[45] and Fgfr2 mutants (Y. Lu, C. Deng, and J. C. Cross, unpublished observations) is instructive. When labyrinth development is examined on successive days from E9.5 to 11.5, one observes that, rather than being arrested at a specific stage of development, chorioallantoic morphogenesis proceeds, but is very slow compared to wild type. This implies that, for *Fgfr2* and *Grb2* at least, these genes regulate the progression, but not the initiation of morphogenesis.

CONCLUSION

Development of the chorioallantoic structure of the placenta involves a number of intricate cell signaling events, associated first with regulation of chorioallantoic attachment through expression of α4 integrin, subsequently with maintenance of Gcm1 expression for initiation of morphogenesis, and finally with stimulating the progression of morphogenesis. The large number of mutant mice that show phenotypes in these events provides exceptional opportunities for exploring signaling pathways in a clinically important biological setting.

ACKNOWLEDGMENTS

The work performed in the Cross lab was supported by operating grants from the Canadian Institutes of Health Research (CIHR) and an establishment grant from the Alberta Heritage Foundation for Medical Research (AHFMR). J. C. Cross is an Investigator of the CIHR and a Senior Scholar of the AHFMR.

REFERENCES

1. Cross, J.C., M. Hemberger, Y. Lu *et al.* 2002. Trophoblast functions, angiogenesis, and remodeling of the maternal vasculature in the placenta. Mol. Cell. Endocrinol. **187:** 207–212.
2. Cross, J.C., Z. Werb & S.J. Fisher. 1994. Implantation and the placenta: key pieces of the development puzzle. Science **266:** 1508–1518.
3. Cross, J.C. 2000. Genetic insights into trophoblast differentiation and placental morphogenesis. Semin. Cell Dev. Biol. **11:** 105–113.
4. Kingdom, J.C.P. & P. Kaufmann. 1997. Oxygen and placental villous development: origins of fetal hypoxia. Placenta **18:** 613–621.
5. Rossant, J. & J.C. Cross. 2001. Placental development: lessons from mouse mutants. Nat. Rev. Genet. **2:** 538–548.
6. Kwee, L., H.S. Baldwin, H.M. Shen *et al.* 1995. Defective development of the embryonic and extraembryonic circulatory systems in vascular cell adhesion molecule (VCAM-1) deficient mice. Development **121:** 489–503.
7. Gurtner, G.C., V. Davis, H. Li *et al.* 1995. Targeted disruption of the murine VCAM1 gene: essential role of VCAM1 in chorioallantoic fusion and placentation. Genes Dev. **9:** 1–14.
8. Yang, J.T., H. Rayburn & R.O. Hynes. 1995. Cell adhesion events mediated by alpha-4 integrins are essential in placental and cardiac development. Development **121:** 549–560.
9. Downs, K.M. 2002. Early placental ontogeny in the mouse. Placenta **23:** 116–131.
10. Downs, K.M. & R.L. Gardner. 1995. An investigation into early placental ontogeny: allantoic attachment to the chorion is selective and developmentally regulated. Development **121:** 407–416.
11. Parr, B.A., V.A. Cornish, M.I. Cybulsky *et al.* 2001. Wnt7b regulates placental development in mice. Dev. Biol. **237:** 324–332.
12. Galceran, J., I. Farinas, M.J. Depew *et al.* 1999. Wnt3a–/–-like phenotype and limb deficiency in Lef1(–/–)Tcf1(–/–) mice. Genes Dev. **13:** 709–717.

13. SOLLOWAY, M.J. & E.J. ROBERTSON. 1999. Early embryonic lethality in Bmp5;Bmp7 double mutant mice suggests functional redundancy within the 60A subgroup. Development **126:** 1753–1768.
14. XU, X., M. WEINSTEIN, C. LI *et al.* 1998. Fibroblast growth factor receptor 2 (FGFR2)–mediated reciprocal regulation loop between FGF8 and FGF10 is essential for limb induction. Development **125:** 753–765.
15. HUNTER, P.J., B.J. SWANSON, M.A. HAENDEL *et al.* 1999. *Mrj* encodes a DnaJ-related co-chaperone that is essential for murine placental development. Development **126:** 1247–1258.
16. CYR, D.M., T. LANGER & M.G. DOUGLAS. 1994. DnaJ-like proteins: molecular chaperones and specific regulators of Hsp70. Trends Biochem. Sci. **19:** 176–181.
17. IZAWA, I., M. NISHIZAWA, K. OHTAKARA *et al.* 2000. Identification of Mrj, a DnaJ/Hsp40 family protein, as a keratin 8/18 filament regulatory protein. J. Biol. Chem. **275:** 34521–34527.
18. CHUANG, J.Z., H. ZHOU, M. ZHU *et al.* 2002. Characterization of a brain-enriched chaperone, MRJ, that inhibits Huntingtin aggregation and toxicity independently. J. Biol. Chem. **277:** 19831–19838.
19. ALTSHULLER, Y., N.G. COPELAND, D.J. GILBERT *et al.* 1996. *Gcm1*, a mammalian homolog of *Drosophila glial cells missing*. FEBS Lett. **393:** 201–204.
20. AKIYAMA, Y., T. HOSOYA, A.M. POOLE *et al.* 1996. The Gcm-motif: a novel DNA-binding motif conserved in *Drosophila* and mammals. Proc. Natl. Acad. Sci. USA **93:** 14912–14916.
21. BASYUK, E., J.C. CROSS, J. CORBIN *et al.* 1999. The murine *Gcm1* gene is expressed in a subset of placental trophoblast cells. Dev. Dyn. **214:** 303–311.
22. JANATPOUR, M.J., M.F. UTSET, J.C. CROSS *et al.* 1999. A repertoire of differentially expressed transcription factors that offers insight into mechanisms of human cytotrophoblast differentiation. Dev. Genet. **25:** 146–157.
23. SCHREIBER, J., E. RIETHMACHER-SONNENBERG, D. RIETHMACHER *et al.* 2000. Placental failure in mice lacking the mammalian homolog of glial cells missing, GCMa. Mol. Cell. Biol. **20:** 2466–2474.
24. ANSON-CARTWRIGHT, L., K. DAWSON, D. HOLMYARD *et al.* 2000. The glial cells missing-1 protein is essential for branching morphogenesis in the chorioallantoic placenta. Nat. Genet. **25:** 311–314.
25. NAIT-OUMESMAR, B., A.B. COPPERMAN & R.A. LAZZARINI. 2000. Placental expression and chromosomal localization of the human GCM1 gene. J. Histochem. Cytochem. **48:** 915–922.
26. YAMADA, K., H. OGAWA *et al.* 1999. A GCM motif protein is involved in placenta-specific expression of human aromatase gene. J. Biol. Chem. **274:** 32279–32286.
27. KIM, J., B.W. JONES, C. ZOCK *et al.* 1998. Isolation and characterization of mammalian homologs of the *Drosophila* gene glial cells missing. Proc. Natl. Acad. Sci. USA **95:** 12364–12369.
28. STECCA, B., B. NAIT-OUMESMAR, K.A. KELLEY *et al.* 2002. Gcm1 expression defines three stages of chorio-allantoic interaction during placental development. Mech. Dev. **115:** 27–34.
29. NAKAYAMA, H., Y. LIU, S. STIFANI *et al.* 1997. Developmental restriction of *Mash-2* expression in trophoblast correlates with potential activation of the NOTCH-2 pathway. Dev. Genet. **21:** 21–30.
30. LI, Y. & R.R. BEHRINGER. 1998. Esx1 is an X-chromosome-imprinted regulator of placental development and fetal growth. Nat. Genet. **20:** 309–311.
31. SCHORPP-KISTNER, M., Z.Q. WANG, P. ANGEL *et al.* 1999. JunB is essential for mammalian placentation. EMBO J. **18:** 934–948.
32. CASTELLUCC, M., G. KOSANKE, F. VERDENELLI *et al.* 2000. Villous sprouting: fundamental mechanisms of human placental development. Hum. Reprod. Update **6:** 485–494.
33. WARBURTON, D., M. SCHWARZ, D. TEFFT *et al.* 2000. The molecular basis of lung morphogenesis. Mech. Dev. **92:** 55–81.
34. BURROW, C.R. 2000. Regulatory molecules in kidney development. Pediatr. Nephrol. **14:** 240–253.

35. Yamaguchi, T.P., A. Bradley, A.P. McMahon *et al.* 1999. A Wnt5a pathway underlies outgrowth of multiple structures in the vertebrate embryo. Development **126:** 1211–1223.
36. Monkley, S.J., S.J. Delaney, D.J. Pennisi *et al.* 1996. Targeted disruption of the Wnt2 gene results in placentation defects. Development **122:** 3343–3353.
37. Ishikawa, T., Y. Tamai, A.M. Zorn *et al.* 2001. Mouse Wnt receptor gene Fzd5 is essential for yolk sac and placental angiogenesis. Development **128:** 25–33.
38. Hemberger, M. & J.C. Cross. 2001. Genes governing placental development. Trends Endocrinol. Metab. **12:** 162–168.
39. Ohmichi, H., U. Koshimizu, K. Matsumoto *et al.* 1998. Hepatocyte growth factor (HGF) acts as a mesenchyme-derived morphogenic factor during fetal lung development. Development **125:** 1315–1324.
40. Schmidt, C., F. Bladt, S. Goedecke *et al.* 1995. Scatter factor/hepatocyte growth factor is essential for liver development. Nature **373:** 699–702.
41. Schaeper, U., N.H. Gehring, K.P. Fuchs *et al.* 2000. Coupling of Gab1 to c-Met, Grb2, and Shp2 mediates biological responses. J. Cell Biol. **149:** 1419–1432.
42. Skaer, H. 1997. Morphogenesis: FGF branches out. Curr. Biol. **7:** R238–R241.
43. Cheng, A.M., T.M. Saxton, R. Sakai *et al.* 1998. Mammalian Grb2 regulates multiple steps in embryonic development, lineage commitment, and malignant transformation. Cell **95:** 793–803.
44. Qian, X., L. Esteban, W.C. Vass *et al.* 2000. The Sos1 and Sos2 Ras-specific exchange factors: differences in placental expression and signaling properties. EMBO J. **19:** 642–654.
45. Saxton, T.M., A.M. Cheng, S.H. Ong *et al.* 2001. Gene dosage–dependent functions for phosphotyrosine-Grb2 signaling during mammalian tissue morphogenesis. Curr. Biol. **11:** 662–670.

Tissue Involution and the Acute Phase Response

MARIT NILSEN-HAMILTON,[a] QUANSHENG LIU,[b] JOEL RYON,[a]
LEE BENDICKSON,[a] PIERIG LEPONT,[c] AND QING CHANG[a]

[a]*Department of Biochemistry, Biophysics, and Molecular Biology, Iowa State University, Ames, Iowa, USA*

[b]*Department of Immunology, Shanghai Medical College, Fudan University, Shanghai, China*

[c]*Center for Molecular Studies, University of Cincinnati, Cincinnati, Ohio, USA*

ABSTRACT: After their roles in reproduction are completed, the mass of the uterus and the mammary gland decrease rapidly by the process of involution that involves an ordered series of events including apoptosis, neutrophil entry, the release of degradative enzymes, and phagocytosis of cellular debris. The acute phase proteins are produced by the liver and other tissues in response to inflammation or a toxic challenge. Uterocalin (SIP24/24p3) is one of these proteins. During involution, the mammary gland and uterus express high levels of uterocalin that reach an average of 0.2–0.5% of the total extractable protein at its peak. Uterocalin and its orthologues have been demonstrated *in vitro* to (1) bind certain fatty acids and (2) specifically induce apoptosis in neutrophils and other leukocytes. The period of uterocalin expression during involution is consistent with the hypothesis that one of its physiological roles is to induce apoptosis of invading neutrophils and delay the entry of neutrophils into the tissue until the second phase of involution. Interestingly, it has been shown that uterocalin expression remains higher in primiparous gland than in virgin glands after the pregnant glands have completely involuted. This observation and the known protective effect of early pregnancy on later development of breast cancer suggest that the ability of uterocalin to induce apoptosis in neutrophils might also decrease oxidative and carcinogenic activity in the gland and result in a lower mutation rate and thus a lower probability of cancer in the primiparous gland.

KEYWORDS: uterus; mammary gland; apoptosis; uterocalin; 24p3

TISSUE REMODELING

During reproduction in the female mammal, several tissues remodel extensively in a cyclical manner, most notably the mammary gland and the uterus. Remodeling of these tissues is initiated by systemic signals generated during reproductive activities

Address for correspondence: Marit Nilsen-Hamilton, Department of Biochemistry, Biophysics, and Molecular Biology, 3206 Molecular Biology Building, Iowa State University, Ames, IA 50011. Voice: 515-294-9996; fax: 515-294-0453.
marit@iastate.edu

Ann. N.Y. Acad. Sci. 995: 94–108 (2003).

such as the ovarian cycle and pregnancy. However, many events in remodeling are regulated by signals generated within the tissue. Epithelial-mesenchymal interactions are involved in these events, such as the branching morphogenesis that occurs during some stages in remodeling of the mammary gland. Local signals are also likely to control entry of immune cells, including phagocytes, which contribute to resolving involution of tissues such as the uterus and the mammary gland.

Epithelial cells that interface with the external environment also play a role in the innate immune response as the first line of defense against microbial challenge. During involution of the uterus and mammary gland, this challenge is particularly intense due to the external exposure of the tissue surfaces. Postpartum uterine infection (metritis) and postweaning mammary gland infection (mastitis) are reproductive risks. Microbial accessibility to these tissues is highest during reproduction, and the involuting tissue environment is a rich nutrient source for microbial growth. Therefore, these tissues might be expected to possess enhanced capabilities of combating microbial infections, particularly during the period of their involution.

INVOLUTION DURING REPRODUCTION

The uterus and the mammary gland undergo extensive proliferation and differentiation during pregnancy. For both tissues, recovery to a normal state after pregnancy involves involution that occurs in the uterus after birth and in the mammary gland after weaning. Tissue structure and mass change dramatically and rapidly during involution.

Careful studies of the mammary gland revealed two distinct stages of involution.[1,2] Apoptosis dominates the first phase of involution in the mammary gland. During the second phase of involution, the mammary alveoli collapse and the periductal stroma produces stromelysin and plasminogen activator. This increase in proteolytic activity in the gland is followed by an influx of macrophages that are believed responsible for scavenging the remaining cellular debris. Simultaneously, adipocytes begin to populate the gland. In this manner, the mouse mammary gland recovers its pre-pregnant morphology by 21 days after weaning.[2]

Like the mammary gland, the uterus also progresses through two stages of involution that involve an early increase in apoptosis followed by increased protease expression.[3,4] During the time period of postpartum days 1 through 6, the wet weight of the uterus decreases 8-fold from an average 1.1 g to an average 0.15 g of tissue per uterus (FIG. 1). The major part of this loss in mass is due to a loss of cellular content because it could be also measured as loss of DNA content of the tissue. Postpartum involution of the uterus results in the rapid loss of additional cellular content that had accumulated over 5 gestational days in the mouse and thus returns the uterus to its condition prior to pregnancy.[5]

Tissue remodeling, such as occurs during involution, that involves apoptosis and tissue destruction prior to resorption also results in the production of signals that activate the innate immune response. A fundamental question regarding tissues such as the uterus and the mammary gland is how these tissues, which harbor a large number of phagocytes during involution and certain other reproductive stages, control the inflammatory response. Oxidative damage during inflammation can result in permanent DNA damage leading to cancer. How do these tissues protect themselves

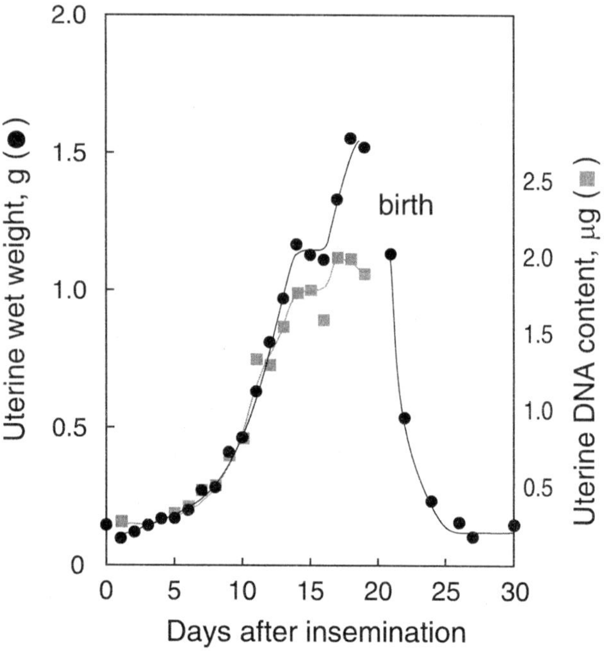

FIGURE 1. Growth and involution of the pregnant uterus. Uteri were dissected from mice at various stages of pregnancy or postpartum, rinsed, external fluid removed, and then weighed. The average weight per uterus is plotted as a function of days after insemination, with birth set at day 20 for postpartum timing. Each data point is the average of at least three mice, except for the day 2 sample (one value). Data for total DNA per uterus is from Fang *et al.*[5] Uterine weight (●); uterine DNA content (■).

from excessive acute damage by activated phagocytes, particularly during periods of tissue remodeling such as occurs during involution?

THE ACUTE PHASE RESPONSE (APR)

The acute phase response was initially defined by the proteins that appear rapidly (within 24 h) in the bloodstream in response to an introduced systemic infection or toxin. These proteins are produced as the result of altered gene expression in the liver and are referred to as the acute phase proteins. The acute phase proteins are part of the innate immune response, and several of them have been shown to modulate the cellular immune response.[6]

Uterocalin is an acute phase protein that exhibits one of the largest early increases in gene expression in the liver on toxic challenge.[7] Uterocalin is also expressed in response to a lipopolysaccharide challenge.[8] Lipopolysaccharides (LPS) are an important feature by which gram-negative bacteria are recognized as infectious agents. FIGURE 2 shows the results of experiments in which mice were challenged

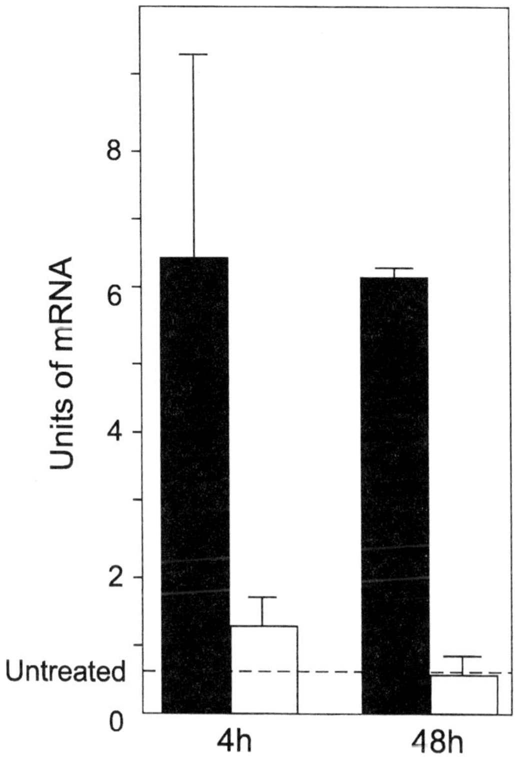

FIGURE 2. Regulation of uterocalin expression in the lung by LPS. Mice were anesthetized and treated intranasally for 4 or 48 h with either PBS (*white bars*) or LPS in PBS (*black bars*) made up in nano-pure water. Total RNA was then isolated and analyzed by Northern blot. The membrane was first hybridized with a ^{32}P-labeled probe to uterocalin/24p3 and then with a ^{32}P-labeled probe to rat GPDH. The results show the ratio of cpm hybridized with the uterocalin mRNA over the cpm hybridized with GPDH mRNA. The data have been normalized to GPDH and a uterocalin standard. The level of uterocalin expression in untreated mice is shown as a dashed line. At least three mice were used to obtain each value.

with inhaled LPS. Uterocalin was expressed by the lungs within 4 h of challenge and expression was even higher after 24 h. The chicken orthologue of uterocalin, ex-FABP (extracellular fatty acid binding protein), is also induced by the inflammatory signals, LPS and interleukin-6.[9]

Originally called SIP24 (superinducible protein, M_r 24,000), later named 24p3 and then uterocalin, the gene is now identified as *Lcn2* in the mouse genome.[10–12] The protein was renamed uterocalin because the highest levels of gene expression *in vivo* were found in the postpartum uterus. However, this new name may have been given prematurely as it was soon discovered that the mammary gland expresses even higher levels of uterocalin than the uterus or the liver.[13] Although a more appropriate name for this protein must await resolution of its physiological role, the name uterocalin is more relevant to its role *in vivo* than either SIP24 or 24p3.

The human equivalent of uterocalin is the protein that is best known as neutrophil gelatinase associated lipocalin (NGAL). The name NGAL came from the observation that a small fraction of the protein is covalently associated with gelatinase (mmp9).[14–17] NGAL is 71% identical in amino acid sequence with mouse uterocalin. As for uterocalin, there is evidence that NGAL is an acute phase protein.[18,19] NGAL is only expressed at the myelocyte/metamyelocyte stage in the course of neutrophil maturation.[20] Once synthesized, the protein is stored in the azurophilic granules until it is released when the neutrophil is activated.[20,21] Consistent with its homology in protein sequence, uterocalin is also detected in mouse neutrophils stained with anti-uterocalin antibodies (T. Bratt, personal communication).

There are many reports of elevated NGAL expression associated with human disease. As expected, based on its identification as an acute phase protein, elevated serum NGAL has been observed associated with several diseases with an inflammatory component, including atherosclerosis, inflammatory bowel disease, and ischemic cerebrovascular diseases.[22–26] NGAL is also expressed by the intestinal epithelium in inflammatory bowel disease, in primary breast cancers, and by a variety of human ovarian cancer cell lines.[16,23,24,27] A more extensive discussion of lipocalins, including NGAL, and cancer can be found in a review by Bratt and colleagues.[28]

In addition to their expression during adult life, uterocalin and its orthologues are expressed during fetal development in both fetal and extraembryonic tissues. The rat protein (NRL) is expressed in prehypertrophic chondrocytes and in forming skeletal muscle.[29] Similarly, in the developing chicken, ex-FABP is expressed by chondrocytes and myocytes.[30,31] Uterocalin expression also occurs in the midgestational mouse placenta and peaks between days 9 and 10 postinsemination (Liu and Nilsen-Hamilton, unpublished).

UTEROCALIN EXPRESSION BY THE UTERINE EPITHELIUM

The postpartum involuting uterus is another major site of uterocalin expression. On day 19 in mouse gestation, the uterocalin mRNA level in the uterus begins to rise rapidly. By postpartum day 1, it reaches its peak level, which is about 400-fold over the midgestational level. The protein can then be detected easily in the uterus by Western blot.[12]

Despite the massive increase in uterocalin expression in the postpartum uterus, the uterocalin mRNA level in the maternal liver does not increase postpartum (FIG. 3), and uterocalin protein is not detected in the postpartum bloodstream.[12] By contrast, during the acute phase response, the levels of uterocalin mRNA increase dramatically in the liver and the protein is released systemically.[7] This suggests that, during involution, there is local control of uterocalin expression in the uterus by one or more agents that do not circulate in sufficient quantities to stimulate postpartum maternal liver expression. Thus, uterocalin levels in the placenta, uterus, and maternal liver seem to be under separate, and probably local, regulation during pregnancy.

The results from immunohistochemical stains for uterocalin in late-gestational uterine tissue provided an explanation for why uterocalin is not secreted by the uterus into the bloodstream. Uterocalin is localized mainly in the uterine luminal epithelium of gestational day 19 animals.[12] Judging from the distribution of utero-

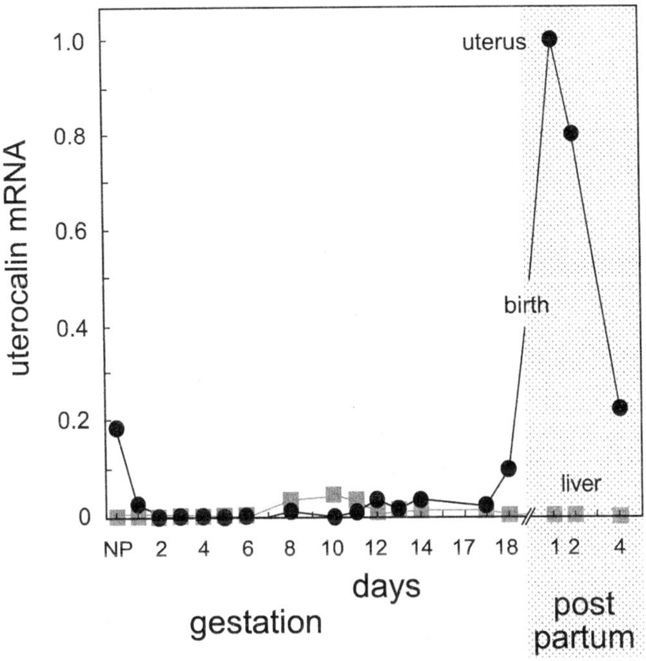

FIGURE 3. Expression of uterocalin mRNA in the uterus during reproduction. Northern blot analysis of uterine and maternal liver RNA from female CF-1 mice at various stages of pregnancy and around birth. Total RNAs were extracted from the uteri and resolved by agarose gel electrophoresis (15 μg/lane). The RNAs were transferred to nylon membranes and hybridized sequentially with ^{32}P-labeled uterocalin/24p3 and rat 18S rRNA cDNA probes. The amount of probe hybridized to uterocalin mRNA was normalized to the amount of 18S probe hybridized to the 18S rRNA in the same lane: ●, mammary gland; ■, liver. The data were derived from two independent analyses. Each point is the mean of 2–7 independent values. Adapted with permission from ref. 13, with additional data included.

calin in the epithelial cells, it appears as though the protein is secreted luminally in the direction of the placenta and fetus. However, without results from *in situ* hybridization studies, we cannot exclude the possibility that uterocalin is taken up by the epithelium and synthesized by another cell type. Interestingly, most of the uterocalin was observed to be inside the cells of the prepartum uterus, whereas it was found as a layer on the luminal epithelial surface of the postpartum uterus.[12] Quantitative Western blot analysis showed that uterocalin was about 0.15% of the total detergent-extracted protein obtained from 1-day postpartum uteri.

UTEROCALIN EXPRESSION IN THE MAMMARY GLAND AND BY MAMMARY EPITHELIAL CELLS

Uterocalin expression also begins to increase in the mammary gland in late gestation.[13] The increase parallels that in the uterus through day 1 postpartum. By

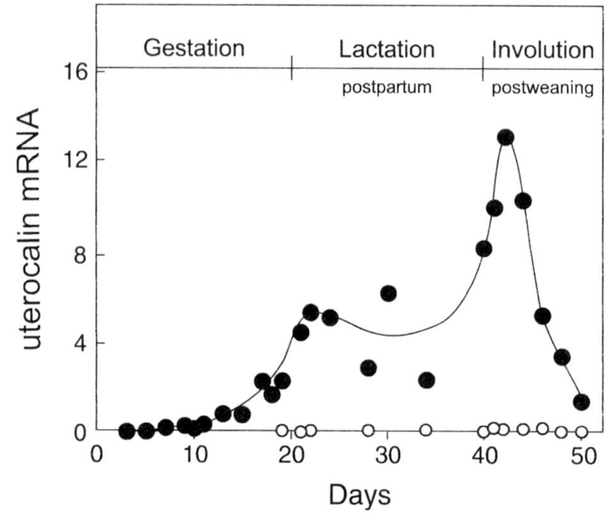

FIGURE 4. Expression of uterocalin in the mammary gland during reproduction. Female CF-1 mice were sacrificed at various times during gestation, lactation, and post-weaning. Mammary gland and liver tissues were collected, and total RNAs were isolated and analyzed by Northern blot. The membranes were probed first with a ^{32}P-labeled uterocalin probe and second with a ^{32}P-labeled rat GPDH probe as a control for loading. The values for uterocalin RNA were normalized to those for GPDH in the same sample. The amount of uterocalin present in the mammary gland and liver is plotted as a function of reproductive stage. For the mammary gland, most points are the average of 3 values, each from different mice. Exceptions are d11 (2 mice) and PW1 (5 mice). For the liver, all points are the average of 2 mice. ●, mammary gland; ○, liver. Adapted with permission from ref. 13.

this time, the levels of uterocalin expression in the mammary gland and uterus are about equal and similar to the level in the acute phase liver. Whereas, by day 2 post-partum, the uterine uterocalin level has begun to decline, the mammary uterocalin level remains high throughout lactation. Immediately postweaning, there is a peak of uterocalin expression in the mammary gland (FIG. 4).

Immunocytochemistry showed that uterocalin is mainly distributed diffusely throughout the mammary gland with some concentration over the epithelium. This is not an unusual distribution for a secreted protein. However, with these studies, we were unable to locate the protein to a single cell type. Quantitative Western blot analysis showed that the mammary gland contains a large quantity of uterocalin during lactation. In midlactation, uterocalin constitutes about 0.2% of the detergent-extracted protein and, at its peak during involution, the level of uterocalin reaches about 0.5% of the detergent-extracted protein.

Weaning of the young induces apoptosis and involution in the mammary gland.[32] The level of uterocalin expression also increases immediately after weaning, whether weaning is forced or with natural timing. Apoptosis is initiated in the mammary gland within hours of weaning.[32] Thus, the increase in uterocalin expression is correlated in time with apoptosis in the involuting mammary gland and the involuting uterus.

To determine whether mammary epithelial cells produce uterocalin, we tested the ability of the HC11 mammary subclone of the COMMA-1D mouse mammary epithelial cell line to produce uterocalin.[33] This cell line was chosen because it has been well studied and has many of the characteristics of secretory alveolar mammary epithelial cells. RNA was extracted from quiescent HC11 cells and tested by Northern blot analysis to detect uterocalin mRNA. The results showed that these cells express at least as much uterocalin mRNA as a lactating mammary gland when both were normalized to the glyceraldehyde-3-phosphate dehydrogenase (G3PDH) mRNA.

UTEROCALIN EXPRESSION DURING APOPTOSIS IN THE FIRST PHASE OF INVOLUTION

The period of high uterocalin expression in the mammary gland corresponds to a previously defined initial phase of involution when the levels of expression of the SGP2 (clusterin) and TIMP (tissue inhibitor of metalloproteinases) genes increase. Whereas TIMP protects against metalloproteinase activity, the SGP2 protein, otherwise called apolipoprotein J or clusterin, causes cell aggregation that protects cells against damage by oxidative stress.[34] SGP2 is produced by bystander cells rather than by the cells undergoing apoptosis and may help to prepare the bystander cells, which are not destined for death, to endure the inflammatory activities of invading phagocytes during the second phase of involution.[35]

The first phase of mammary gland involution is followed by a second phase when the level of expression of the stromelysin-1 increases. This second phase is marked by an increase in metalloproteases such as stromelysin-1 and plasminogen activator that are products of the mammary stromal cells.[1,36]

The uterus follows a similar time course of involution as the mammary gland with respect to the expression of uterocalin, SGP2, and stromelysin-1. Immediately postpartum, the uterus enters the first phase of involution during which apoptosis increases as does uterocalin and SGP2 expression.[13] The parallel decline in uterine weight and uterocalin expression suggests that the cells that are undergoing apoptosis may be those that are producing uterocalin.

MOLECULAR FUNCTIONS OF UTEROCALIN AND NGAL

Various functions have been ascribed to uterocalin and its orthologues. Several functions are related to those of other members of the lipocalin protein family to which it belongs. Lipocalins form a basketlike structure known as a β-barrel that is created by a series of 9 antiparallel beta-sheets. The cavity formed by this structure usually binds fatty acids or other hydrophobic molecules.

The earliest reported ligand for uterocalin and orthologues was fMLP (*N*-formylmethionine-leucine-phenylalanine), which was found cross-linked to NGAL after light activation of a photoactivatable fMLP derivative.[37] However, more recent studies showed that fMLP does not bind stoichiometrically and does not bind in the NGAL pocket.[28] The search for uterocalin ligands led to the identification of cholesteroyl oleate, oleic acid, and retinoic acid as ligands for uterocalin using an assay that involved measuring the quenching of a Trp located in the uterocalin pocket.[38] However, the reported stoichiometry for retinoic acid was very low and no stoichiometry was reported for cholesteroyl oleate or oleic acid. Fatty acid ligands, including arachidonic acid and oleic acid, have also been reported for the chicken orthologue, ex-FABP.[39] In this study, the stoichiometry of ex-FABP binding to linoleic acid was shown to be close to 1. Thus, these long-chain, unsaturated fatty acids appear to be bona fide ligands of ex-FABP that probably bind in the hydrophobic pocket of this lipocalin. More recently, as a result of solving the crystal structure of NGAL, another potential group of ligands have been identified. These are catecholate-type ferric siderophores that are released by bacteria to sequester the iron necessary for their proliferation.

Thus, two quite different types of ligands have been identified for uterocalin and its orthologues. Both are consistent with some of the biological functions ascribed to these proteins. Some fatty acid ligands identified for ex-FABP are precursors of prostaglandins. These potential ligands are more consistent with the hypothesis that uterocalin regulates cellular functions such as apoptosis.[40,41] The identification of a siderophore as an NGAL ligand is more consistent with the hypothesis that the high level of uterocalin expression during the acute phase response combats bacterial infection. High levels of uterocalin expression in the involuting mammary gland and uterus might function to capture iron-loaded siderophores and thus limit bacterial growth. This latter proposed function is also consistent with the external direction of uterocalin secretion by epithelial cells of the uterus and mammary gland. In the uterus at birth, uterocalin is found on the luminal surface of the epithelium as the gland begins to involute.[12] Secretion into the milk demonstrates that the mammary epithelium also secretes uterocalin externally and into the compartment where bacterial infection is most likely to occur.[13]

CELLULAR FUNCTIONS OF UTEROCALIN

Although the function of uterocalin has not been demonstrated for most tissues in which it is expressed, it has been demonstrated that uterocalin induces apoptosis in cultured leukocyte lines and primary human neutrophils.[41] Interleukin-3 (IL-3) suppresses expression of uterocalin by FL5.12 pro-B cells. Thus, this effect explains the observation that removal of IL-3 from the culture medium of these cells results in their death by apoptosis.[41] These results also suggest that the association of uterocalin expression with apoptosis in the mammary gland and the uterus could be because uterocalin induces apoptosis of the epithelium in these involuting tissues.

MECHANISMS OF REGULATION OF UTEROCALIN GENE EXPRESSION

The tissue specificity and large variations in uterocalin expression *in vivo* point to some likely specific mechanisms of gene regulation. Studies of uterocalin gene expression in cell culture also provide some clues. In fibroblasts, uterocalin expression is stimulated by fibroblast growth factor (FGF), epidermal growth factor (EGF), tumor necrosis factor-α (TNFα), interleukin-6 (synergistically with TNFα), and dexamethasone.[7,10,42–44] These hormones and growth factors are expressed at various times during gestation and lactation. Thus, there are many candidate regulators of uterocalin expression *in vivo*. Other growth factors and cytokines have no effect or inhibit uterocalin expression. For example, insulin does not alter expression, whereas interferon-γ inhibits uterocalin expression.[10,42] Insulin-like growth factor (IGF-1) also inhibits uterocalin expression by neutrophils and may mediate suppression of uterocalin expression by IL-3.[41]

Of the growth factors known to regulate uterocalin in cell culture, EGF might be expected to regulate uterocalin expression in the mammary gland because this growth factor has been shown to be important for growth and development of epithelial cells in the mammary gland. However, uterocalin expression in HC11 epithelial cells is not regulated by EGF (Ryon and Nilsen-Hamilton, unpublished); nor is uterocalin expression regulated by prolactin, a hormone that, with EGF, regulates many genes in mammary epithelial cells. β-Casein, a gene that is regulated by prolactin, is synthesized and secreted into the milk along with uterocalin during lactation. However, the patterns of uterocalin and β-casein expression diverge drastically on involution, when β-casein expression decreases and uterocalin expression increases sharply.[1] Thus, it seems unlikely that uterocalin is regulated *in vivo* by either EGF or prolactin.

We have also tested the possibility that EGF and its relatives, amphiregulin and TGFα, might regulate the expression of uterocalin in the uterus. Uterocalin levels were measured in postpartum uterine tissues of triple knockout mice and their normal counterparts obtained from David Lee (unpublished results).[45] The results showed no difference in expression of uterocalin between normal tissues and knockout tissues in which the three EGF-family growth factors (EGF, TGFα, and amphiregulin) are not expressed (Ryon and Nilsen-Hamilton, unpublished data).

Uterocalin expression is regulated by glucocorticoids *in vivo* and in cultured fibroblasts and HC11 epithelial cells.[12,13,42] This regulation is likely mediated by one or both glucocorticoid response elements that are located in the proximal uterocalin promoter.[43] However, it is unlikely that glucocorticoids are responsible for the increase in uterocalin expression during involution because glucocorticoids delay the onset of involution and inhibit apoptosis in the mammary epithelium.[1,46]

Prostaglandin F2α (PGF2α) induces uterocalin expression in 3T3 cells and is a likely candidate for the regulation of uterocalin in the uterus and the mammary gland.[10] Like uterocalin, the level of PGF2α in the blood of several species falls below the nonpregnant levels in early gestation. There is then a transient midgestational rise in PGF2α and an increase again at parturition.[47] PGF2α continues to be produced in the mammary gland during lactation and is found in breast milk.[47] The apparent coincidence of uterocalin expression and the production of PGF2α in the uterus and the mammary gland suggests that PGF2α could be involved in the action or regulation of uterocalin.

The transcription factor NF-κB is very likely to be an *in vivo* regulator of uterocalin gene expression. This is suggested by the fact that uterocalin gene expression is regulated *in vivo* by LPS (FIG. 2). Uterocalin expression is also stimulated in cultured cells by LPS and polyIC.[8,44] Both LPS and polyIC act through toll-like receptors for which NF-κB is the primary downstream transcription factor.[48,49] Also, the NGAL proximal promoter includes a consensus site for NF-κB binding.[50]

INFLAMMATION, UTEROCALIN, AND CANCER

Chronic inflammation is a well-established risk factor for cancer.[51–53] The risk of colon cancer is decreased by about 50% with regular ingestion of nonsteroidal anti-inflammatory drugs (NSAIDs) such as aspirin.[54] The association between inflammation and cancer risk is also seen in human diseases in which chronic inflammation is often associated with higher cancer risk.[52,55] Inflammatory conditions or surgical treatments that alter exposure of tissues to inflammatory stimuli alter the risk of cancer in an inverse relationship.[51,53,56,57] As well, diseases and gene deletions in mice and humans that result in the loss of anti-inflammatory proteins demonstrate higher cancer rates.[58–60]

It is well established that parity (full-term pregnancy) conveys protection against cancers of the breast and other reproductive tissues. This correlation has been demonstrated in many studies using rodent models and from epidemiologic analyses of human populations. Full-term pregnancy seems necessary for the protection against breast cancer.[61,62] Protection by parity is a combined effect of host and local tissue effects.[63] A recent microarray analysis of gene expression in parous and virgin rat and mouse mammary glands revealed several differences in gene expression in the mammary glands from parous compared with virgin female mice and rats.[64] Uterocalin is one of the small subset of genes for which expression remained elevated beyond the involution period in the primiparous animal compared with the virgin animal.

Uterocalin stimulates apoptosis of neutrophils, but not of macrophages.[41] Whereas neutrophils are the quintessential acute phase responders, macrophages play a subsequent, but extremely important, role in the inflammatory response; this role is to engulf apoptotic neutrophils and to subdue the inflammatory response.[65] Activated neutrophils, while essential for combating microbial infections, also produce numerous oxidative and carcinogenic products. Carcinogens such as DMBA must be metabolized (bioactivated) to become carcinogenic *in vivo*.[66] The ability of activated neutrophils to metabolize and produce active carcinogenic species has been proposed as an explanation of why epithelial cells in the breast are sensitive targets for chemical carcinogenesis despite their own limited bioactivation capacity.[67] Thus, it is reasonable to suppose that the extent of DNA damage and mutation in a tissue increases in parallel with the average number of activated neutrophils present over time.

Being elevated in the mammary glands of parous animals compared with virgin animals, and having a function that would suppress inflammation, uterocalin is an excellent candidate for the long-sought gene that contributes to long-term protection of the mammary gland from breast cancer.

ACKNOWLEDGMENTS

This work was funded in part by Grant No. BC9964884 from the Department of Defense and by Iowa State University College of Agriculture.

REFERENCES

1. Lund, L.R., J. Romer, N. Thomasset *et al.* 1996. Two distinct phases of apoptosis in mammary gland involution: proteinase-independent and -dependent pathways. Development **122:** 181–193.

2. WAGNER, K.U., C.A. BOULANGER, M.D. HENRY *et al.* 2002. An adjunct mammary epithelial cell population in parous females: its role in functional adaptation and tissue renewal. Development **129:** 1377–1386.
3. TAKAMOTO, N., P.C. LEPPERT & S.Y. YU. 1998. Cell death and proliferation and its relation to collagen degradation in uterine involution of rat. Connect. Tissue Res. **37:** 163–175.
4. BALBIN, M., A. FUEYO, V. KNAUPER *et al.* 1998. Collagenase 2 (MMP-8) expression in murine tissue-remodeling processes: analysis of its potential role in postpartum involution of the uterus. J. Biol. Chem. **273:** 23959–23968.
5. FANG, Y., P. LEPONT, J. FASSETT *et al.* 1999. Signaling between the placenta and the uterus involving the mitogen-regulated protein/proliferins. Endocrinology **140:** 5239–5274.
6. SUFFREDINI, A.F., G. FANTUZZI, R. BADOLATO *et al.* 1999. New insights into the biology of the acute phase response. J. Clin. Immunol. **19:** 203–214.
7. LIU, Q. & M. NILSEN-HAMILTON. 1995. Identification of a new acute phase protein. J. Biol. Chem. **270:** 22565–22570.
8. MEHEUS, L.A., L.M. FRANSEN, J.G. RAYMACKERS *et al.* 1993. Identification by microsequencing of lipopolysaccharide-induced proteins secreted by mouse macrophages. J. Immunol. **151:** 1535–1547.
9. CERMELLI, S., B. ZEREGA, M. CARLEVARO *et al.* 2000. Extracellular fatty acid binding protein (ex-FABP) modulation by inflammatory agents: "physiological" acute phase response in endochondral bone formation. Eur. J. Cell Biol. **79:** 155–164.
10. NILSEN-HAMILTON, M., R.T. HAMILTON & G.A. ADAMS. 1982. Rapid selective stimulation by growth factors of the incorporation by BALB/C 3T3 cells of [35S]methionine into a glycoprotein and five superinducible proteins. Biochem. Biophys. Res. Commun. **108:** 158–166.
11. HRABA-RENEVEY, S., H. TURLER, M. KRESS *et al.* 1989. SV40-induced expression of mouse gene 24p3 involves a post-transcriptional mechanism. Oncogene **4:** 601–608.
12. LIU, Q., J. RYON & M. NILSEN-HAMILTON. 1997. Uterocalin: a mouse acute phase protein expressed in the uterus around birth. Mol. Reprod. Dev. **46:** 507–514.
13. RYON, J., L. BENDICKSON & M. NILSEN-HAMILTON. 2002. High expression in involuting reproductive tissues of uterocalin/24p3, a lipocalin and acute phase protein. Biochem. J. **I367:** 271–277.
14. KJELDSEN, L., A.H. JOHNSEN, H. SENGELOV *et al.* 1993. Isolation and primary structure of NGAL, a novel protein associated with human neutrophil gelatinase. J. Biol. Chem. **268:** 10425–10432.
15. KJELDSEN, L., D.F. BAINTON, H. SENGELOV *et al.* 1994. Identification of neutrophil gelatinase-associated lipocalin as a novel matrix protein of specific granules in human neutrophils. Blood **83:** 799–807.
16. BARTSCH, S. & H. TSCHESCHE. 1995. Cloning and expression of human neutrophil lipocalin cDNA derived from bone marrow and ovarian cancer cells. FEBS Lett. **357:** 255–259.
17. TRIEBEL, S., J. BLASER, H. REINKE *et al.* 1992. A 25 kDa alpha-2-microglobulin-related protein is a component of the 125 kDa form of human gelatinase. FEBS Lett. **314:** 386–388.
18. ERSSON, A., M. WALLES, K. OHLSSON *et al.* 2002. Chronic hyperbaric exposure activates proinflammatory mediators in humans. J. Appl. Physiol. **92:** 2375–2380.
19. LIN, Y., M.W. RAJALA, J.P. BERGER *et al.* 2001. Hyperglycemia-induced production of acute phase reactants in adipose tissue. J. Biol. Chem. **276:** 42077–42083.
20. BORREGAARD, N., M. SEHESTED, B.S. NIELSEN *et al.* 1995. Biosynthesis of granule proteins in normal human bone marrow cells: gelatinase is a marker of terminal neutrophil differentiation. Blood **85:** 812–817.
21. LE CABEC, V., J.B. COWLAND, J. CALAFAT *et al.* 1996. Targeting of proteins to granule subsets is determined by timing and not by sorting: the specific granule protein NGAL is localized to azurophil granules when expressed in HL-60 cells. Proc. Natl. Acad. Sci. USA **93:** 6454–6457.
22. ELNEIHOUM, A.M., P. FALKE, L. AXELSSON *et al.* 1996. Leukocyte activation detected by increased plasma levels of inflammatory mediators in patients with ischemic cerebrovascular diseases. Stroke **27:** 1734–1738.

23. NIELSEN, O.H., P. GIONCHETTI, M. AINSWORTH *et al.* 1999. Rectal dialysate and fecal concentrations of neutrophil gelatinase–associated lipocalin, interleukin-8, and tumor necrosis factor-alpha in ulcerative colitis. Am. J. Gastroenterol. **94:** 2923–2928.
24. LAWRANCE, I.C., C. FIOCCHI & S. CHAKRAVARTI. 2001. Ulcerative colitis and Crohn's disease: distinctive gene expression profiles and novel susceptibility candidate genes. Hum. Mol. Genet. **10:** 445–456.
25. FORSBLAD, J., A. GOTTSATER, K. PERSSON *et al.* 2002. Clinical manifestations of atherosclerosis in an elderly population are related to plasma neopterin, NGAL, and endothelin-1, but not to *Chlamydia pneumoniae* serology. Int. Angiol. **21:** 173–179.
26. XU, S. & P. VENGE. 2000. Lipocalins as biochemical markers of disease. Biochim. Biophys. Acta **1482:** 298–307.
27. STOESZ, S.P., A. FRIEDL, J.D. HAAG *et al.* 1998. Heterogeneous expression of the lipocalin NGAL in primary breast cancers. Int. J. Cancer **79:** 565–572.
28. BRATT, T., S. OHLSON & N. BORREGAARD. 1999. Interactions between neutrophil gelatinase–associated lipocalin and natural lipophilic ligands. Biochim. Biophys. Acta **1472:** 262–269.
29. ZEREGA, B., S. CERMELLI, B. MICHELIS *et al.* 2000. Expression of NRL/NGAL (neu-related lipocalin/neutrophil gelatinase–associated lipocalin) during mammalian embryonic development and in inflammation. Eur. J. Cell Biol. **79:** 165–172.
30. DESCALZI CANCEDDA, F., B. DOZIN, B. ZEREGA *et al.* 2000. Ex-FABP: a fatty acid binding lipocalin developmentally regulated in chicken endochondral bone formation and myogenesis. Biochim. Biophys. Acta **1482:** 127–135.
31. GENTILI, C., S. CERMELLI, C. TACCHETTI *et al.* 1998. Expression of the extracellular fatty acid binding protein (ex-FABP) during muscle fiber formation *in vivo* and *in vitro*. Exp. Cell Res. **242:** 410–418.
32. LI, M., X. LIU, G. ROBINSON *et al.* 1997. Mammary-derived signals activate programmed cell death during the first stage of mammary gland involution. Proc. Natl. Acad. Sci. USA **94:** 3425–3430.
33. BALL, R.K., R.R. FRIIS, C.A. SCHOENENBERGER *et al.* 1988. Prolactin regulation of beta-casein gene expression and of a cytosolic 120-kd protein in a cloned mouse mammary epithelial cell line. EMBO J. **7:** 2089–2095.
34. SCHWOCHAU, G.B., K.A. NATH & M.E. ROSENBERG. 1998. Clusterin protects against oxidative stress *in vitro* through aggregative and nonaggregative properties. Kidney Int. **53:** 1647–1653.
35. BACH, U.C., M. BAIERSDORFER, G. KLOCK *et al.* 2001. Apoptotic cell debris and phosphatidylserine-containing lipid vesicles induce apolipoprotein J (clusterin) gene expression in vital fibroblasts. Exp. Cell Res. **265:** 11–20.
36. FREIJE, J.M., I. DIEZ-ITZA, M. BALBIN *et al.* 1994. Molecular cloning and expression of collagenase-3, a novel human matrix metalloproteinase produced by breast carcinomas. J. Biol. Chem. **269:** 16766–16773.
37. ALLEN, R.A., R.W. ERICKSON & A.J. JESAITIS. 1989. Identification of a human neutrophil protein of Mr 24,000 that binds *N*-formyl peptides: co-sedimentation with specific granules. Biochim. Biophys. Acta **991:** 123–133.
38. CHU, S., H. LIN, H. HUANG *et al.* 1998. The hydrophobic pocket of 24p3 protein from mouse uterine luminal fluid: fatty acid and retinol binding activity and predicted structural similarity to lipocalins. J. Pept. Res. **52:** 390–397.
39. CANCEDDA, F.D., M. MALPELI, C. GENTILI *et al.* 1996. The developmentally regulated avian Ch21 lipocalin is an extracellular fatty acid–binding protein. J. Biol. Chem. **271:** 20163–20169.
40. PERSENGIEV, S.P., L.R. DEVIREDDY & M.R. GREEN. 2002. Inhibition of apoptosis by ATFx: a novel role for a member of the ATF/CREB family of mammalian bZIP transcription factors. Genes Dev. **16:** 1806–1814.
41. DEVIREDDY, L.R., J.G. TEODORO, F.A. RICHARD *et al.* 2001. Induction of apoptosis by a secreted lipocalin that is transcriptionally regulated by IL-3 deprivation. Science **293:** 829–834.
42. DAVIS, T.R., L. TABATABAI, K. BRUNS *et al.* 1991. Basic fibroblast growth factor induces 3T3 fibroblasts to synthesize and secrete a cyclophilin-like protein and beta-2-microglobulin. Biochim. Biophys. Acta **1095:** 145–152.

43. GARAY-ROJAS, E., M. HARPER, S. HRABA-RENEVEY *et al.* 1996. An apparent autocrine mechanism amplifies the dexamethasone- and retinoic acid–induced expression of mouse lipocalin-encoding gene 24p3. Gene **170:** 173–180.
44. HAMILTON, R.T., M. NILSEN-HAMILTON & G. ADAMS. 1985. Superinduction by cycloheximide of mitogen-induced secreted proteins produced by Balb/c 3T3 cells. J. Cell. Physiol. **123:** 201–208.
45. LUETTEKE, N.C., T.H. QIU, S.E. FENTON *et al.* 1999. Targeted inactivation of the EGF and amphiregulin genes reveals distinct roles for EGF receptor ligands in mouse mammary gland development. Development **126:** 2739–2750.
46. BERG, M.N., A.M. DHARMARAJAN & B.J. WADDELL. 2002. Glucocorticoids and progesterone prevent apoptosis in the lactating rat mammary gland. Endocrinology **143:** 222–227.
47. POYSER, N.L. 1981. Prostaglandin production by the uterus during early pregnancy. Acta Vet. Scand. Suppl. **77:** 241–249.
48. IMLER, J.L. & J.A. HOFFMANN. 2000. Toll and toll-like proteins: an ancient family of receptors signaling infection. Rev. Immunogenet. **2:** 294–304.
49. O'NEILL, L.A. 2002. Wanted: a molecular basis for specificity in toll-like receptor signal transduction. Mol. Cell **10:** 969–971.
50. COWLAND, J.B. & N. BORREGAARD. 1997. Molecular characterization and pattern of tissue expression of the gene for neutrophil gelatinase–associated lipocalin from humans. Genomics **45:** 17–23.
51. KUPER, H., H.O. ADAMI & D. TRICHOPOULOS. 2000. Infections as a major preventable cause of human cancer. J. Intern. Med. **248:** 171–183.
52. ERNST, P.B. & B.D. GOLD. 2000. The disease spectrum of *Helicobacter pylori*: the immunopathogenesis of gastroduodenal ulcer and gastric cancer. Annu. Rev. Microbiol. **54:** 615–640.
53. NESS, R.B., J.A. GRISSO, C. COTTREAU *et al.* 2000. Factors related to inflammation of the ovarian epithelium and risk of ovarian cancer. Epidemiology **11:** 111–117.
54. COLLET, J.P., C. SHARPE, E. BELZILE *et al.* 1999. Colorectal cancer prevention by non-steroidal anti-inflammatory drugs: effects of dosage and timing. Br. J. Cancer **81:** 62–68.
55. BERNATSKY, S., A. CLARKE & R. RAMSEY-GOLDMAN. 2002. Malignancy and systemic lupus erythematosus. Curr. Rheumatol. Rep. **4:** 351–358.
56. DE MARZO, A.M., V.L. MARCHI, J.I. EPSTEIN *et al.* 1999. Proliferative inflammatory atrophy of the prostate: implications for prostatic carcinogenesis. Am. J. Pathol. **155:** 1985–1992.
57. ERNST, P. 1999. Review article: the role of inflammation in the pathogenesis of gastric cancer. Aliment. Pharmacol. Ther. **13**(suppl. 1)**:** 13–18.
58. SHORTER, R.G. 1983. Risks of intestinal cancer in Crohn's disease. Dis. Colon Rectum **26:** 686–689.
59. LEVIN, B. 1992. Ulcerative colitis and colon cancer: biology and surveillance. J. Cell. Biochem. Suppl. **16G:** 47–50.
60. ZHANG, Z., G.C. KUNDU, D. PANDA *et al.* 1999. Loss of transformed phenotype in cancer cells by overexpression of the uteroglobin gene. Proc. Natl. Acad. Sci. USA **96:** 3963–3968.
61. NEWCOMB, P.A., B.E. STORER, M.P. LONGNECKER *et al.* 1994. Lactation and a reduced risk of premenopausal breast cancer. N. Engl. J. Med. **330:** 81–87.
62. RUSSO, J. & I.H. RUSSO. 1980. Susceptibility of the mammary gland to carcinogenesis: II. Pregnancy interruption as a risk factor in tumor incidence. Am. J. Pathol. **100:** 497–512.
63. ABRAMS, T.J., R.C. GUZMAN, S.M. SWANSON *et al.* 1998. Changes in the parous rat mammary gland environment are involved in parity-associated protection against mammary carcinogenesis. Anticancer Res. **18:** 4115–4121.
64. D'CRUZ, C.M., S.E. MOODY, S.R. MASTER *et al.* 2002. Persistent parity-induced changes in growth factors, TGF-beta3, and differentiation in the rodent mammary gland. Mol. Endocrinol. **16:** 2034–2051.
65. SAVILL, J.S., A.H. WYLLIE, J.E. HENSON *et al.* 1989. Macrophage phagocytosis of aging neutrophils in inflammation: programmed cell death in the neutrophil leads to its recognition by macrophages. J. Clin. Invest. **83:** 865–875.

66. DiGiovanni, J. & M.R. Juchau. 1980. Biotransformation and bioactivation of 7,12-dimethylbenz[a]anthracene (7,12-DMBA). Drug Metab. Rev. **11:** 61–101.
67. Josephy, P.D. & B.L. Coomber. 1998. The 1996 Veylien Henderson Award of the Society of Toxicology of Canada. Current concepts: neutrophils and the activation of carcinogens in the breast and other organs. Can. J. Physiol. Pharmacol. **76:** 693–700.

How Proteases Regulate Bone Morphogenesis

NATHALIE ORTEGA, DANIELLE BEHONICK, DOMINIQUE STICKENS, AND ZENA WERB

Department of Anatomy, University of California, San Francisco, California 94143, USA

ABSTRACT: Matrix metalloproteinases (MMPs) degrade most components of the extracellular matrix (ECM), as well as many non-ECM molecules. MMPs participate in (1) degradation of ECM to allow cell migration; (2) alteration of the ECM microenvironment resulting in alteration in cellular behavior; (3) modulation of biologically active molecules by direct cleavage or release from ECM stores; (4) regulation of the activity of other proteases; and (5) cell attachment, proliferation, differentiation, and apoptosis. We have sought to understand the role of MMPs during development and tissue repair in transgenic mice. Endochondral bone formation presents a particularly interesting developmental challenge. During this process, an avascular tissue (cartilage) is converted into one of the most highly vascularized tissues (bone) in the vertebrate body. Ossification begins with invasion of the calcified hypertrophic cartilage by capillaries. Apoptosis of the terminal hypertrophic chondrocytes, degradation of the cartilage matrix, and deposition of bone matrix by osteoblasts accompanies neovascularization of the growth plate. Remodeling of ECM results in a cavity filled with vascular channels containing hematopoietic cells. Our results reveal that MMP9, MMP13, and vascular endothelial growth factor are key regulators for the remodeling of the skeletal tissues. They coordinate not only matrix degradation, but also the recruitment and differentiation of endothelial cells, osteoclasts, chondroclasts, and osteoprogenitors.

KEYWORDS: matrix metalloproteinases; endochondral ossification; angiogenesis; VEGF; MMP9

INTRODUCTION

The skeleton develops by one of two processes: plate bones are formed by intramembranous ossification, while skeletal long bones are made from a cartilage model. In higher vertebrates, the cartilage of the appendicular skeleton is replaced by bone in a process called endochondral ossification.[1] During this process, the epiphyseal growth plate undergoes morphogenesis. A region of resting chondrocytes differentiates into a zone of proliferating chondrocytes that then hypertrophies and finally undergoes apoptotic cell death. In this area of bone formation, blood vessel invasion from the metaphysis coincides with degradation of the extracellular matrix (ECM) and apoptosis of hypertrophic chondrocytes. Following the invasion of the capillaries,

Address for correspondence: Zena Werb, Department of Anatomy, HSW 1320/1321, University of California, 513 Parnassus Avenue, San Francisco, CA 94143-0452. Voice: 415-476-4622; fax: 415-476-4565.
zena@itsa.ucsf.edu

Ann. N.Y. Acad. Sci. 995: 109–116 (2003). © 2003 New York Academy of Sciences.

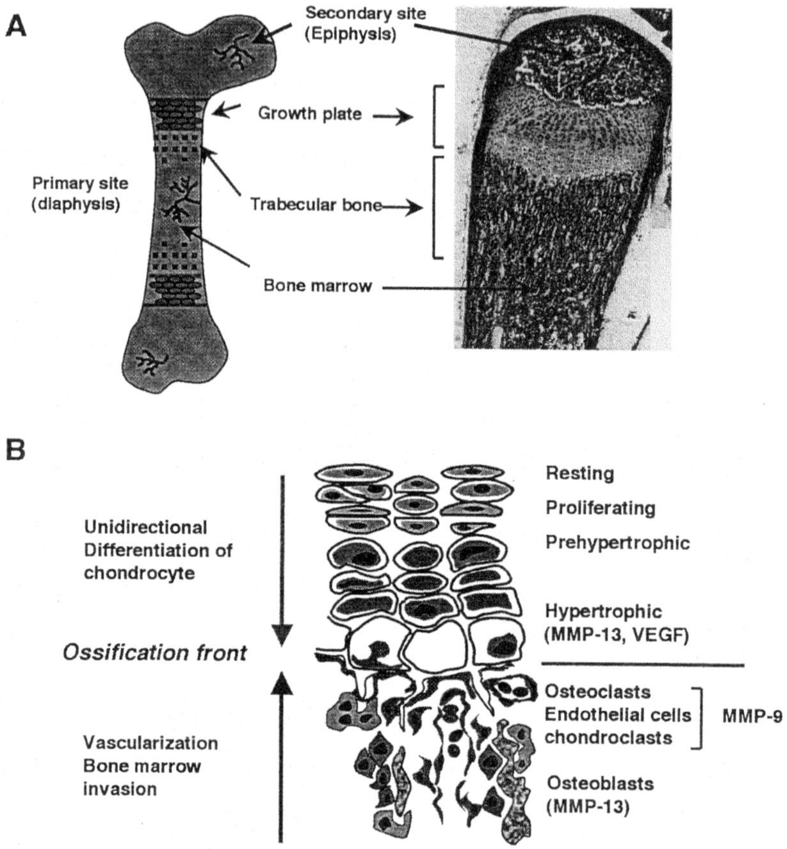

FIGURE 1. Endochondral ossification process. **(A)** Endochondral ossification is a process by which the majority of the bones form in vertebrates. During embryogenesis, cartilaginous skeleton derives from condensation of mesenchyme. In long bones, the cartilaginous templates consist of a central diaphysis and two epiphyses surrounded by a perichondrium. Within the diaphysis, chondrocytes gradually become hypertrophic and the matrix calcifies. The perichondrium around this area acquires osteogenic properties, and mesenchymal progenitors and blood vessels invade this calcified matrix, establishing the primary site of ossification. Osteoprogenitor cells differentiate into osteoblasts and use the calcified matrix (longitudinal septa) as a support for deposition of bone matrix, giving rise to trabecular bone. Later in development, secondary sites develop in the epiphyses. Mesenchymal progenitors differentiate into osteoblasts and hematopoietic cells. The bone marrow cavity extends into the diaphysis and is limited on each extremity by the growth plates. **(B)** The growth plates are critical sites regulating the growth of the long bones through a highly controlled and unidirectional differentiation of chondrocytes. The distal part of the growth plate contains resting chondrocytes. Some of these cells receive signals to enter the cell cycle, resulting in layers of proliferating chondrocytes. These chondrocytes give rise to mature, or prehypertrophic, chondrocytes, which then differentiate into hypertrophic chondrocytes. Chondrocytes degenerate and neovessels, osteoblasts, and hematopoietic cells invade the empty lacunae. An angiogenic switch occurs along the growth plate with an increase in the expression of the

osteogenic progenitors are recruited and trabecular bone is deposited. A new structure, the growth plate, is formed, from which the skeletal bones elongate. Endochondral ossification occurs at two distinct sites in the vertebrate long bone—the so-called primary and secondary sites of ossification. Bone development is not synchronized between these sites; the process initiates first at the primary site (diaphysis) and later ossification takes place at the secondary site (epiphysis) (FIG. 1).

The formation of bone from cartilage is a complex process. During this process, cartilage, an avascular tissue, is converted into bone, which, at 40% endothelial cells, is one of the most highly vascularized tissues in the vertebrate body. The ECM is completely remodeled in that the switch from resting cartilage to prehypertrophic cartilage is also accompanied by a switch in the synthesis of ECM proteins. Whereas proliferating chondrocytes express and secrete collagen type II, hypertrophic chondrocytes express and secrete collagen type X, and osteoblasts, responsible for depositing the "bone matrix", express and secrete collagen type I. As such, it is clear that the ECM and its remodeling play a significant role in the sequential events involved in endochondral ossification.

MATRIX METALLOPROTEINASE FAMILY

The matrix metalloproteinases (MMPs) are a family of ECM-degrading enzymes that share common functional domains and activation mechanisms.[2] These enzymes have an essential Zn ion in the catalytic domain. They are synthesized as secreted or transmembrane proenzymes and processed to an active form by the removal of an amino-terminal propeptide. A cysteine residue on the propeptide forms the fourth ligand of the Zn ion and thereby maintains the enzyme in an inactive conformation. This link with the Zn ion, which must be broken for enzyme activation, is usually disrupted by one or more proteolytic cleavages that remove the propeptide from the enzyme.

MMPs are endopeptidases that are extracellular regulators of cell growth, migration, and ECM remodeling.[3] They are produced by a variety of cell types including epithelial cells, fibroblasts, and inflammatory cells. The secreted MMPs include collagenases (MMP1, MMP8, MMP13), gelatinases (MMP2, MMP9), stromelysins (MMP3, MMP10, MMP11), and other MMPs (MMP7, MMP12). To date, six membrane-type MMPs have been identified (MMP14 [also called MT1-MMP], MMPs-15, 16, 17, 24, 25). With the exception of MMP17 and 24, these enzymes are

major angiogenic factor, VEGF, in hypertrophic chondrocytes, whereas different cell types below the front of ossification express receptors for the factor. This switch is also dependent on a strong remodeling of the matrix coordinated with the vascularization. Two critical enzymes involved in that process are MMP9 and MMP13, expressed by osteoclasts, chondroclasts, and endothelial cells for MMP9 and hypertrophic chondrocytes and osteoblasts for MMP13. The VEGF system and MMP9 interfere in certain pathways implicated in the regulation of endochondral ossification.

single-pass type I membrane proteins with an extracellular N-terminus and a short cytoplasmic C-terminal domain. Membrane-type MMPs are activated intracellularly in the secretory pathway by furin-like enzymes.[4,5] The MMPs are involved in binding various molecules in their exosites (distant from their catalytic sites), including inhibitors and substrates. The different exosites confer specificity for interaction of each MMP with different protein substrates. MMPs are regulated at multiple levels including gene expression, spatial localization, zymogen activation, and inhibition by the naturally occurring tissue inhibitors (e.g., tissue inhibitors of metalloproteinases [TIMPs] 1–4, α2-macroglobulin, and Reck[6,7]).

MMPs have been implicated in a number of physiological and pathological processes including both normal and tumor angiogenesis.[8–10] Angiogenesis requires locally restricted extracellular proteolysis achieved by a finely tuned balance of MMPs and MMP inhibitors and the assembly of components of one or more protease systems on the cell surface. MMP2, MMP9 and MT1-MMP are of particular interest for angiogenesis.[11–15] MT1-MMP, in particular, is endowed with two characteristics important for locally restricted proteolysis during endothelial migration: it associates with the plasma membrane and can catalytically activate the precursor of MMP2, allowing for amplification of the degradation process.[16]

MMPs IN ENDOCHONDRAL OSSIFICATION

At least three MMPs are highly expressed during endochondral ossification. When neovascularization of the cartilage anlage begins, MT1-MMP and MMP9 are expressed in the preosteoclasts and other chondroclastic cells of unknown origin.[14,15] MMP13 is expressed in the terminal hypertrophic chondrocytes and in the newly recruited osteoblasts.[13] The critical role of these proteases in osteogenesis can be seen by treating mice of 1–2 weeks of age with a broad-spectrum MMP inhibitor, GM6001. The result is a dramatic expansion of the zone of hypertrophic cartilage in the growth plate at the primary site of ossification. In addition, the secondary ossification sites do not form in the presence of the inhibitor. These results show that MMPs are required for normal bone development.

All three MMPs that are expressed in developing bone have been investigated for their role in endochondral ossification.[13–18] MT1-MMP-deficient transgenic animals do not display any defects in the process of endochondral ossification at the primary site.[14,15] However, MMP9-null transgenic animals have a very pronounced phenotype, displaying substantially more hypertrophic cartilage, impaired endochondral ossification at the primary site, and a delay in formation of the bone marrow (BM) cavity.[13] In addition, the MMP9-null mouse shows decreased trabecular bone formation, chondrocyte apoptosis, vascular invasion, and osteoclastic recruitment. This phenotype is mostly manifest during the developmental stage; a full recovery can be observed after about 4 weeks of age. In older null animals, the only remaining phenotype is long bones that are about 10% shorter than those of wild-type animals.[3]

Together, MMP9 and MMP13 account for most of the protease-dependent steps of endochondral ossification.[18] MMP9 and MMP13 are expressed in nonoverlapping cell types at the front of ossification. Whereas MMP9 is expressed in osteoclasts, endothelial cells, and some other cell types including BM stromal cells, MMP13 is expressed in the hypertrophic chondrocytes and osteoblasts.

MOLECULAR MECHANISMS UNDERLYING THE MMP-NULL PHENOTYPES

The MMP9 protein is present in the hypertrophic cartilage primarily at the front of ossification, although it is also present along the trabeculae and in osteoclasts. There are many processes that these MMPs may mediate in developing bone; examples include matrix proteolysis, vascular invasion, bioavailability of growth factors, and chondrocyte apoptosis. The excessive hypertrophic cartilage in the MMP9-null mouse indicates either a defect in the rate of entry of cells into the proliferating/maturing chondrocyte pool or a defect in the exit from this pool in processes involved in cartilage resorption, chondrocyte apoptosis, vascular invasion, and bone formation. The expression of several markers and experiments of BrdU incorporation indicate that there is no alteration in the rate of entry of the cells into the hypertrophic chondrocyte pool. Therefore, the defect in the MMP9-null mouse is in the rate of exit of chondrocytes from the late hypertrophic pool.

Angiogenic factors are expressed in the developing bone. In particular, vascular endothelial growth factor is made by the terminal hypertrophic chondrocytes. In the MMP9-null mouse, there is an increased amount of terminal hypertrophic chondrocytes and, according to *in situ* hybridization studies, a 10-fold increase in the amount of VEGF mRNA produced. Moreover, in an *in vitro* angiogenesis assay, cartilage explants from MMP9-null mice induce a delayed angiogenic response.[8] These observations lead to the hypothesis that the protease may regulate the availability of VEGF in hypertrophic chondrocytes, such as when the protease remodels extracellular matrix. To understand the molecular role of the VEGF system in the endochondral ossification process, a soluble VEGF receptor, the mFlt-1-Fc containing the terminal 3 Ig repeats of the receptor fused to the Fc portion of IgG, was used to compete for VEGF.[17] Wild-type mice treated with this soluble receptor display a large increase in the hypertrophic cartilage zone and a decrease in bone formation and vessel invasion. Thus, blocking VEGF activity phenocopies the MMP9-null phenotype. Transplanting wild-type bone marrow into the bone of an MMP9-null animal allows for rescue of the null phenotype within a week. Likewise, if we provide exogenous VEGF, the hypertrophic cartilage is rapidly remodeled and the amount of trabecular bone increases, giving a partial rescue of the VEGF-depleted phenotype. VEGF is a chemoattractant for osteoclasts. Moreover, invasion of osteoclasts into the hypertrophic cartilage requires VEGF because blocking VEGF function inhibits this process.[17,18] Taken together, these observations point to a link between MMP9 and VEGF in the coordination of the processes involved in endochondral ossification.

Hypertrophic cartilages from wild-type and MMP9-null animals show strikingly different protein patterns when compared by silver-stained SDS-PAGE. Indeed, many proteins accumulate in the absence of MMP9. These accumulated proteins may be protein precursors or downstream signaling proteins that accumulate as a result of altered proteolysis.

Mice deficient in MMP9 exhibit a delay in osteoclast recruitment in specialized invasion and bone resorption models *in vitro*.[18] MMP9 is specifically required for the invasion of osteoclasts and endothelial cells into the discontinuously mineralized hypertrophic cartilage that fills the core of the diaphysis, whereas other MMPs, principally MMP13, are required for the passage of the cells through unmineralized type I collagen of the nascent bone collar and play a role in resorption of mineralized matrix.

MMP9 stimulates the solubilization of unmineralized cartilage by MMP13, a collagenase highly expressed in hypertrophic cartilage prior to osteoclast invasion.[18]

MMP9 and MMP13 are synergistic in their effects on bone development and remodeling.[18] There are many other molecules in this region, including connective tissue growth factor, and many other cellular processes, including osteoblast and endothelial cell recruitment, that are required for bone development and remodeling. VEGF provides a molecular linkage between the degradation of the cellular matrix, the vascular invasion that occurs, and the hypertrophic chondrocyte apoptosis. Because VEGF receptors are expressed on osteogenic precursors and endothelial cells, it is possible that this developmental process is under very tight regulation involving MMPs and VEGF.

The precise mechanism involving MMP9 in controlling the bioavailability of VEGF in bone development has yet to be elucidated. However, MMP9 has many ECM substrates including collagens,[9,19] aggrecan, laminin, osteonectin, heparin, link protein and nonmatrix proteins including serpins, $\alpha 1$ proteinase inhibitor, some integrins, plasminogen, and latent TGFα. While most of these proteins have been shown to be substrates only *in vitro*, cleavage of one of these may activate cleavage of VEGF *in vivo*.

MMPs IN BONE MARROW AND HEMATOPOIESIS

The bone presents an unusual challenge in that it is, in actuality, one organ (the bone marrow) nested within another (the bony skeleton). While endochondral ossification is producing calcified bone from a cartilaginous template, bone marrow is also being made. In MMP9-null mice, this process is delayed. However, these mice do, at later stages, display normal hematopoietic ability and have red and white blood cell counts comparable to wild-type animals.

Recruitment of hematopoietic stem cells (HSCs), hematopoietic precursors, and endothelial precursors from the bone marrow is required and occurs during a number of physiological and pathological processes. This recruitment is a dynamic process that can be induced by various growth factors including VEGF, but that requires MMPs. In particular, VEGF cannot mobilize endothelial precursors into the circulation of MMP9-null mice[20] or of wild-type mice treated with an MMP inhibitor. Activated MMP9 promotes the release of the stem cell–active cytokine soluble Kit-ligand (sKitL), thereby driving stem, hematopoietic, and endothelial progenitor cells from their quiescent niches to their proliferative niches within the BM.[20,21] Furthermore, histological examination of the BM of VEGF-A-treated mice reveals increased immunoreactive MMP-9 on stromal and hematopoietic cells. Collectively, these data suggest that MMP9 activation is required for the recruitment and mobilization of BM-derived stem, hematopoietic, and endothelial precursors.

CONCLUSIONS

Many functions lie downstream of MMPs. Defects in endochondral ossification due to absence of MMP9 include delay in recruitment of osteoclasts and in vascular invasion, altered apoptosis of the hypertrophic chondrocytes, accumulation of ECM, and a delayed recruitment and/or differentiation of osteoblast precursors. MMP13 synergizes with MMP9, while also affecting bone remodeling more directly. MMP9 also serves as the decisive checkpoint for the rapid recruitment and mobilization of stem, hematopoietic, and endothelial precursors, thereby modulating angiogenesis and hematopoiesis. Determining the exact molecular targets of these MMPs will give us insights into these events and how they are coordinated to regulate normal bone development, repair, and homeostasis, and into what perturbations take place in developmental disorders and adult bone pathologies.

ACKNOWLEDGMENTS

This work was supported by NIH Grant Nos. AR46238 and DE13058.

REFERENCES

1. Karsenty, J. & E.F. Wagner. 2002. Reaching a genetic and molecular understanding of skeletal development. Dev. Cell. **4:** 389–406.
2. Sternlicht, M.D. & Z. Werb. 1999. Matrix metalloproteinases. *In* Guidebook to the Extracellular Matrix, Anchor, and Adhesion Proteins, pp. 519–538. Sambrook & Tooze (Oxford University Press). London.
3. Brinckerhoff, C.E. & L.M. Matrisian. 2002. Matrix metalloproteinases: a tail of a frog that became a prince. Nat. Rev. Mol. Cell. Biol. **3:** 207–214.
4. Sato, H. *et al.* 1996. Activation of a recombinant membrane type 1–matrix metalloproteinase (MT1-MMP) by furin and its interaction with tissue inhibitor of metalloproteinases (TIMP)–2. FEBS Lett. **393:** 101–104.
5. Kang, T. *et al.* 2002. Activation of membrane-type matrix metalloproteinase 3 zymogen by the proprotein convertase furin in the trans-Golgi network. Cancer Res. **62:** 675–681.
6. Sternlicht, M.D. & Z. Werb. 1999. Tissue inhibitors of metalloproteinases. *In* Guidebook to the Extracellular Matrix, Anchor, and Adhesion Proteins, pp. 539–542. Sambrook & Tooze (Oxford University Press). London.
7. Baker, A.H. *et al.* 2002. Metalloproteinase inhibitors: biological actions and therapeutic opportunities. J. Cell Sci. **115:** 3719–3727.
8. Vu, T.H. & Z. Werb. 2000. Matrix metalloproteinases: effectors of development and normal physiology. Genes Dev. **14:** 2123–2133.
9. Egeblad, M. & Z. Werb. 2002. New roles for the matrix metalloproteinases in the progression of cancer. Nat. Rev. Cancer **2:** 163–176.
10. Chang, C. & Z. Werb. 2001. The many faces of metalloproteases: growth signaling, invasion, angiogenesis, and metastasis. Trends Cell. Biol. **11:** S37–S43.
11. Itoh, T. *et al.* 1998. Reduced angiogenesis and tumor progression in gelatinase A–deficient mice. Cancer Res. **58:** 1048–1051.
12. Bergers, G. *et al.* 2000. Matrix metalloproteinase-9 triggers the angiogenic switch during carcinogenesis. Nat. Cell Biol. **2:** 737–744.
13. Vu, T.H. *et al.* 1998. MMP-9/gelatinase B is a key regulator of growth plate angiogenesis and apoptosis of hypertrophic chondrocytes. Cell **93:** 411–422.
14. Zhou, Z. *et al.* 2000. Impaired endochondral ossification and angiogenesis in mice deficient in membrane-type matrix metalloproteinase I. Proc. Natl. Acad. Sci. USA **97:** 4052–4057.
15. Holmbeck, K. *et al.* 1999. MT1-MMP-deficient mice develop dwarfism, osteopenia, arthritis, and connective tissue disease due to inadequate collagen turnover. Cell **99:** 81–92.
16. Sato, H. *et al.* 1994. A matrix metalloproteinase expressed on the surface of invasive tumour cells. Nature **370:** 61–65.

17. GERBER, H-P. *et al.* 1999. VEGF couples hypertrophic cartilage remodeling, ossification, and angiogenesis during endochondral bone formation. Nat. Med. **5:** 623–628.
18. ENGSIG, M.T. *et al.* 2000. Matrix metalloproteinase 9 and vascular endothelial growth factor are essential for osteoclast recruitment into developing long bones. J. Cell Biol. **151:** 879–890.
19. OPDENAKKER, G. *et al.* 2001. Gelatinase B: a tuner and amplifier of immune functions. Trends Immunol. **22:** 571–579.
20. HEISSIG, B. *et al.* 2002. Mobilization of stem cells from the bone marrow niche requires MMP-9-mediated release of Kit ligand. Cell **109:** 625–637.
21. HATTORI, K. *et al.* 2002. Placental growth factor reconstitutes hematopoiesis by recruiting VEGFR1+ stem cells from bone-marrow microenvironment. Nat. Med. **8:** 841–849.

Cell Migration and Programmed Cell Death of Drosophila Germ Cells

CLARK R. COFFMAN

Department of Zoology and Genetics, Iowa State University, Ames, Iowa 50011, USA

ABSTRACT: **Cell migration and programmed cell death are essential components of animal development and homeostasis, and the germ cells of Drosophila provide a simple genetic system to study the molecular mechanisms that govern these important cellular processes. Detailed descriptions of germ cell migration in Drosophila were accomplished long ago, but most genetic and molecular analyses of the process have occurred within the past 10 years. A few of the genes required for germ cell migration have been identified, and a very interesting picture is emerging. However, a process as complex as cell migration must involve the functions of many more molecules. In addition, cell migration and cell death mechanisms are often linked, as it is important to eliminate cells that are misplaced and could present a danger to the organism. In Drosophila, genes involved in germ cell migration can also affect programmed cell death. Currently, very little is known about how germ cells ectopic to the gonads are eliminated. To date, only four genes have been reported with roles in germ cell death, and three of these have additional functions in germ cell pathfinding. The nature of the cell death program has not been elucidated. Here, I provide a brief review of Drosophila germ cell migration and programmed cell death at both the descriptive and molecular levels. Many questions remain to be answered, but advances made in recent years are providing useful insights into these critical biological phenomena.**

KEYWORDS: **apoptosis; autophagy; cell movement; development; *Drosophila melanogaster***

INTRODUCTION

During animal development, many cells must migrate to their ultimate destinations in order to function. In the adult organism, migrating cells are essential for wound healing, immune system function, and angiogenesis.[1–4] Often, cell death mechanisms are coupled to cell migration to eliminate cells that fail to migrate appropriately, present a danger to the animal, or are in excess.[5–15] The precise regu-

Address for correspondence: Clark R. Coffman, Department of Zoology and Genetics, Iowa State University, 3258 Molecular Biology Building, Ames, IA 50011-3260. Voice: 515-294-3911; fax: 515-294-6755.
ccoffman@iastate.edu

Ann. N.Y. Acad. Sci. 995: 117–126 (2003). © 2003 New York Academy of Sciences.

lation of cell movements and cell death must involve a complex array of intercellular and intracellular signaling events that result in changes in gene expression, cellular architecture, and cell physiology. Deciphering these mechanisms is complicated by the fact that the role played by some components varies depending upon the cellular context at the time of signaling.

The germ cells of *Drosophila melanogaster* provide a simple model to study the basic mechanisms regulating cell migration and cell death that is amenable to genetic, developmental, and molecular approaches.[16–31] In the early embryo, there are relatively few germ cells, and their movements have been well documented using morphological and molecular markers. Germ cells ectopic to the gonads or that migrate inappropriately are eliminated with great efficiency. Recent studies have revealed that both attractive and repulsive guidance cues are involved.[30,32–36] Interestingly, some of the molecules identified are lipid-modifying enzymes. This raises the possibility that the signals carry lipid modifications or that the signaling molecules may be lipids. In addition, some of the genes with roles in cell migration also function in cell death.[30,31]

A DESCRIPTION OF DROSOPHILA GERM CELL DEVELOPMENT

The primordial germ cells (PGCs) are the first cells that differentiate in the Drosophila embryo.[16,17,26,37–39] After approximately 2 hours of development, the presumptive germ cells form at the posterior pole. At this point, the PGCs are segregated from the somatic cells of the embryo. The germ cells then undergo up to 2 additional rounds of mitoses resulting in ~35 germ cells, on average, by the cellular blastoderm stage.[20] The PGCs then cease mitosis until they have been incorporated into the developing embryonic gonad.[16]

Through a series of stereotyped movements, the germ cells must reenter the embryo, migrate to the presumptive gonad region, and coalesce with the mesoderm-derived somatic gonadal precursor cells to form the gonad. Detailed observations of these movements have been made by several authors.[16–19,22–24,26,40] FIGURE 1 illustrates some of these germ cell movements as visualized using a germ cell–specific marker.[41] During gastrulation and germ band extension, the germ cells are displaced dorsally and come to lie in a pocket of invaginated epithelial cells at the center of the embryo, the posterior midgut primordium (PMG) (FIG. 1B). At ~4.5 hours, the presumptive germ cells become amoeboid and reenter the embryo by crossing the PMG epithelium (FIG. 1C).[16,23,24,42] These movements involve changes in the actin cytoskeleton of both the germ cells and the PMG epithelium, as well as alterations in the adherence junctions of the PMG epithelial layer. Studies by Jaglarz and Howard, utilizing heterochronic transplantation of germ cells, indicate that the timing of germ cell exit from the PMG is controlled by the somatic cells of the embryo.[40] The presumptive germ cells remain associated with the PMG epithelial cells after transiting them. They then move laterally, forming left and right populations of cells. This is followed by migration to and association with the somatic gonad precursor cells.[43–45] The germ cells and associated somatic gonad precursor cells then migrate together and coalesce in abdominal segment 5, the site of the mature gonad (FIG. 1D). The migration of the germ cells is complete by about 10 hours of development.[26]

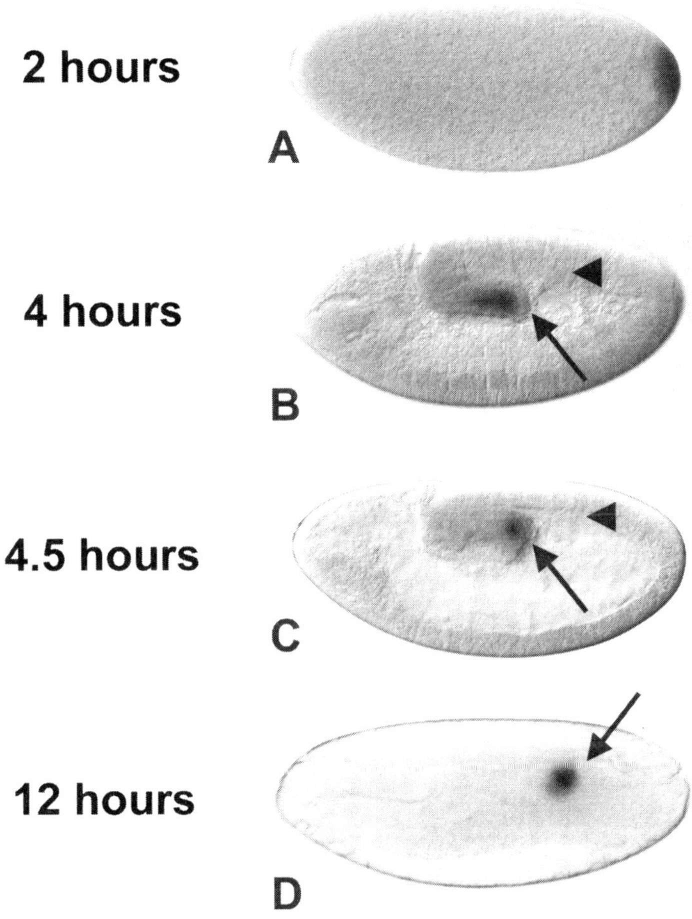

FIGURE 1. Germ cell migration in Drosophila. In all panels, lateral views are shown. Anterior is to the left, and dorsal is at the top. Germ cells are labeled using a *fat facets–lacZ* transgene and are stained using X-Gal. **(A)** The germ cells form at the posterior of the embryo at around 2 hours of development and are the first cells to differentiate. The somatic cells of the embryo will form later from the nonstaining cells at the left. **(B)** An approximately 4-hour-old embryo is shown. As gastrulation begins, the germ band extends and the embryo folds back on itself. The germ cells are still at the posterior, but now lie within the epithelium of the posterior midgut primordium (PMG). The arrows in B and C denote the PMG epithelial cells. **(C)** The germ cells become amoeboid and migrate through the PMG epithelium. They separate into right and left populations while still in contact with the PMG. They then move through several mesoderm cell types to associate with the somatic gonad precursor cells that are forming in the region labeled by the arrowheads in B and C. **(D)** The germ cells coalesce with the somatic gonad precursor cells as the germ band retracts. Posterior is again at the extreme right. A condensed gonad primordium that includes both germ cells and somatic gonad cells is denoted by the arrow.

IDENTIFICATION OF SOME MOLECULES REQUIRED FOR GERM CELL MIGRATION

Genetic studies exploring the mechanisms of germ cell migration have been reported from several labs.[23,24,27,30–36,40,44–50] These studies have provided intriguing insights into the regulation of germ cell migration and have revealed multiple signaling mechanisms within and between cells involving many cell types.

Embryonically Expressed Attractive and Repulsive Signals

In a large-scale screen of the third chromosome, Lehmann and colleagues identified 25 zygotically required genes that affect germ cell migration.[34] Most of these disrupt embryonic patterning or the specification of the somatic gonad precursor cells. Among the genes identified were *hedgehog* (*hh*), *HMG-CoA reductase* (*hmgcr*), and *thread* (*th*). Hh and HMGCR proteins are both expressed in the somatic gonad precursor cells and appear to have roles in germ cell pathfinding.[35,36]

Hh, an extracellular signaling molecule, may attract germ cells to the somatic gonad precursor cells. When expressed in ectopic locations, Hh can cause germ cells to migrate towards the cells expressing it.[36] In the receiving cell, transmission of the Hh signal is often mediated by Patched and Smoothened. Consistent with the hypothesis that Hh signaling is involved in germ cell migration, *patched* and *smoothened* mutants exhibit germ cell migration defects.[36] However, direct evidence of a germ cell–autonomous requirement for *patched* or *smoothened* function has not been shown.

HMGCR is an intracellular protein that regulates an early step in isoprenoid synthesis.[51] Like Hh, forced expression of HMGCR in ectopic locations will cause germ cells to move toward the cells expressing HMGCR.[35] It has been proposed that HMGCR is involved in a modification of Hh.[36] However, the relationship between the attractive signal and a possible lipid modification of the signaling molecule is still untested. Alternatively, HMGCR may generate an independent, lipid-based signal from the somatic gonad that attracts the germ cells. Lysophosphatidic acid and sphingosine 1-phosphate have been shown to mediate cell migration in other systems.[6,11–13]

Thread (also known as DIAP1) is an inhibitor of apoptosis protein (IAP). IAPs have been shown to inhibit the activity of caspases, key components of the apoptosis and autophagic cell death pathways.[52–55] It will be interesting to see what role, if any, *thread* has in germ cell pathfinding and programmed cell death.

In other screens, *wunen* (*wun*) and *wunen2* (*wun2*) have been found to be necessary for the generation of a signal that repels germ cells away from the PMG primordium.[30,32,33] These genes perform redundant functions. Single mutants do not exhibit a germ cell migration defect, while removal of both genes results in a severe disruption of germ cell migration. The *wun* genes encode type 2 phosphatidic acid phosphatases. These findings provide additional evidence that a lipid-based signaling mechanism or some lipid modification of another protein may be necessary for germ cell migration. Wun and Wun2 may also have functions in germ cell death.

Maternally Contributed Gene Products

To date, only three maternally required genes have been identified with roles in germ cell migration. These are *nanos* (*nos*), *polar granule component* (*pgc*), and *scattershot* (*sctt*).[31,46,49,56] Germ cells from females lacking *nos* do not migrate to the gonad.[46,49] Instead, they cluster just after exiting the PMG. Using germ cell transplants, Kobayashi and colleagues demonstrated that the function of *nos* is required within the germ cells for normal migration.[46] In wild-type germ cells, cell division ceases once gastrulation begins,[16] and RNA polymerase II transcription within the germ cells is repressed.[57–59] Germ cells in embryos from *nos* mutant females prematurely initiate transcription and cell divisions.[46,48,50] Nos is a zinc finger protein that can act as a translational repressor.[60,61] However, the downstream targets of Nos regulation that are involved in germ cell migration remain to be identified.

The product of the *polar granule component* gene (*pgc*), an apparently untranslatable RNA, is localized to the polar granules within the germ cells.[56] Expression of antisense *pgc* RNA causes a decrease in the number of germ cells that reach the gonad and an increase in the percentage of agametic ovaries. The levels of *nos* RNA are decreased in *pgc* antisense–expressing embryos, and it has not been determined whether *pgc* has functions in germ cell pathfinding that are independent of *nos*.

In *sctt* mutants, the germ cells form normally and initiate migratory movements, but the germ cells disperse throughout the posterior half of the embryo and very few, if any, reach the gonad (FIG. 2B).[31] This migration defect can be rescued with a paternally supplied wild-type copy of the gene (FIG. 2C). Interestingly, germ cells ectopic to the gonad fail to die in these rescued embryos.

GERM CELL DEATH

The germ cells of Drosophila are an excellent model for studying programmed cell death. In the past 60 years, several researchers have observed that only about 50% of the germ cells that originally form are incorporated into the somatic gonad.[16,17,20,37,62] In addition, very few germ cells are observed outside the normal migratory path.[21,22,30,31] This suggests that errant germ cells are rapidly eliminated. Primordial germ cells that have been labeled using horseradish-peroxidase or radioactive thymidine do not transdifferentiate.[20,62] Thus, the fate of the germ cells is restricted to the germ cell lineage. Collectively, these studies demonstrate that germ cells ectopic to the gonad die and that there must be a mechanism regulating germ cell survival within the embryo. Programmed cell deaths via apoptosis or autophagy have been reported in Drosophila.[54,63–67] However, the molecular mechanism of germ cell death remains virtually unexplored.

Regulators of Germ Cell Death

There are only four genes that have demonstrated roles in the programmed death of Drosophila germ cells. These are *outsiders* (*out*), *sctt*, *wun*, and *wun2*.[30–33] The *wun* and *wun2* genes function redundantly. In *out* mutant embryos, a wild-type number of germ cells are incorporated into the gonads (FIG. 2D).[31] This indicates that germ cell pathfinding is not impaired by loss of *out* function. However, many germ cells ectopic to the gonads that should undergo programmed cell death persist and continue to express germ cell markers. The function of the *out* gene is required zygotically.

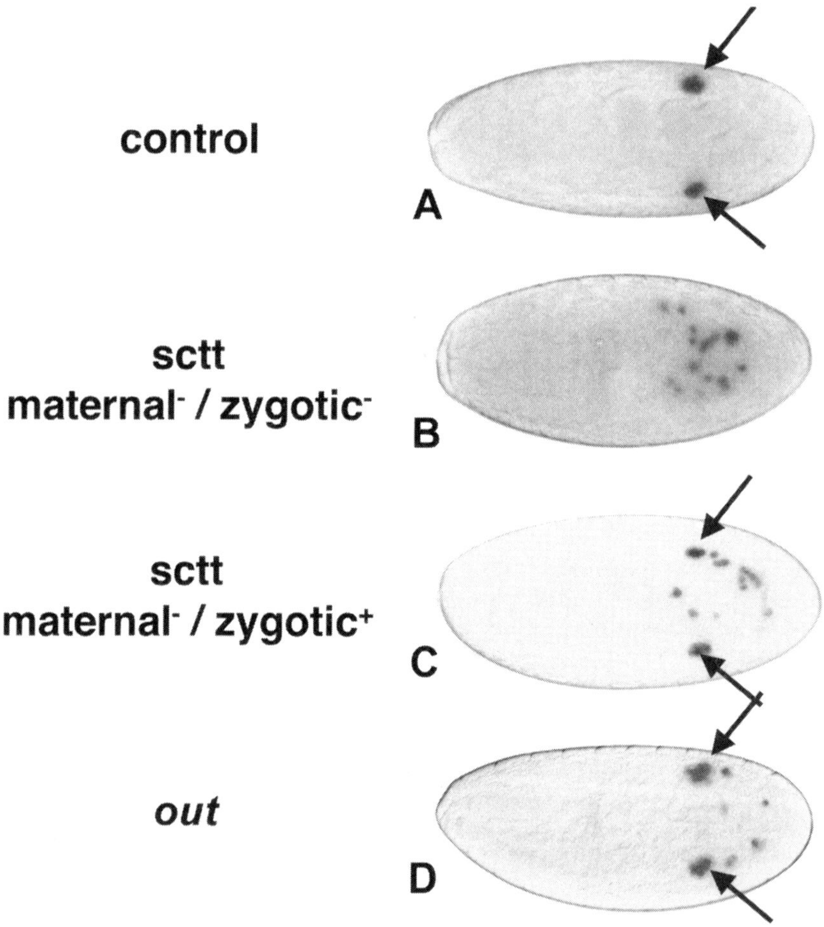

FIGURE 2. Examples of embryos displaying defects in germ cell migration and/or programmed cell death. Dorsal views are shown in all panels. Anterior is to the left, and right is at the top. All embryos are 12–15 hours old. Arrows denote the developing gonads. **(A)** A control embryo showing the wild-type Fat-facets–β-Gal germ cell staining pattern. Only those germ cells that successfully migrate to and become incorporated into the gonads continue to stain for this germ cell marker. **(B)** An embryo from a homozygous *sctt* mutant mother that also lacks zygotic *sctt* function is shown. Note that the germ cells are scattered throughout the posterior of the embryo and that no germ cells are clustered within the presumptive gonad region. Both germ cell migration and the death of germ cells ectopic to the gonads are disrupted in these embryos. The somatic gonad forms normally. **(C)** An embryo from a homozygous *sctt* mutant mother that received a paternal wild-type copy of the gene shows complete rescue of the germ cell migration defect. A wild-type number of germ cells become incorporated into the gonads. However, the germ cells ectopic to the gonads are not eliminated. **(D)** In *out* mutant embryos, germ cell migration is normal, but germ cells outside of the gonads continue to express germ cell characteristics.

The *scattershot* (*sctt*) gene, when mutated, disrupts both germ cell pathfinding and programmed cell death.[31] Both phenotypes require that the mother be homozygous mutant. However, the migration and cell death functions can be genetically uncoupled. A paternally supplied wild-type copy of the gene can rescue the cell migration, but not the cell death phenotype (FIG. 2C).

Wun and Wun2 are phosphatidic acid phosphatases that are required to generate a signal that repels the germ cells away from the PMG.[30,32,33] If the functions of *wun* and *wun2* are disrupted, the germ cells exit the PMG and disperse throughout the posterior of the embryo. The distribution of germ cells in *wun* mutants is very similar to that observed in *sctt* mutants. Interestingly, overexpression of *wun* or *wun2* causes a decrease in germ cell number.[30] Thus, *wun* and *wun2* act to repel germ cells away from the PMG and to activate a program of cell death within the germ cells.

In Drosophila, programmed cell death can occur via different mechanisms. Apoptosis and autophagy have been reported,[54,63,65–67] and there is evidence indicating that non-caspase-mediated cell death programs may also exist.[64] While apoptosis and autophagy have distinct morphologies, they utilize some of the same molecular machinery. Therefore, these alternative forms of programmed cell death cannot always be distinguished using molecular markers. The nature of germ cell death in Drosophila will be elucidated in future studies.

OUTSTANDING ISSUES AND QUESTIONS

The screens for germ cell migration mutants have been very successful. We now know some of the genes required for normal germ cell migration in *Drosophila*, but their roles remain enigmatic and many questions remain unanswered. Germ cell migration no doubt requires a myriad of molecules. What other genes are necessary for germ cell guidance/pathfinding? It has been shown that both repulsive and attractive signals are involved. For each signal, the active signaling molecule or molecules must be synthesized and made available in a form that can be recognized by the receiving cell. The signal must be received and interpreted, and finally the receiving cell must change its physiology to respond to the signal correctly. A considerable amount of work remains before we fully understand this complex process.

What are the genes involved in the programmed cell death of germ cells? The existence of mutants that disrupt the death of germ cells ectopic to the gonads argues that this process occurs via an active, developmentally regulated mechanism. However, it is still unclear whether apoptosis, autophagy, or some other mechanism is involved.

Are there other genes with roles in both germ cell migration and programmed cell death? In other organisms and in other contexts, cell migration and cell survival are often linked.[5–15] Establishing the relationships between migration cues and cell survival/death signals will require that we identify and characterize more of the molecules involved. Forward genetic screens are a powerful means to identify required functions. It will also be informative to examine conserved candidate genes that are predicted to regulate cell pathfinding or cell death.

REFERENCES

1. LAUFFENBURGER, D.A. & A.F. HORWITZ. 1996. Cell migration: a physically integrated molecular process. Cell **84:** 359–369.

2. Jacobson, M.D., M. Weil *et al.* 1997. Programmed cell death in animal development. Cell **88:** 347–354.
3. Holder, N. & R. Klein. 1999. Eph receptors and ephrins: effectors of morphogenesis. Development **126:** 2033–2044.
4. Meier, P., A. Finch *et al.* 2000. Apoptosis in development. Nature **407:** 796–801.
5. Fleischman, R.A. 1993. From white spots to stem cells: the role of the Kit receptor in mammalian development. Trends Genet. **8:** 285–290.
6. Goetzl, E.J. & S. An. 1998. Diversity of cellular receptors and functions for the lysophospholipid growth factors lysophosphatidic acid and sphingosine 1-phosphate. FASEB J. **12:** 1589–1598.
7. Kunisada, T., H. Yoshida *et al.* 1998. Transgene expression of steel factor in the basal layer of epidermis promotes survival, proliferation, differentiation, and migration of melanocyte precursors. Development **125:** 2915–2923.
8. Wakamatsu, Y., M. Mochii *et al.* 1998. Avian neural crest–derived neurogenic precursors undergo apoptosis on the lateral migration pathway. Development **125:** 4205–4213.
9. Ashman, L.K. 1999. The biology of stem cell factor and its receptor C-kit. Int. J. Biochem. Cell Biol. **31:** 1037–1051.
10. De Felici, M. 2000. Regulation of primordial germ cell development in the mouse. Int. J. Dev. Biol. **44:** 575–580.
11. Goetzl, E.J., H. Lee *et al.* 2000. Mechanisms of lysolipid phosphate effects on cellular survival and proliferation. Ann. N.Y. Acad. Sci. **905:** 177–187.
12. Swarthout, J.T. & H.W. Walling. 2000. Lysophosphatidic acid: receptors, signaling, and survival. Cell Mol. Life Sci. **57:** 1978–1985.
13. Fukushima, N., I. Ishii *et al.* 2001. Lysophospholipid receptors. Annu. Rev. Pharmacol. Toxicol. **41:** 507–534.
14. Kierszenbaum, A.L. & L.L. Tres. 2001. Primordial germ cell–somatic cell partnership: a balancing cell signaling act. Mol. Reprod. Dev. **60:** 277–280.
15. Miller, F.D. & D.R. Kaplan. 2001. Neurotrophin signalling pathways regulating neuronal apoptosis. Cell Mol. Life Sci. **58:** 1045–1053.
16. Sonnenblick, B.P. 1941. Germ cell movements and sex differentiation of the gonads in the Drosophila embryo. Proc. Natl. Acad. Sci. USA **27:** 484–489.
17. Sonnenblick, B.P. 1950. The early embryology of *Drosophila melanogaster*. *In* Biology of Drosophila, pp. 62–167. Wiley. New York.
18. Counce, S.J. 1963. Developmental morphology of polar granules in Drosophila including observations on pole cell behavior and distribution during embryogenesis. J. Morphol. **112:** 129–145.
19. Fullilove, S.L. & A.G. Jacobson. 1978 Embryonic development: descriptive. *In* The Genetics and Biology of Drosophila, pp. 105–227. Academic Press. New York.
20. Underwood, E.M., J.H. Caulton *et al.* 1980. Developmental fate of pole cells in *Drosophila melanogaster*. Dev. Biol. **77:** 303–314.
21. Hay, B., L. Ackerman *et al.* 1988. Identification of a component of Drosophila polar granules. Development **103:** 625–640.
22. Lasko, P.F. & M. Ashburner. 1990. Posterior localization of Vasa protein correlates with, but is not sufficient for, pole cell development. Genes Dev. **4:** 905–921.
23. Warrior, R. 1994. Primordial germ cell migration and the assembly of the Drosophila embryonic gonad. Dev. Biol. **166:** 180–194.
24. Jaglarz, M.K. & K.R. Howard. 1995. The active migration of Drosophila primordial germ cells. Development **121:** 3495–3503.
25. Williamson, A. & R. Lehmann. 1996. Germ cell development in Drosophila. Annu. Rev. Cell Dev. Biol. **12:** 365–391.
26. Campos-Ortega, J. & V. Hartenstein. 1997. The Embryonic Development of *Drosophila melanogaster*. Second edition. Springer-Verlag. New York/Berlin.
27. Moore, L.A., H.T. Broihier *et al.* 1998. Gonadal mesoderm and fat body initially follow a common developmental path in Drosophila. Development **125:** 837–844.
28. Wylie, C. 1999. Germ cells. Cell **96:** 165–174.
29. Wylie, C. 2000. Germ cells. Curr. Opin. Genet. Dev. **10:** 410–413.
30. Starz-Gaiano, M., N.K. Cho *et al.* 2001. Spatially restricted activity of a Drosophila lipid phosphatase guides migrating germ cells. Development **128:** 983–991.

31. COFFMAN, C.R., R.C. STROHM *et al.* 2002. Identification of X-linked genes required for migration and programmed cell death of *Drosophila melanogaster* germ cells. Genetics **162:** 273–284.
32. ZHANG, N., J.P. ZHANG *et al.* 1996. Identification and genetic analysis of *wunen*, a gene guiding *Drosophila melanogaster* germ cell migration. Genetics **143:** 1231–1241.
33. ZHANG, N., J. ZHANG *et al.* 1997. The Drosophila protein Wunen repels migrating germ cells. Nature **385:** 64–67.
34. MOORE, L.A., H.T. BROIHIER *et al.* 1998. Identification of genes controlling germ cell migration and embryonic gonad formation in Drosophila. Development **125:** 667–678.
35. VAN DOREN, M., H.T. BROIHIER *et al.* 1998. HMG-CoA reductase guides migrating primordial germ cells. Nature **396:** 466–469.
36. DESHPANDE, G., L. SWANHART *et al.* 2001. Hedgehog signaling in germ cell migration. Cell **106:** 759–769.
37. RABINOWITZ, M. 1941. Studies on the cytology and early embryology of the egg of *Drosophila melanogaster*. J. Morphol. **69:** 1–49.
38. TURNER, F.R. & A.P. MAHOWALD. 1976. Scanning electron microscopy of Drosophila embryogenesis. 1. The structure of the egg envelopes and the formation of the cellular blastoderm. Dev. Biol. **50:** 95–108.
39. FOE, V.E. & B.M. ALBERTS. 1983. Studies of nuclear and cytoplasmic behaviour during the five mitotic cycles that precede gastrulation in Drosophila embryogenesis. J. Cell Sci. **61:** 31–70.
40. JAGLARZ, M.K. & K.R. HOWARD. 1994. Primordial germ cell migration in *Drosophila melanogaster* is controlled by somatic tissue. Development **120:** 83–89.
41. FISCHER-VIZE, J.A., G.M. RUBIN *et al.* 1992. The *fat facets* gene is required for Drosophila eye and embryo development. Development **116:** 985–1000.
42. CALLAINI, G., M.G. RIPARBELLI *et al.* 1995. Pole cell migration through the gut wall of the Drosophila embryo: analysis of cell interactions. Dev. Biol. **170:** 365–375.
43. BROOKMAN, J.J., A.T. TOOSY *et al.* 1992. The 412 retrotransposon and the development of gonadal mesoderm in Drosophila. Development **116:** 1185–1192.
44. BOYLE, M. & S. DINARDO. 1995. Specification, migration, and assembly of the somatic cells of the Drosophila gonad. Development **121:** 1815–1825.
45. BROIHIER, H.T., L.A. MOORE *et al.* 1998. *zfh-1* is required for germ cell migration and gonadal mesoderm development in Drosophila. Development **125:** 655–666.
46. KOBAYASHI, S., M. YAMADA *et al.* 1996. Essential role of the posterior morphogen *nanos* for germline development in Drosophila. Nature **380:** 708–711.
47. BOYLE, M., N. BONINI *et al.* 1997. Expression and function of *clift* in the development of somatic gonadal precursors within the Drosophila mesoderm. Development **124:** 971–982.
48. ASAOKA, M., H. SANO *et al.* 1998. Maternal Nanos regulates zygotic gene expression in germline progenitors of *Drosophila melanogaster*. Mech. Dev. **78:** 153–158.
49. FORBES, A. & R. LEHMANN. 1998. Nanos and Pumilio have critical roles in the development and function of Drosophila germline stem cells. Development **125:** 679–690.
50. DESHPANDE, G., G. CALHOUN *et al.* 1999. Novel functions of *nanos* in downregulating mitosis and transcription during the development of the Drosophila germline. Cell **99:** 271–281.
51. GOLDSTEIN, J.L. & M.S. BROWN. 1990. Regulation of the mevalonate pathway. Nature **343:** 425–430.
52. HAY, B.A., D.A. WASSARMAN *et al.* 1995. Drosophila homologs of baculovirus inhibitor of apoptosis proteins function to block cell death. Cell **83:** 1253–1262.
53. DEVERAUX, Q.L. & J.C. REED. 1999. IAP family proteins—suppressors of apoptosis. Genes Dev. **13:** 239–252.
54. LEE, C.Y. & E.H. BAEHRECKE. 2001. Steroid regulation of autophagic programmed cell death during development. Development **128:** 1443–1455.
55. THUMMEL, C.S. 2001. Steroid-triggered death by autophagy. Bioessays **23:** 677–682.
56. NAKAMURA, A., R. AMIKURA *et al.* 1996. Requirement for a noncoding RNA in Drosophila polar granules for germ cell establishment. Science **274:** 2075–2079.
57. ZALOKAR, M. 1976. Autoradiographic study of protein and RNA formation during early development of Drosophila eggs. Dev. Biol. **49:** 425–437.

58. Seydoux, G. & M.A. Dunn. 1997. Transcriptionally repressed germ cells lack a subpopulation of phosphorylated RNA polymerase II in early embryos of *Caenorhabditis elegans* and *Drosophila melanogaster*. Development **124:** 2191–2201.
59. Van Doren, M., A.L. Williamson *et al.* 1998. Regulation of zygotic gene expression in Drosophila primordial germ cells. Curr. Biol. **8:** 243–246.
60. Tautz, D. 1988. Regulation of the Drosophila segmentation gene *hunchback* by two maternal morphogenetic centres. Nature **332:** 281–284.
61. Wharton, R.P. & G. Struhl. 1991. RNA regulatory elements mediate control of Drosophila body pattern by the posterior morphogen Nanos. Cell **67:** 955–967.
62. Technau, G.M. & J.A. Campos-Ortega. 1986. Lineage analysis of transplanted individual cells in embryos of *Drosophila melanogaster* III. Commitment and proliferative capabilities of pole cells and midgut progenitors. Roux's Arch. Dev. Biol. **195:** 489–498.
63. Abrams, J.M., K. White *et al.* 1993. Programmed cell death during Drosophila embryogenesis. Development **117:** 29–43.
64. Foley, K. & L. Cooley. 1998. Apoptosis in late stage Drosophila nurse cells does not require genes within the H99 deficiency. Development **125:** 1075–1082.
65. Abrams, J.M. 1999. An emerging blueprint for apoptosis in Drosophila. Trends Cell Biol. **9:** 435–440.
66. Bangs, P. & K. White. 2000. Regulation and execution of apoptosis during Drosophila development. Dev. Dyn. **218:** 68–79.
67. Lee, C.Y. & E.H. Baehrecke. 2000. Genetic regulation of programmed cell death in Drosophila. Cell Res. **10:** 193–204.

Differentiation and Morphological Integration of Neural Progenitor Cells Transplanted into the Developing Mammalian Eye

D. S. SAKAGUCHI,[a,b,c] S. J. VAN HOFFELEN,[a,c] AND M. J. YOUNG[d]

[a]*Department of Zoology and Genetics,* [b]*Department of Biomedical Sciences, and* [c]*Neuroscience Program, Iowa State University, Ames, Iowa 50011, USA*

[d]*Schepens Eye Research Institute, Department of Ophthalmology, Harvard Medical School, Boston, Massachusetts 02114, USA*

ABSTRACT: Transplantation of neural stem/progenitor cells has been proposed as a novel approach for the replacement and repair of damaged CNS tissues. We have evaluated the influence of the host cellular microenvironment upon the survival, differentiation, and integration of neural progenitor cells transplanted into the CNS. Using this approach, we have investigated the fate of neural progenitor cells *in vivo* following transplantation into the developing mammalian eye. Murine brain progenitor cells (mBPCs) isolated from neonatal mice expressing the green fluorescent protein (GFP) transgene were transplanted into the eyes of Brazilian opossums (*Monodelphis domestica*). *Monodelphis* pups are born in an extremely immature, fetal-like state. The eyes of neonatal pups provide a fetal-like environment in which to study cellular interactions between host tissues and transplanted neural progenitor cells. mBPCs were transplanted by intraocular injection in hosts ranging in age from 5 days postnatal to adult. The transplanted cells were easily identified because of their GFP fluorescence. Extensive survival, differentiation, and morphological integration of mBPCs within the host tissue was observed. We found that the younger retinas provided a more supportive environment for the morphological integration of the transplanted mBPCs. Cells with morphologies characteristic of specific retinal cell types were observed. Moreover, some transplanted mBPCs were labeled with antibodies characteristic of specific neural/retinal phenotypes. These results suggest that the host environment strongly influences progenitor cell differentiation and that transplantation of neural progenitor cells may be a useful approach aimed at treating degeneration and pathology of the CNS.

KEYWORDS: transplantation; stem cells; retinal cell differentiation; retinal development; xenotransplantation; tissue engineering

INTRODUCTION

As a potential treatment for neurodegenerative diseases, stem cells have been proposed as a unique source of transplantable cells to replace damaged and diseased

Address for correspondence: D. S. Sakaguchi, Department of Zoology and Genetics, Iowa State University, 339 Science II, Ames, IA 50011. Voice: 515-294-3112; fax: 515-284-8457.
dssakagu@iastate.edu

Ann. N.Y. Acad. Sci. 995: 127–139 (2003).

neurons and glia of the central nervous system (CNS). Stem cells, which are multipotential progenitors that can give rise to a variety of cell types, are capable of self-renewal and have been isolated from many different tissues from embryonic, fetal, and postnatal donors. Neural stem cells and progenitor cells have been successfully isolated from various regions of the CNS such as the hippocampus, subventricular zone, spinal cord ependyma, and retina.[1–7] In general, neural stem cells can expand in serum-free medium when cultured with high levels of growth factors such as epidermal growth factor (EGF) and/or basic fibroblast growth factor (FGF-2). They are defined by their ability to differentiate into cells of all neural lineages (neurons, oligodendroglia, and astroglia), to self renew (to give rise to new neural stem cells with similar potential), and to populate developing and/or degenerating CNS regions.[8] Neural stem cells possess unique characteristics that make them ideal vectors for CNS repair. They can be clonally expanded in culture, providing a renewable supply of material for cell-based transplantation procedures. In addition, they may be genetically engineered to express exogenous genes for neurotransmitters and neurotrophic factors that may facilitate neuron survival and regeneration. Novel therapeutic strategies are being developed to take advantage of the ability of neural stem cells to proliferate in culture and to survive following transplantation into the mammalian brain, where they may integrate and stably express foreign genes or alternatively repopulate the damaged or diseased nervous system. It is important to note that studies examining the transplantation of "neural stem cells" are in fact grafting mixed populations of cells, which may contain some "true" neural stem cells, but which also contain cells that are more differentiated. These cells are termed neural progenitor cells or precursor cells.

With respect to the mammalian visual system, the death of specific populations of cells within the retina is associated with blinding diseases of the eye such as age-related macular degeneration, glaucoma, and retinitis pigmentosa. Transplantation of stem cells to replace lost cells or to act as support cells to facilitate rescue may become a practical approach to treating blinding diseases of the visual system in the future.

A number of animal studies have tested the viability of cell replacement in the retina using neural stem cells and progenitor cells transplanted either into the subretinal space or within the vitreal chamber.[9–13] Adult hippocampal progenitor cells transplanted into the retina of normal neonatal rats were capable of morphologically integrating into the host tissue. In contrast, migration of the adult hippocampal progenitor cells did not occur when grafted into normal adult recipients.[9] Recent results show that grafting of adult hippocampal progenitor cells into normal adult eyes will give rise to neuronal integration only when the retina is damaged at the time of transplantation.[10,14,15] Recently, the neuronal differentiation and morphological integration of adult hippocampal progenitor cells in the dystrophic retina of the adult Royal College of Surgeons (RCS) rat was investigated.[10] This study showed that neuronal integration was enhanced by the degenerating environment in the host eye, supporting the hypothesis that neural progenitors can be influenced by injury cues in the mature CNS.[10] In some cases, cells incorporated into the host retina even displayed morphologies similar to some retinal cell types, although the grafted cells failed to express retinal-specific proteins. To address issues associated with stem cell plasticity, it is essential to investigate how the different host environments influence stem/progenitor cell survival, differentiation, and integration (FIG. 1).

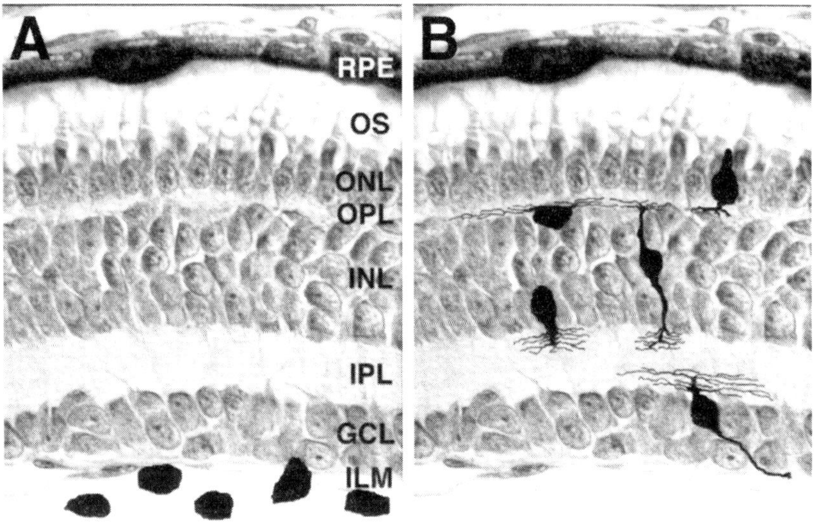

FIGURE 1. Schematic representation illustrating some important issues related to the migration, differentiation, and integration of stem/progenitor cells transplanted into the eye. **(A)** In this diagram, stem cells were inserted into the eye via intraocular transplantation. Five "transplanted" cells have aligned themselves along the neural retina in the vicinity of the ILM (black-filled cells). **(B)** Several weeks later, these "transplanted" cells have migrated, differentiated, and integrated into the host retina. Investigating how the host microenvironment influences these processes is essential for a better understanding of the biology of stem cells and progenitor cells. Abbreviations: RPE, retinal pigment epithelium; OS, outer segments; ONL, outer nuclear layer; OPL, outer plexiform layer; INL, inner nuclear layer; IPL, inner plexiform layer; GCL, ganglion cell layer; ILM, inner limiting membrane.

THE RETINA IS AN IDEAL CNS STRUCTURE TO EXAMINE STEM CELL DIFFERENTIATION AND INTEGRATION

The retina is an ideal *in vivo* experimental model system to investigate tissue engineering and the integration of transplanted stem/progenitor cells in the CNS. The retina is a part of the CNS as it is embryonically derived from the neural tube and it forms as an out-pocketing from the wall of the diencephalon. In addition, the development of the mammalian retina has been well characterized. Architecturally, it is a highly organized structure (FIG. 1) that is easily accessible for experimental manipulation and whose circuitry is relatively well defined. Furthermore, all vertebrate retinas are organized in the same fashion and are composed of five basic neuronal cell types: photoreceptors, whose cell bodies are located in the outer nuclear layer (ONL); ganglion cells, whose cell bodies form the ganglion cell layer (GCL); horizontal cells; bipolar cells; and amacrine cells, with their cell bodies in the inner nuclear layer (INL) (FIG. 1). In addition, two types of glia are also present within most mammalian retinas (Müller cells, which are the intrinsic retinal glia; and astrocytes that migrate in through the optic nerve). The nuclear layers are separated by synaptic layers, the outer and inner plexiform layers (OPL and IPL, respectively) (FIG. 1). The neurons elaborate their input (dendrites) and output (axons) processes within specific sublamina of these plexiform layers. The different cell types can be reliably identified based upon their position and morphology within the retina.

MONODELPHIS DOMESTICA: A NOVEL *IN VIVO* MODEL SYSTEM TO INVESTIGATE STEM CELL AND PROGENITOR CELL PLASTICITY

The retina of the gray short-tailed Brazilian opossum, *Monodelphis domestica*, possesses several advantages over other mammalian retinas and therefore has served as a unique host environment for recent transplant studies from our laboratory. The Brazilian opossum is a small pouchless marsupial that breeds well under laboratory conditions and whose young are born in an extremely immature, fetal-like state after a 14-day gestation (compared to approximately 19- and 21-day gestation for mice

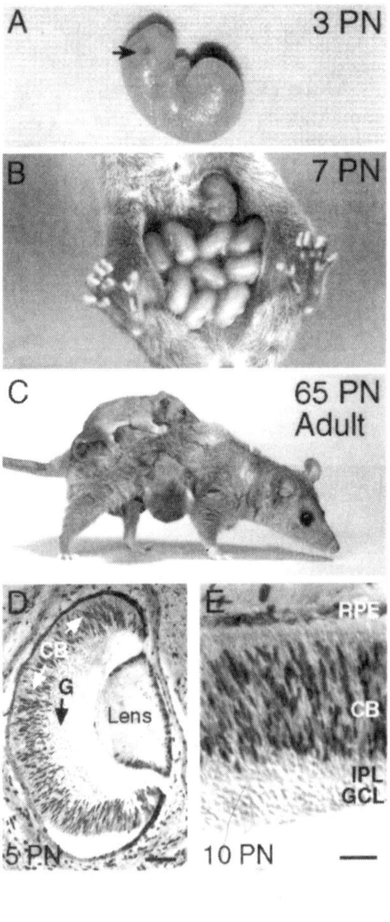

FIGURE 2. The Brazilian opossum as an *in vivo* experimental model system. **(A)** A 3 PN *Monodelphis* pup is approximately 1 cm in length. The pigmented eye is clearly visible beneath the skin (arrow). **(B)** Ventral view of a female with ten 7 PN pups attached to her nipples. **(C)** Adult female opossum (about the size of a hamster) with three 65 PN pups grasping onto their mother. **(D, E)** Neurogenesis within the developing *Monodelphis* retina. Images showing BrdU-IR within the postnatal retina. Pups received subcutaneous injections of BrdU at 5 and 10 PN and were sacrificed 2 hours later. Extensive BrdU incorporation was observed during early postnatal development. Abbreviations: CB, cytoblast layer; G and GCL, ganglion cell layer; IPL, inner plexiform layer; RPE, retinal pigment epithelium. Scale bars for D and E = 20 μm.

and rats, respectively).[16] Litters vary in size from 3 to 13 pups (FIG. 2). At birth (designated as 1 day postnatal: 1 PN), the retina is at a developmentally embryonic state and undergoes a protracted period of neurogenesis.[16] Eye opening occurs around 35 PN, pups are weaned at 60–65 PN, and the animals reach reproductive maturity at about 6 months of age. Their lack of a pouch and their fetal-like nature at birth circumvents the need for *in utero* experimental procedures that would be required on more conventional mammalian model systems such as mice, rats, rabbits, ferrets, or cats. Thus, their immaturity at birth and their protracted period of postnatal development make marsupials, like *Monodelphis*, excellent models for the study of visual system development and plasticity.[17–22]

Bromodeoxyuridine (BrdU) birth-dating analysis has revealed that the vast majority of retinal cytogenesis occurs postnatally in *Monodelphis*. FIGURE 2 illustrates that extensive BrdU incorporation, indicative of DNA synthesis, occurs during early postnatal life. At birth, the *Monodelphis* retina is a relatively undifferentiated neuroepithelium, with the first postmitotic cells, the earliest differentiating ganglion cells, located in the dorsocentral retina.[23] At 5 PN, the outer half of the retina, the cytoblast region, displays extensive BrdU incorporation (FIG. 2D). The ganglion cell layer has formed, although the other retinal layers are not yet obvious (FIG. 2D). By 10 PN, the nascent IPL is present, separating the GCL from the mitotically active cytoblast layer (FIG. 2E). Retinogenesis continues to 25 PN in *Monodelphis*. The first 15 days of postnatal life in *Monodelphis* development roughly corresponds with late embryonic development in rodents. For example, the CNS in *Monodelphis* at postnatal day 16 is approximately equivalent to that of a postnatal day 1 rat.[16,24]

IN VITRO DIFFERENTIATION OF MURINE BRAIN PROGENITOR CELLS (mBPCs)

An *in vitro* analysis of stem cells and progenitor cells is essential for a better understanding of their proliferation and differentiation potentials. The cells used in this analysis are termed murine brain progenitor cells (mBPCs) and were isolated from the brains of newborn enhanced green fluorescent protein (GFP)–expressing transgenic mice [TgN(β-act-eGFP)04Obs][25] as reported by Shatos and colleagues.[5]

The mBPCs were maintained as neurospheres by culturing in 20 ng/mL of epidermal growth factor (EGF). To initiate differentiation, the neurospheres were collected, dissociated, and plated onto an extracellular matrix (ECM) substrate in the absence of the growth factor. After 72 h under these differentiation conditions, the cells were found to express microtubule-associated protein (MAP2); class III beta tubulin and calretinin, which are markers associated with neurons; as well as glial fibrillary acidic protein (GFAP), which is a marker associated with glial cells (FIG. 3). This *in vitro* analysis revealed that the mBPCs, when cultured under differentiation conditions, possessed the ability to generate cells expressing both neuronal and glial markers.

IN VIVO ANALYSIS OF mBPCs FOLLOWING TRANSPLANTATION: SURVIVAL, DIFFERENTIATION, AND INTEGRATION

To test the hypothesis that the age of the host microenvironment influences neural stem cell and progenitor cell differentiation and integration, we transplanted the

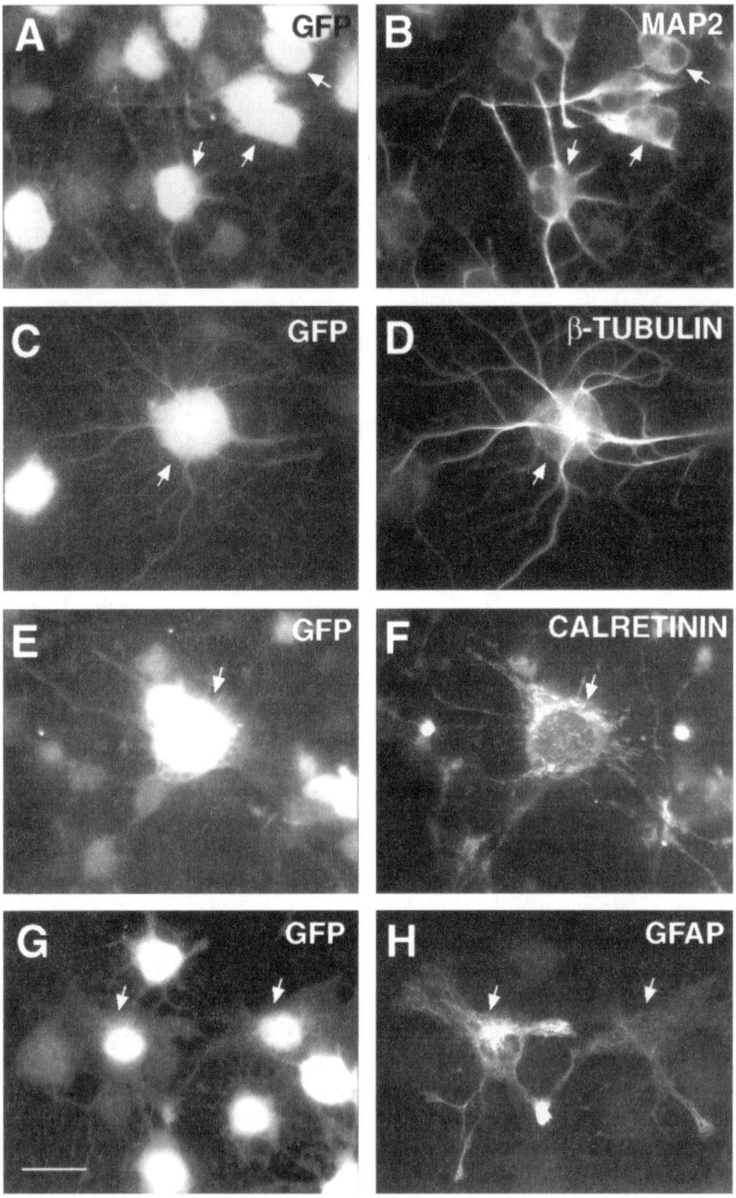

FIGURE 3. Expression of neuronal and glial markers by mBPCs *in vitro*. **(A, C, E, G)** Fluorescent images illustrating GFP-expressing mBPCs, and immunoreactivity for MAP2 **(B)**, class III beta tubulin **(D)**, calretinin **(F)**, and GFAP **(H)**. The arrows point to examples of GFP-expressing mBPCs that also expressed neural markers. Scale bar = 20 μm.

mBPCs into differently aged host retinas. The host animals were anesthetized for all transplantation procedures. mBPCs were transplanted by intraocular injection into the vitreal chamber of the eye through the caudal pole of the globe. Then, 1–2 μL of cell suspension containing ~50,000 cells/μL was pressure-injected using a beveled glass micropipette. Hosts were permitted to survive for 1, 2, or 4 weeks post-transplant, at which time the eye tissue was prepared for analysis. By using "fetal-like" hosts of 10 PN and younger, as well as maturing (30–35 PN) and mature (older than 35 PN) host animals, we have been able to investigate the influence of the host cellular microenvironment on the mBPCs *in vivo*. All of our procedures using animals were carried out in accordance with guidelines, had the approval of the Iowa State University Committee on Animal Care, and conform to the ARVO Statement on the Use of Animals in Ophthalmic Research.

Our initial studies verified that the mBPCs were capable of surviving following xenografting (FIG. 4). Indeed, the transplanted cells survived, morphologically differentiated, and integrated into host tissue (FIG. 4). Furthermore, the host tissue appeared relatively normal and there were no obvious signs of disruption of the laminar organization of the retina (FIG. 4).

However, we observed dramatic differences when mBPCs were grafted into hosts of different ages (FIG. 5). Extensive morphological differentiation and integration were only observed after transplantation into the youngest (5–10 PN) host retinas at 4 weeks postgrafting. Although cells were capable of surviving, little integration was observed in the older host retinas (35, 79 PN and adult hosts) (FIG. 5). Transplanted cells were often found in the posterior segment of the eye along the inner limiting membrane (ILM). Since virtually no integration of transplanted cells was observed in the older host eyes, we focused our efforts on characterizing the integration of the transplanted cells into the developing environment of the eye (FIGS. 5 and 6). Transplanted cells in the youngest hosts often displayed differentiated morphologies

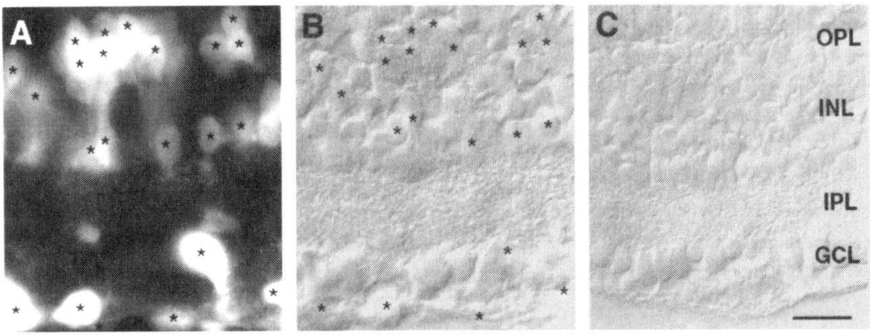

FIGURE 4. The architectural integrity of the retina was maintained after incorporation of mBPCs. **(A)** GFP-expressing cells integrated throughout the 10 PN host retina after 4 weeks posttransplant. **(B)** Differential interference contrast (DIC) microscopy image of the retina illustrated in A and an age-matched, control retina **(C)**. Stars in A and B correspond to examples of the same GFP-expressing transplanted somata. All tissue sections were obtained from 38 PN animals [i.e., 10 PN host plus 4 weeks (28 days) survival post-transplant]. Abbreviations: OPL, outer plexiform layer; INL, inner nuclear layer; IPL, inner plexiform layer; GCL, ganglion cell layer. Scale bar = 20 μm.

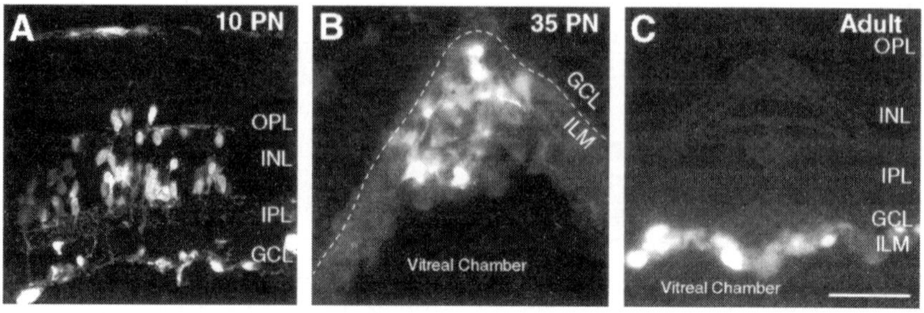

FIGURE 5. Transplantation of mBPCs into host eyes of different ages. Fluorescent images of mBPCs in sections of retina at 4 weeks posttransplant. GFP-expressing cells were observed within the eye at all ages. **(A)** GFP-expressing cells were found integrated throughout the 10 PN host retina, mBPCs were observed within all layers of the retina, and GFP-expressing cells were observed in the ONL, INL, and GCL, with processes in the OPL and IPL. **(B)** In an older host of 35 PN, mBPCs were found within the vitreal chamber, along the ILM, and processes were found extending into the inner retina, but somata were not. **(C)** In an adult host, isolated GFP-expressing processes were seen integrated within the inner retina; however, GFP-expressing somata were mainly found within the vitreal chamber or adjacent to the ILM. Abbreviations: OPL, outer plexiform layer; INL, inner nuclear layer; IPL, inner plexiform layer; GCL, ganglion cell layer; ILM, inner limiting membrane. Scale bar = 20 μm.

characteristic of specific retinal cell types (FIG. 6). Some GFP+ cells possessed morphologies similar to horizontal, bipolar, amacrine, and ganglion cells (FIG. 6). A striking observation was the ability of the grafted cells to respect the architectural organization of the retina. Transplanted GFP-expressing cell bodies were generally localized to the nuclear layers of the retina, while the GFP+ processes were restricted to the plexiform layers. Furthermore, within the IPL, there often appeared to be a segregation of the GFP+ processes within specific sublamina. The mammalian IPL can be subdivided into an OFF and ON sublamina both anatomically and physiologically (FIG. 6). In many cases, GFP+ processes were observed ramifying specifically within one sublamina (FIG. 6). Taken together, these results suggest that mBPCs transplanted into a developing retina were capable of migrating into specific layers. Moreover, the growing processes of the transplanted cells were capable of recognizing specific molecular cues located within the host tissue microenvironment.

The results described here suggest that grafted stem cells are capable of responding to cues in the developing CNS. A number of other studies have indicated that grafted stem cells are also able to respond to cues in the diseased adult CNS. Indeed, many studies show that the transplanted cells integrate and differentiate much better in dystrophic recipients than in normal controls.[10] While there are many possible explanations for this result, one could hypothesize that the diseased CNS in some manner recapitulates development, expressing factors normally only expressed during embryonic growth. The identification of these factors, and the understanding of their impact upon stem cell differentiation, is an important area of study in neurobiology. The use of recipients such as the Brazilian opossum provides one with an excellent model system to investigate the role of these factors for neural transplantation.

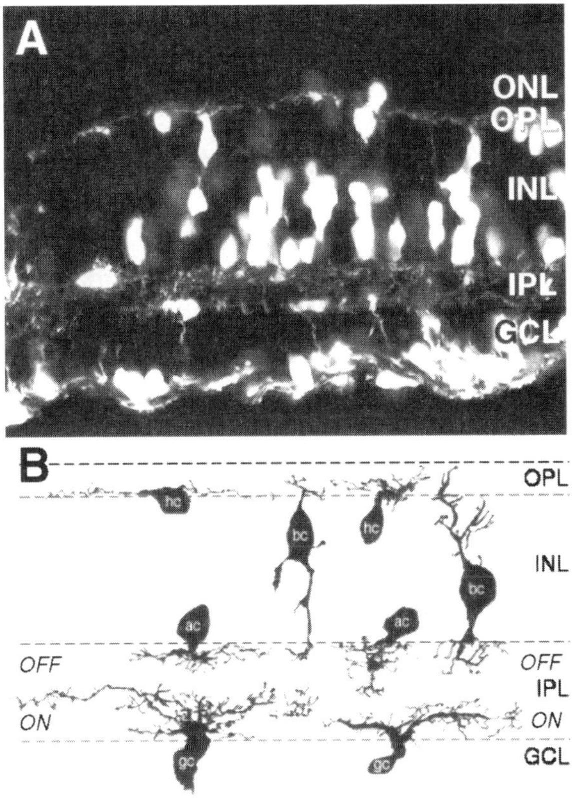

FIGURE 6. Morphological differentiation and integration of mBPCs into the 10 PN retina at 4 weeks posttransplant. GFP-expressing cells appeared intertwined between host cells in an organized extensive network in images captured using confocal microscopy. **(A)** Fluorescence images illustrating that GFP-expressing cells were found throughout the host retina. GFP+ somata were located within nuclear layers and displayed processes extending within the plexiform layers. Within the IPL, processes often appeared to organize within specific sublamina. **(B)** Transplanted GFP-expressing mBPCs were capable of developing morphologies similar to retinal cell types. Reconstructions of GFP-expressing cells morphologically integrated into the developing retina. Each cell was reconstructed from a series of confocal images. Cells exhibiting morphologies similar to retinal neurons are displayed together in a single field. Abbreviations: hc, horizontal-like cell; bc, bipolar-like cell; ac, amacrine-like cell; gc, retinal ganglion-like cells (or displaced amacrine-like cell); OPL, outer plexiform layer; INL, inner nuclear layer; IPL, inner plexiform layer; GCL, ganglion cell layer; OFF, the OFF sublamina of the IPL; ON, the ON sublamina of the IPL. The dashed lines demarcate the approximate boundaries of the plexiform layers.

PHENOTYPIC DIFFERENTIATION OF TRANSPLANTED mBPCs

To investigate the molecular differentiation of the transplanted cells, we used a panel of specific antibodies in an immunohistochemical analysis. A subset of GFP+ cells were observed expressing MAP2, calretinin, and recoverin (FIG. 7). The ex-

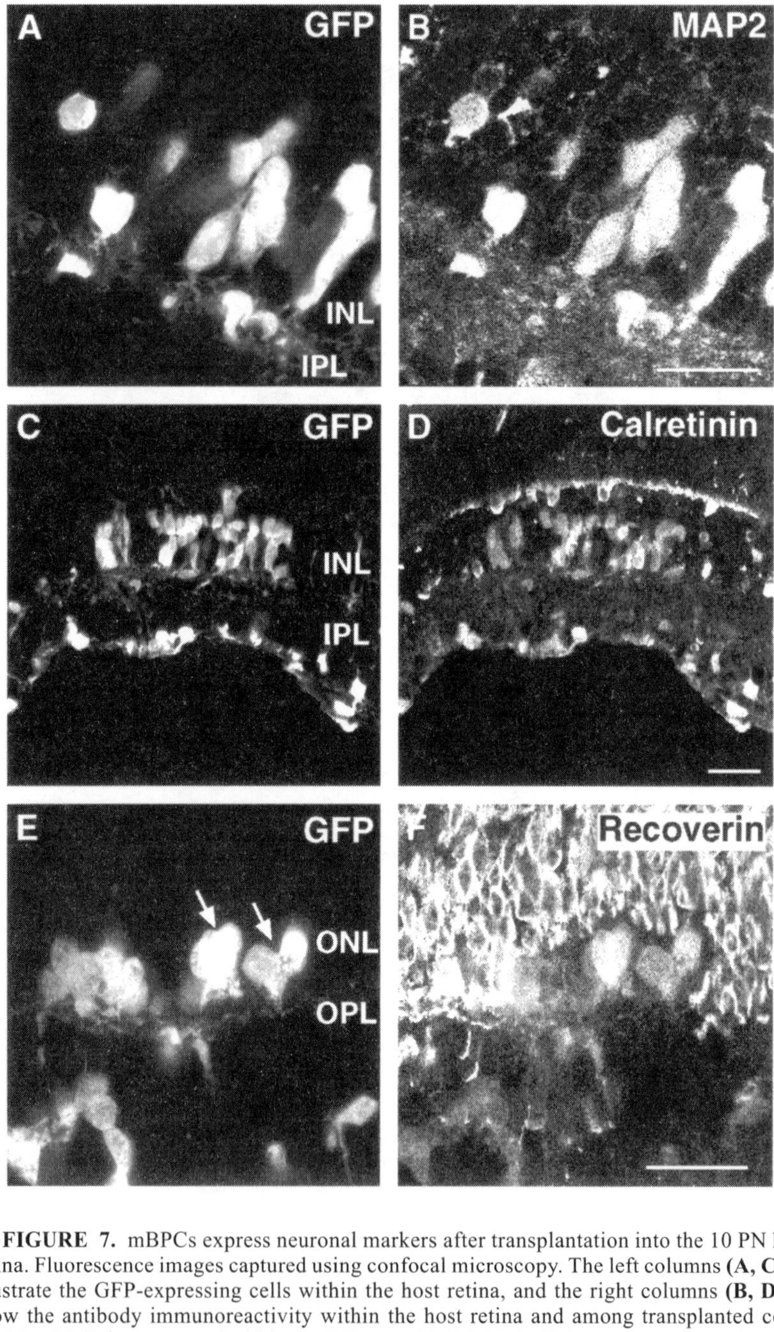

FIGURE 7. mBPCs express neuronal markers after transplantation into the 10 PN host retina. Fluorescence images captured using confocal microscopy. The left columns **(A, C, E)** illustrate the GFP-expressing cells within the host retina, and the right columns **(B, D, F)** show the antibody immunoreactivity within the host retina and among transplanted cells. **(A, B)** mBPCs incorporated within the INL were MAP2-IR (cells incorporated into the outer layers of the retina were not MAP2-IR). **(C, D)** Calretinin-IR was observed among GFP-expressing cells in the inner retina. Cells were located among calretinin-IR amacrine and

pression of these particular proteins by transplanted cells appears to be specific since not all GFP+ cells expressed these markers. This suggests that subpopulations of the mBPCs may exist that possess particular cell fate potentials with respect to their differentiation capabilities. Alternatively, the position of the cells within the host retina may strongly influence the fate of the transplanted cells. For example, MAP2 expression is normally restricted to the inner two-thirds of the retina. GFP+ cells that were also MAP2-immunoreactive (-IR) were in general also restricted to the inner two-thirds of the retina. Recoverin-IR within the retina is restricted primarily to photoreceptor cells in the ONL (although some bipolar cells also stain with anti-recoverin antibodies). GFP+ cells that were also recoverin-IR were only observed within cell bodies in the ONL. These preliminary findings suggest that there may be regional differences within the host environment that influence the phenotypic differentiation of the mBPCs following transplantation. Our results demonstrate that mBPCs were capable of survival following xenografting even in the absence of immunosuppression. This may in part be due to the relative purity of the mBPCs, which lack antigen-presenting cells and passenger leukocytes that would be present in conventional grafts of neural tissue, as well as to the low level of MHC expression exhibited by CNS stem cells.[26]

SUMMARY

Neural stem cells and progenitor cells clearly possess a remarkable degree of plasticity. The data presented here demonstrate that mBPCs can survive intravitreal transplantation and incorporate within the intact, developing retina. We have shown for the first time that the developing microenvironment may be critical in determining the fate of transplanted progenitor cells. Four weeks after transplantation into a developing environment, murine-derived neural progenitors integrated within the neural retina, respected the architectural organization, and adopted retinal-like morphologies and phenotypes.

ACKNOWLEDGMENTS

We wish to thank A. Dizhoor, Wayne State University, Detroit, MI, for generously providing the antirecoverin antibody for this project; M. H. West Greenlee for comments on the manuscript; and M. Shatos for assistance with the mBPCs. We also thank the Iowa State University Laboratory Animal Resources personnel for animal care. This work was partially supported by grants from the NIH (No. 09595 to

---

ganglion cells. No calretinin-IR mBPCs were found in the vitreous or outer retina at this age. **(E, F)** Recoverin-IR was observed among GFP-expressing cells located within the ONL (arrows). mBPCs were found to express recoverin only when located within the outer retina. Arrows highlight examples of colocalization in E and F. Abbreviations: ONL, outer nuclear layer; OPL, outer plexiform layer; INL, inner nuclear layer; IPL, inner plexiform layer. Scale bars = 20 μm.

M. J. Young), the Minda de Gunzburg Research Center for Retinal Transplantation (to M. J. Young), the Carver Trust (No. 701-17-52 to D. S. Sakaguchi), the Iowa State University Biotechnology Foundation (No. 102-47-45 to D. S. Sakaguchi), the NSF (No. IBN-9311054 to D. S. Sakaguchi), and The Glaucoma Foundation (to D. S. Sakaguchi). This manuscript, designated by Iowa State University as J-19811 of the Iowa Agriculture and Home Economics Experiment Station, Ames, IA, project number 3205, was supported by Hatch Act and State of Iowa Funds.

REFERENCES

1. GAGE, F.H., J. RAY & L.J. FISHER. 1995. Isolation, characterization, and use of stem cells from the CNS. Annu. Rev. Neurosci. **18:** 159–192.
2. PALMER, T.D., J. TAKAHASHI & F.H. GAGE. 1997. The adult rat hippocampus contains primordial neural stem cells. Mol. Cell. Neurosci. **8**(6)**:** 389–404.
3. SHIHABUDDIN, L.S., J. RAY & F.H. GAGE. 1997. FGF-2 is sufficient to isolate progenitors found in the adult mammalian spinal cord. Exp. Neurol. **148**(2)**:** 577–586.
4. WEISS, S., C. DUNNE, J. HEWSON *et al.* 1996. Multipotent CNS stem cells are present in the adult mammalian spinal cord and ventricular neuroaxis. J. Neurosci. **16**(23)**:** 7599–7609.
5. SHATOS, M.A., K. MIZUMOTO, H. MIZUMOTO *et al.* 2001. Multipotent stem cells from the brain and retina of green mice. J. Reg. Med. **2:** 13–15.
6. CHACKO, D.M., J.A. ROGERS, J.E. TURNER *et al.* 2000. Survival and differentiation of cultured retinal progenitors transplanted in the subretinal space of the rat. Biochem. Biophys. Res. Commun. **268**(3)**:** 842–846.
7. TROPEPE, V., B.L. COLES, B.J. CHIASSON *et al.* 2000. Retinal stem cells in the adult mammalian eye. Science **287**(5460)**:** 2032–2036.
8. GAGE, F.H. 2000. Mammalian neural stem cells. Science **287**(5457)**:** 1433–1438.
9. TAKAHASHI, M., T.D. PALMER, J. TAKAHASHI *et al.* 1998. Widespread integration and survival of adult-derived neural progenitor cells in the developing optic retina. Mol. Cell. Neurosci. **12**(6)**:** 340–348.
10. YOUNG, M.J., J. RAY, S.J. WHITELEY *et al.* 2000. Neuronal differentiation and morphological integration of hippocampal progenitor cells transplanted to the retina of immature and mature dystrophic rats. Mol. Cell. Neurosci. **16**(3)**:** 197–205.
11. PRESSMAR, S., M. ADER, G. RICHARD *et al.* 2001. The fate of heterotopically grafted neural precursor cells in the normal and dystrophic adult mouse retina. Invest. Ophthalmol. Visual Sci. **42**(13)**:** 3311–3319.
12. WARFVINGE, K., C. KAMME, U. ENGLUND *et al.* 2001. Retinal integration of grafts of brain-derived precursor cell lines implanted subretinally into adult, normal rats. Exp. Neurol. **169**(1)**:** 1–12.
13. WOJCIECHOWSKI, A.B., U. ENGLUND, C. LUNDBERG *et al.* 2002. Subretinal transplantation of brain-derived precursor cells to young RCS rats promotes photoreceptor cell survival. Exp. Eye Res. **75**(1)**:** 23–37.
14. NISHIDA, A., M. TAKAHASHI, H. TANIHARA *et al.* 2000. Incorporation and differentiation of hippocampus-derived neural stem cells transplanted in injured adult rat retina. Invest. Ophthalmol. Visual Sci. **41**(13)**:** 4268–4274.
15. KURIMOTO, Y., H. SHIBUKI, Y. KANEKO *et al.* 2001. Transplantation of adult rat hippocampus-derived neural stem cells into retina injured by transient ischemia. Neurosci. Lett. **306**(1–2)**:** 57–60.
16. KUEHL-KOVARIK, M.C., D.S. SAKAGUCHI, J. IQBAL *et al.* 1995. The gray short-tailed opossum: a novel model for mammalian development. Lab. Anim. **24:** 24–29.
17. CAVALCANTE, L.A. 1985. Postnatal neurogenesis and the formation of neural connections in the visual system of a marsupial. Dev. Neurobiol. Mammals **59:** 1–29.
18. DUNLOP, S.A., W.M. ROSS & L.D. BEAZLEY. 1994. The retinal ganglion cell layer and optic nerve in a marsupial, the honey possum (*Tarsipes rostratus*). Brain. Behav. Evol. **44**(6)**:** 307–323.

19. DUNLOP, S.A., L.B. TEE, R.D. LUND *et al.* 1997. Development of primary visual projections occurs entirely postnatally in the fat-tailed dunnart, a marsupial mouse, *Sminthopsis crassicaudata*. J. Comp. Neurol. **384**(1)**:** 26–40.
20. HARMAN, A.M., K.J. SANDERSON & L.D. BEAZLEY. 1992. Biphasic retinal neurogenesis in the brush-tailed possum, *Trichosurus vulpecula*: further evidence for the mechanisms involved in formation of ganglion cell density gradients. J. Comp. Neurol. **325:** 595–606.
21. MACLAREN, R.E. & J.S. TAYLOR. 1995. A critical period for axon regrowth through a lesion in the developing mammalian retina. Eur. J. Neurosci. **7**(10)**:** 2111–2118.
22. MARK, R.F., T. FREEMAN, Y. DING *et al.* 1993. Two stages in the development of a mammalian retinocollicular projection. Neuroreport **5**(2)**:** 117–120.
23. WEST GREENLEE, M.H., J.J. SWANSON, J.J. SIMON *et al.* 1996. Postnatal development and the differential expression of presynaptic terminal–associated proteins in the developing retina of the Brazilian opossum, *Monodelphis domestica*. Dev. Brain. Res. **96:** 159–172.
24. SAUNDERS, N.R., E. ADAM, M. READER *et al.* 1989. *Monodelphis domestica* (grey short-tailed opossum): an accessible model for studies of early neocortical development. Anat. Embryol. (Berl.) **180**(3)**:** 227–236.
25. OKABE, M., M. IKAWA, K. KOMINAMI *et al.* 1997. “Green mice” as a source of ubiquitous green cells. FEBS Lett. **407**(3)**:** 313–319.
26. KLASSEN, H., M.R. SCHWARTZ, A.H. BAILEY *et al.* 2001. Surface markers expressed by multipotent human and mouse neural progenitor cells include tetraspanins and non-protein epitopes. Neurosci. Lett. **312**(3)**:** 180–182.

Role of Collagen Type II and Perlecan in Skeletal Development

ERIKA GUSTAFSSON,[a] ATTILA ASZÓDI,[a,b] NATHALIE ORTEGA,[c] ERNST B. HUNZIKER,[d] HANS-WERNER DENKER,[e] ZENA WERB,[c] AND REINHARD FÄSSLER[a,b]

[a]*Department of Experimental Pathology, Lund University, 22185 Lund, Sweden*

[b]*Department of Molecular Medicine, Max-Planck-Institute for Biochemistry, 82152 Martinsried, Germany*

[c]*Department of Anatomy, University of California, San Francisco, California 94143, USA*

[d]*ITI Research Institute for Dental and Skeletal Biology, University of Bern, 3010 Bern, Switzerland*

[e]*Institute for Anatomy, Universitätsklinikum Essen, 45122 Essen, Germany*

ABSTRACT: The cartilage extracellular matrix is composed of a dense collagen network that entraps a range of other specialized proteins important for the proper formation and function of the tissue. Loss of two abundant cartilage components, type II collagen and perlecan, has drastic effects on skeletal development. Both collagen II and perlecan mutants have severe and lethal chondrodysplasia characterized by disorganized growth plate, lack of collagen network, defective endochondral bone formation, and abnormal intervertebral disk development. To test whether the reduced collagen density in the perlecan-null cartilage is due to enhanced activity of collagen-degrading proteinases, we have analyzed gelatinase expression and activity in the mutant tissue. Immunohistochemical analysis revealed a weak, but clear, expansion of MMP-9 deposition into the hypertrophic zone of the perlecan-null growth plate. However, *in situ* and SDS-PAGE zymography showed that the activity of gelatinases (MMP-2 and MMP-9) is not altered in perlecan-null cartilage, suggesting that they are not primarily linked to the reduced fibrillar network observed in the mutant. Likewise, intercrossing of perlecan mutants onto an MMP-9-null background could not rescue the ultrastructural abnormalities of the perlecan-deficient cartilage.

KEYWORDS: chondrodysplasia; collagen type II; extracellular matrix; matrix metalloproteinases; perlecan

INTRODUCTION

The vertebrate skeleton forms through two distinct processes: intramembranous (most of the cranial bones and part of the clavicula) and endochondral ossification

Address for correspondence: Erika Gustafsson, Department of Experimental Pathology, Lund University Hospital, SE-22185 Lund, Sweden.

Ann. N.Y. Acad. Sci. 995: 140–150 (2003). © 2003 New York Academy of Sciences.

(bones of the axial and appendicular skeleton). During intramembranous ossification, mesenchymal cells differentiate directly into bone-producing osteoblasts; in contrast, during endochondral ossification (EO), mesenchymal cells differentiate into chondrocytes and bones develop via cartilaginous templates. EO involves the establishment of the growth plate in which chondrocytes undergo sequential differentiation steps of proliferation, maturation, and hypertrophy. Chondrocyte hypertrophy is accompanied by matrix calcification, vascular invasion, degradation of the cartilage matrix, and bone deposition.

Chondrocytes secrete a specialized extracellular matrix (ECM) that provides the structural integrity of the tissue and regulates cellular differentiation influencing the distribution and presentation of growth factors and signaling molecules. Cartilage ECM is composed of a dense collagen fibrillar network that entraps various proteoglycans and noncollagenous glycoproteins. The major proteoglycan component is aggrecan, which generates an osmotic swelling pressure due to its high negative charge. The hydrated matrix is essential to resist compression forces, while the three-dimensional collagen meshwork provides the strength and resilience of the cartilage.

Collagen type II is the most abundant component of the cartilage ECM and interacts with the quantitatively minor collagens IX and XI to form heterotypic fibrils. Collagen II mutations in vertebrae reveal an essential structural and developmental role of the fibrillar network. Mutations in humans lead to a clinically variable spectrum of chondrodysplasias ranging from mild to lethal phenotype.[1] Mice lacking type II collagen develop a lethal chondrodysplasia characterized by abnormally shaped bones resulting from a complete absence of EO and by the lack of intervertebral disk (IVD) formation.[2,3]

Another prominent cartilage component is the proteoglycan perlecan, also present in basement membranes where it plays a role in formation and/or mechanical stabilization of these structures. Although abundant in the ECM, perlecan's importance for cartilage development was not clear until perlecan-null mice were created. Mice carrying a null mutation in the perlecan gene die perinatally at two stages.[4] Most embryos (70%) die at embryonic day (E) 10–12 from defective basement membranes in the heart.[4] The fraction of mutants surviving this crisis suffer from severe skeletal defects characterized by bent initial cartilaginous anlage of long bones, disorganized growth plates, reduced and abnormal collagen network in the hypertrophic matrix, and impaired endochondral ossification.[4,5] The severe cartilage defect and the recently discovered malformation of the great arteries of the heart[6] lead to lethality at birth. Humans lacking functional perlecan protein succumb to Dyssegmental Dysplasia of the Silverman-Handmaker type, which is a syndrome characterized by a severe lethal chondrodysplasia resembling that of the null mice.[7] Mutations in the perlecan gene generating truncated perlecan variants lead to chondrodystrophic myotonia called Schwartz-Jampel syndrome.[8]

The reduced amount and shortened collagen fibrils in the cartilage of perlecan-null mice suggest that perlecan is important in maintaining the collagen network. One possible explanation could be that perlecan protects the cartilage ECM by binding and regulating the activity of matrix metalloproteinases (MMPs). Several collagenases (MMP-1, MMP-8, and MMP-13) and gelatinases (MMP-2 and MMP-9) have been implicated in type II collagen degradation and/or EO.[9–13] This study aims to investigate the possible involvement of collagen-degrading gelatinases in the ECM disorganization of perlecan-deficient cartilage.

METHODS

Mice

The generation of mice carrying inactivated collagen type II, perlecan, or MMP-9 gene were described earlier.[3,4,13]

Staining of Skeletons, Histology, Electron Microscopy, and Immunohistochemistry

Stainings of skeletons, histology, electron microscopy, and immunohistochemistry were performed as described earlier.[2] Primary antibodies used were rabbit antimouse perlecan against domain V (diluted 1:1000, obtained from Rupert Timpl, Max-Planck-Institute for Biochemistry, Martinsried, Germany) and sheep antihuman MMP-9 (diluted 1:1000, obtained from Gillian Murphy, School of Biological Sciences, Norwich, England). Secondary antibodies used were biotinylated goat antirabbit Ig and biotinylated goat antisheep Ig (both from Vector Laboratories).

In Situ *Gelatin Zymography*

This method was modified from a protocol described by Denker.[14] Clean glass microscope slides were defatted by incubation in ether:ethanol (1:1) for 10–20 min and air-dried. Then, 200 µL of 5% gelatin solution (dissolved in deionized water at 50°C) was spread on each slide. After polymerization of the gelatin film in a horizontal position on a cool surface, the slides were fixed in a solution containing 10% formaldehyde and 45% ethanol for 1 h at room temperature. After washing for 20 min in running deionized water, the slides were air-dried and stored at –20°C.

After thawing, the slides were incubated in 7% NH_4OH at room temperature just until the gelatin film started to detach from the slide. After detachment, the gelatin film was stretched out on the slide and air-dried in a vertical position. The slides were soaked in 0.1 M phosphate buffer (pH 7.2) for 5 min and then air-dried. Freshly cut frozen sections (10–20 µm) of E17.5 tibia were placed in direct contact with the gelatinized slides and incubated for 4–12 h at 37°C in a humidified chamber. The reaction was stopped by fixation in saturated aqueous $HgCl_2$ (Merck, cat. no. 4419) for 10 min at 4°C. After careful washing in running tap water, the gelatin films were stained with 0.5% toluidine blue in borate buffer (pH 10.0) for 30 min at 4°C and then briefly washed in running tap water and deionized water. After air-drying, the slides were dehydrated in graded ethanol, cleared in xylene, and mounted with Entellan (Merck). Sites of gelatinase activity were indicated as unstained or slightly stained halos in the gelatin film.

SDS-PAGE Gelatin Zymography

Growth plate cartilage explants from E17.5 wild-type and perlecan-null long bones were cultured in serum-free DMEM medium for 4 days. Supernatants were analyzed for gelatinase activity by gel zymography in a 10% SDS-polyacrylamide gel containing 0.1% gelatin (Merck, Darmstadt, Germany, cat. no. 4070) according to the standard protocol.[15]

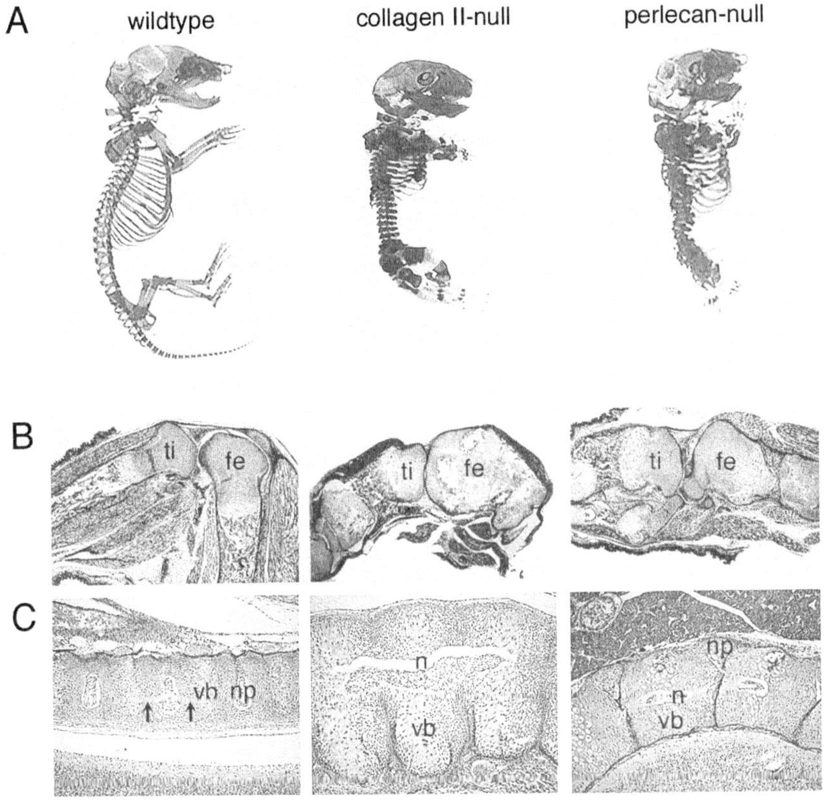

FIGURE 1. Skeletal abnormalities in embryos lacking collagen type II and perlecan, respectively. **(A)** Lateral view of alcian blue (cartilage)/alizarin red (bone)–stained skeletons of E18.5 wild-type, collagen II–, and perlecan-null embryos. Loss of both collagen type II and perlecan results in disproportionate dwarfism with short spine, vertebrae, and long bones, narrow thorax, and malformed ribs. **(B)** Comparative histological analysis of developing E16.5 limbs from wild-type, collagen II–, and perlecan-null embryos reveals the structural disorganization of the long bones in both collagen II– and perlecan-null embryos. In collagen II–null mutants, the epiphyseal cartilages of femur and tibia are enlarged and frequently contained holes. Holes are also present in the growth plate of perlecan-nulls, especially in the hypertrophic and prehypertrophic zones. Both collagen II– and perlecan-null long bones have disorganized growth plates, shorter metaphysis, and defective endochondral bone formation. **(C)** Comparative histological analysis of the developing vertebral columns from E16.5 wild-type, collagen II–, and perlecan-null embryos. In wild-type embryos, the notochord gradually expanded at the IVD areas to form the nucleus pulposus and vanished within the vertebral bodies. The arrows show remnants of the notochord. In collagen II–null embryos, the rod-like structure of the notochord remained relatively unchanged, without any notable differential expansion or regression. In perlecan-null embryos, the annulus are not identifiable and the nucleus pulposus are formed, but located at the periphery of the disk. Note that the notochord still contains cells. Abbreviations: ti, tibia; fe, femur; n, notochord; vb, vertebral body; np, nucleus pulposus.

RESULTS AND DISCUSSION

Collagen Type II– and Perlecan-Null Mice Develop a Severe Chondrodysplasia with Mechanically Weak Cartilage

Skeletal stainings of wild-type embryos with alcian blue (for cartilaginous tissues) and alizarin red (for mineralized bone) showed the structure of a normal skeleton at E18.5 (FIG. 1A). In contrast, both the collagen II– and perlecan-null embryos displayed similar defects with a severely affected endoskeleton (FIG. 1A). Both mutants were dwarfed and all bones formed by EO were malformed. Long bones and the trunk were shortened, the thorax was narrowed, and the size and shape of vertebral bodies and ribs were abnormal when compared to wild-type controls (FIG. 1A). Histological investigation of paraffin-embedded knees of E16.5 embryos revealed the structural disorganization of the long bones in both collagen II– and perlecan-null embryos (FIG. 1B). In collagen II mutants, the epiphyseal cartilages of femur and tibia were enlarged and frequently contained holes (FIG. 1B). Tissue enlargement and the formation of holes or cracks were also obvious in the perlecan-nulls, especially in the hypertrophic and prehypertrophic zones of the growth plate (FIG. 1B). Such holes were not present in plastic-embedded sections (data not shown), suggesting a mechanically weak mutant cartilage that is damaged during routine tissue processing for histology. Both collagen II– and perlecan-null long bones lacked well-organized growth plates, although cells resembling hypertrophic chondrocytes were visible adjacent to the ossification zone, suggesting that mutant chondrocytes undergo terminal differentiation. In both mutants, the metaphysis was greatly shortened, the endochondral bone formation was absent, and bone cavities were formed via protrusions of the periosteal bone (FIG. 1B).

Histological analyses of the developing trunk region at E16.5 also revealed similar defects in collagen II– and perlecan-null embryos (FIG. 1C). In both mutants, the vertebral bodies were increased in size compared with wild type, and the vertebral columns displayed abnormal IVD development. In wild-type mice at E16.5, the IVDs, which separate the neighboring vertebral bodies, consist of the nucleus pulposus in the center and the annulus fibrosus at the periphery (FIG. 1C). During normal IVD development, the annulus is derived from the intervertebral mesenchyme, whereas the nucleus pulposus is derived from the notochord in a process where notochordal cells gradually expand at the intervertebral areas and vanish within the vertebral bodies (FIG. 1C). In collagen II–null embryos, the IVDs were absent, and the rod-like notochord remained unchanged without any detectable expansion or regression (FIG. 1C). The perlecan-null embryos showed a slightly milder IVD phenotype. While the annulus were not identifiable, the nucleus pulposus were formed, but located at the periphery of the disk (FIG. 1C). In the vertebral bodies, we frequently observed that the notochord still contained cells, indicating that the removal of the notochordal cells was incomplete in the perlecan-null mutants (FIG. 1C).

Collagen Type II– and Perlecan-Null Mice Lack a Well-Organized Collagen Fibrillar Network in Cartilage

The almost identical skeletal malformations of the collagen II– and perlecan-null mice were accompanied by similarities in ultrastructural abnormalities. Electron

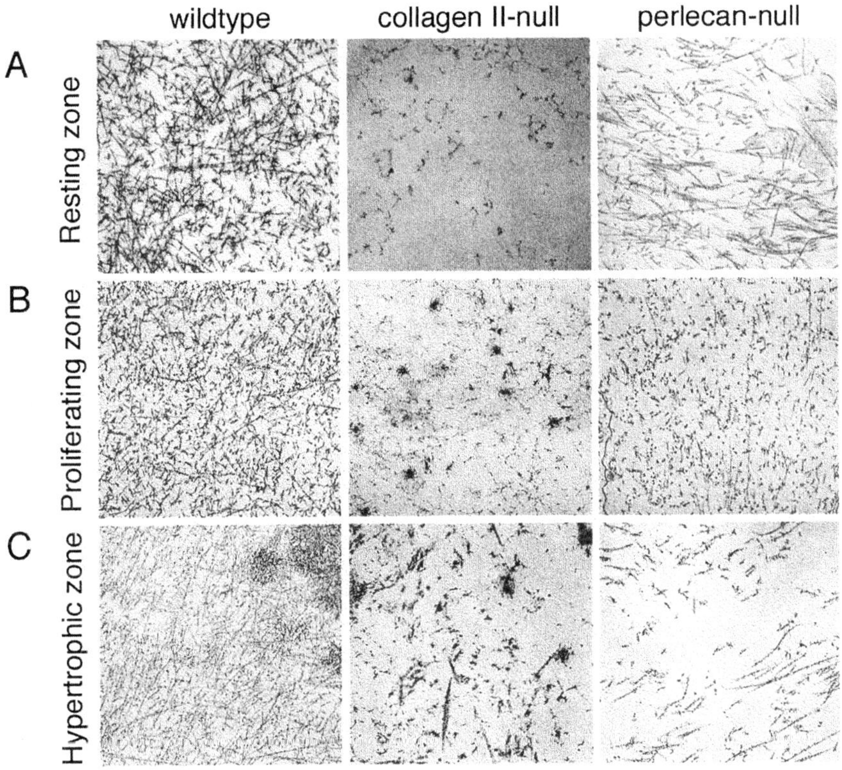

FIGURE 2. Ultrastructure of the matrix in the resting, proliferating, and hypertrophic zones of the cartilage of E17.5 wild-type, collagen II–, and perlecan-null embryos. The wild-type embryo shows a well-organized network of collagen fibrils of uniform diameter and length in all zones examined **(A–C)**. The collagen II–null mutant shows the complete lack of collagen fibrils in all zones **(A–C)**. In the perlecan-null mutant, the fibrils are normal, but less dense in the resting zone compared with wild type **(A)**. The ultrastructural abnormalities are more pronounced in the proliferating zone **(B)** and even more in the hypertrophic zone **(C)**, where the fibrils are shorter in length, have a lower contrast, and a greatly reduced density compared to the wild type.

microscopy of the resting, proliferating, and hypertrophic zones of wild-type growth plate showed a dense network of collagen fibrils with uniform length and diameter (FIG. 2A). The collagen II–null cartilage showed the complete lack of collagen fibrillar network in all zones (FIGS. 2A–2C). The sparse, electron-dense material observed in the mutant ECM was made up of abnormal fibrils formed by the "ectopically" expressed noncartilage collagen types I and III.[2] The perlecan-null cartilage also lacked the well-formed collagen network. In the resting zone, the length and diameter of the fibrils were relatively normal, but the fibrillar density was reduced compared to that of wild type (FIGS. 2A–2C). In the proliferating zone and especially in the hypertrophic zone, the fibrillar density was further reduced and the fibrils were very short, varied in diameter, and did not form a network.

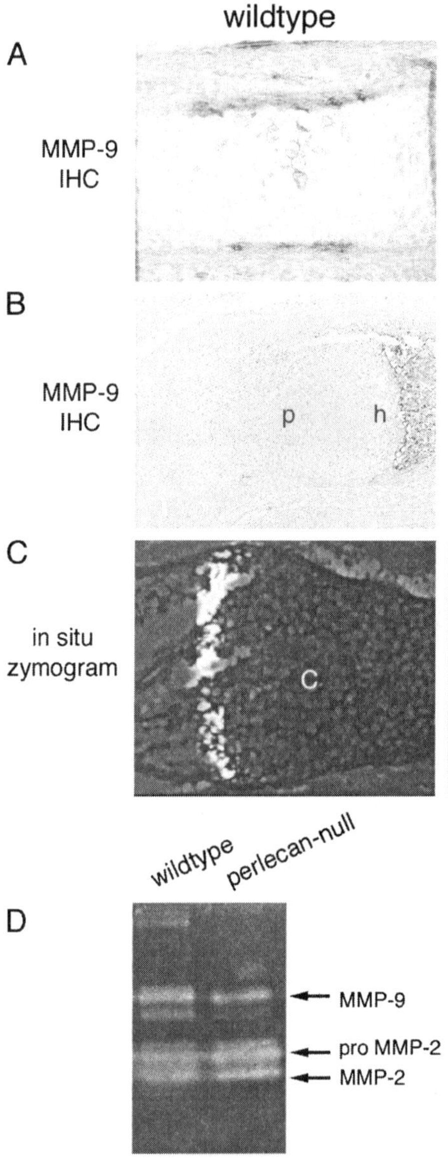

FIGURE 3. Analysis of gelatinase expression and activity in E17.5 wild-type and perlecan-null embryos. **(A)** MMP-9 immunostaining of metatarsals shows that the initial invasion of MMP-9-positive preosteoclasts is delayed in the perlecan-null mice compared to wild type. **(B)** MMP-9 immunostaining of tibias shows a strong immunoreactivity at the cartilage-bone junction and trabecular surfaces in wild type and mutant, and a weak MMP-9 deposition in the mutant hypertrophic zone. Terms: p, proliferative zone; h, hypertrophic zone. **(C)** *In situ* gelatin zymogram of wild-type and perlecan-null tibial sections showing gelatinase activity at the cartilage-bone junction (c = cartilage). **(D)** SDS-PAGE gelatin

Previously, we have suggested that the abnormal collagen matrix in the collagen II–deficient cartilage is unable to constrain proteoglycan-induced osmotic swelling pressure, resulting in tissue enlargement and softening.[2] We also proposed that the fluid accumulation reduces the internal swelling pressure in the cartilage and, as a consequence, the developing vertebral bodies cannot generate the proper "squeezing forces" on the notochord, which initiate the migration of notochordal cells towards the nucleus pulposus of the IVDs. We showed here that the reduced collagen density in the perlecan-null cartilage leads to similar, although less severe, histological abnormalities observed in the collagen II–null mice, further indicating the importance of the three-dimensional collagen network in the mechanical properties of cartilaginous matrices.

Does Abnormal Gelatinase Activity Contribute to the Cartilage Defects of Perlecan Mutants?

The lack of the fibrillar collagen network in the perlecan-null mice suggests that perlecan may have a special function in maintaining the cartilage ECM. Since the hypertrophic zone of long bones has a high activity of collagen-degrading proteases, a possible regulatory function of perlecan could account for this defect. MMP–9, also called gelatinase B, is expressed by osteoclasts and by cells at the osteochondrogenic junctions. It has a pivotal role in endochondral bone formation, regulating growth plate angiogenesis and the apoptosis of hypertrophic chondrocytes.[13] It has been proposed that MMP-9 is important for the degradation of the collagen network at the terminal layer of the hypertrophic zone, allowing the replacement of cartilage by mineralized bone.[16] The finding that the heparan sulfate side chains of perlecan can bind MMP-7 and possibly others such as MMP-1, -2, -9, and -13[17] further strengthens this hypothesis. Therefore, it is possible that perlecan is involved in the storage, stability, or inactivation of MMPs within the growth plate matrix.

To test the possibility that abnormal MMP-9 activity is causing the reduced collagen network in perlecan-deficient cartilage, we first performed immunohistochemical analyses on long bones at E17.5. The initial invasion of MMP-9-positive preosteoclasts was delayed in the perlecan-null mice (FIG. 3A). However, once the cells had entered, strong MMP-9 immunostaining was observed at the cartilage-bone junctions and the trabecular surfaces in both wild-type and perlecan-null embryos (FIG. 3B). We did not observe MMP-9 deposition in other zones of the wild-type growth plate; however, a faint immunoreactivity was detected in the hypertrophic zone of the mutant growth plate. In order to test if this moderate MMP-9 deposition is associated with enhanced MMP-9 activity in the perlecan-null cartilage, we performed *in situ* gelatin zymography assays. Tibial sections from both wild-type and

---

zymograms of conditioned medium from growth plates. Explants from wild-type and perlecan-null long bones were cultured in serum-free DMEM medium for 4 days. Supernatants were analyzed for gelatinase activity by SDS-substrate gel zymography. Bands corresponding to MMP-9, pro-MMP-2, and MMP-2 are visible, but there is no difference in gelatinase activity in supernatants from wild-type and perlecan-null long bones.

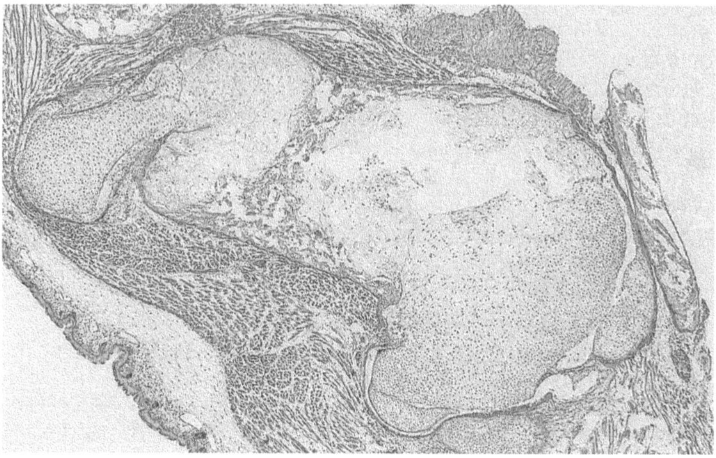

FIGURE 4. Hematoxylin/eosin staining of E17.5 MMP-9 perlecan double-mutant humerus. The double-mutant humerus is short and thick, the metaphyseal bone is very reduced, and the growth plate is severely disorganized. These defects in the bone morphology are identical to those of perlecan-null mice.

perlecan-null E17.5 embryos showed a clearing of the gelatin film only at the cartilage-bone junction (FIG. 3C), indicating that gelatinase activity is not significantly increased in the mutant cartilage. In agreement with the *in situ* data, SDS-PAGE gelatin zymography of conditioned medium from E17.5 epiphyseal and growth plate explant cultures showed similar results (FIG. 3D). Bands corresponding to MMP-9, pro-MMP-2, and MMP-2 were visible, but no differences in gelatinase activity could be detected in supernatants from wild-type and perlecan-null long bones (FIG. 3D). Finally, MMP-9-null mice were intercrossed with the perlecan-null strain to generate MMP-9/perlecan double-mutant embryos. Hematoxylin/eosin staining of E17.5 MMP-9/perlecan double-mutant humerus showed that the mutant humerus was short and thick, the metaphyseal bone was greatly reduced, and the growth plate severely disorganized. The defects were identical to those of perlecan-null mice, suggesting that MMP-9 does not contribute to the primary processes generating the ultrastructural abnormalities of the perlecan-null cartilage matrix (FIG. 4).

CONCLUSIONS

The presence of a correctly organized collagen II–containing fibrillar network is of crucial importance for endochondral bone formation and IVD development. The results from both collagen type II– and perlecan-null mutants support this notion since, in both cases, ultrastructural investigations revealed severe defects in the cartilage fibrillar network resulting in a severe disruption of EO and in impaired IVD development.

The more pronounced loss of the collagen network in the hypertrophic zone of the perlecan mutants leads us to investigate whether collagen-degrading proteases had an increased activity in this region of perlecan-null cartilage. Although we could detect a weak MMP-9 deposition in the hypertrophic matrix, both gel and *in situ* zymography show that the activity of gelatinases (MMP-2 and MMP-9) is not altered in mutant cartilage, suggesting that these proteases are not likely to be involved in generating the defects in the fibrillar network of the perlecan-null cartilage. In agreement, MMP-9-null background cannot rescue the ultrastructural abnormalities of the perlecan-deficient cartilage. One explanation could be that collagenases such as MMP-1, MMP-8, and MMP-13 are involved in these processes.

ACKNOWLEDGMENTS

This work was supported by the National Institutes of Health (Grant No. AR46238), a fellowship from the Osteogenesis Imperfecta Foundation, the Swedish Science Foundation, the Craaford Foundation, the Max Planck Society, and the Fonds der Chemischen Industrie.

REFERENCES

1. Bateman, J.F. 2001. The molecular genetics of inherited cartilage disease. Osteoarthritis Cartilage **9**(suppl. A)**:** S141–S149.
2. Aszódi, A. *et al.* 1998. Collagen II is essential for the removal of the notochord and the formation of intervertebral discs. J. Cell Biol. **143:** 1399–1412.
3. Li, S.W. *et al.* 1995. Transgenic mice with targeted inactivation of the Col2 α1 gene for collagen II develop a skeleton with membranous and periosteal bone, but no endochondral bone. Genes Dev. **9:** 2821–2830.
4. Costell, M. *et al.* 1999. Perlecan maintains the integrity of cartilage and some basement membranes. J. Cell Biol. **147:** 1109–1122.
5. Arikawa-Hirasawa, E. *et al.* 1999. Perlecan is essential for cartilage and cephalic development. Nat. Genet. **23:** 354–358.
6. Costell, M. *et al.* 2002. Abnormal migration of the neural crest and transposition of great arteries in perlecan-null embryos. Circ. Res. **91:** 158–164.
7. Arikawa-Hirasawa, E. *et al.* 2001. Dyssegmental dysplasia, Silverman-Handmaker type, is caused by functional null mutations of the perlecan gene. Nat. Genet. **27:** 431–434.
8. Nicole, S. *et al.* 2000. Perlecan, the major proteoglycan of basement membranes, is altered in patients with Schwartz-Jampel syndrome (chondrodystrophic myotonia). Nat. Genet. **26:** 480–483.
9. Harris, E.D., Jr. & S.M. Krane. 1972. An endopeptidase from rheumatoid synovial tissue culture. Biochim. Biophys. Acta **258:** 566–576.
10. Knauper, V. *et al.* 1996. Biochemical characterization of human collagenase-3. J. Biol. Chem. **271:** 1544–1550.
11. Miller, E.J. *et al.* 1976. Cleavage of type II and III collagens with mammalian collagenase: site of cleavage and primary structure at the NH_2-terminal portion of the smaller fragment released from both collagens. Biochemistry **15:** 787–792.
12. Murphy, G. *et al.* 1982. Partial purification of collagenase and gelatinase from human polymorphonuclear leucocytes: analysis of their actions on soluble and insoluble collagens. Biochem. J. **203:** 209–221.
13. Vu, T.H. *et al.* 1998. MMP-9/gelatinase B is a key regulator of growth plate angiogenesis and apoptosis of hypertrophic chondrocytes. Cell **93:** 411–422.
14. Denker, H.W. 1974. Protease substrate film test. Histochemistry **38:** 331–338.

15. KLEINER, D.E. & W.G. STETLER-STEVENSON. 1994. Quantitative zymography: detection of picogram quantities of gelatinases. Anal. Biochem. **218:** 325–329.
16. LEE, E.R. *et al.* 1999. Active gelatinase B is identified by histozymography in the cartilage resorption sites of developing long bones. Dev. Dyn. **215:** 190–205.
17. YU, W.H. & J.F. WOESSNER, JR. 2000. Heparan sulfate proteoglycans as extracellular docking molecules for matrilysin (matrix metalloproteinase 7). J. Biol. Chem. **275:** 4183–4191.

Remodeling of the Microenvironment by Aggressive Melanoma Tumor Cells

MARY J. C. HENDRIX,[a,b] ELISABETH A. SEFTOR,[a,b] DAWN A. KIRSCHMANN,[a,b] VITO QUARANTA,[c] AND RICHARD E. B. SEFTOR[a,b]

[a]*Department of Anatomy and Cell Biology,* [b]*The Holden Comprehensive Cancer Center, University of Iowa, Iowa City, Iowa 52242-1109, USA*

[c]*Department of Cell Biology, Scripps Research Institute, La Jolla, California 92037, USA*

ABSTRACT: A necessity for development and tumor progression is a blood supply formed by vasculogenic and/or angiogenic events, involving the cooperative interactions of cells with their microenvironment. Based on the recent characterization of vasculogenic mimicry by aggressive melanoma cells, particularly their ability to express VE (vascular endothelial)–cadherin, TIE-1, and EphA2, current studies have focused on the molecular signals deposited by these cells as they remodel their microenvironment. The experimental approach utilizes unique three-dimensional collagen matrices preconditioned by metastatic melanoma cells, which contain laminin 5 γ2 chain–enriched tracks with promigratory cleavage fragments produced by cooperative interactions with specific matrix metalloproteinases (MMPs). The results demonstrate that the collagen matrices preconditioned by the metastatic cells induce poorly aggressive melanoma cells to express, for the first time, key angiogenic/vasculogenic/matrix-remodeling genes. Treatment of aggressive melanoma cells with an MMP inhibitor resulted in the inhibition of vasculogenic mimicry–associated genes in these tumor cells and abrogation of the inductive effects of the preconditioned matrix on poorly aggressive melanoma cells. These observations illustrate the remarkable influence of the microenvironment on the phenotype of melanoma cells and may provide new perspectives on tumor cell plasticity and unique treatment strategies.

KEYWORDS: melanoma; microenvironment; vasculogenic mimicry; laminin; matrix metalloproteinases

INTRODUCTION

The remodeling of the microenvironment during tumor progression recapitulates many of the biological properties germane to embryogenesis, including cell plasticity, migration/invasion, and the formation of a blood supply. For example, during embryonic development, the formation of primary vascular networks occurs by the process of vasculogenesis, which is the *in situ* differentiation of mesodermal progenitor cells (angioblasts or hemangioblasts) to endothelial cells that organize into a primitive

Address for correspondence: Mary J. C. Hendrix, Department of Anatomy and Cell Biology, University of Iowa, 51 Newton Road, 1-100 Bowen Science Building, Iowa City, IA 52242-1109. Voice: 319-335-7755; fax: 319-335-7770.
mary-hendrix@uiowa.edu

Ann. N.Y. Acad. Sci. 995: 151–161 (2003). © 2003 New York Academy of Sciences.

network.[1,2] The subsequent remodeling of the vasculogenic network into a more refined complex of vasculature occurs through angiogenesis, which is the growth of new capillaries from preexisting networks. Similarly, it is widely accepted that, during cancer progression, tumors require a blood supply for growth.[3–5] Based on the molecular profile of aggressive melanoma cells, together with novel *in vitro* observations and correlative histopathology findings, our laboratory and collaborators have introduced the concept of vasculogenic mimicry to describe the plasticity and unique ability of aggressive melanoma tumor cells to express endothelial-associated genes and form tubular structures and patterned networks in three-dimensional culture that mimic embryonic vasculogenic networks.[6–9] In animal and patient tumors, histological analysis has revealed the architecture of the aggressive tumors to consist of laminin-rich networks surrounding spheroids of tumor cells.[9] Many of these networks are lined by tumor cells delineating vascular channels containing red blood cells and plasma, and possibly providing a perfusion mechanism and dissemination route within the tumor compartment that functions either independently of or simultaneously with angiogenesis and vessel co-option. Recent observations have confirmed the presence of a laminin-positive, fluid-conducting meshwork in xenografted cutaneous and primary human uveal melanoma.[10] Tumors exhibiting vasculogenic mimicry have a poor prognosis,[6,11] which underscores the clinical relevance of this new concept. The biological significance of vasculogenic mimicry has been uniquely demonstrated in an ischemic limb model, where human metastatic melanoma cells participated in the neovascularization of circulation-deficient muscle.[12] This study provided new perspectives on tumor cell plasticity and challenged existing dogma regarding the fate and function of aggressive tumor cells in an ischemic microenvironment.

The molecular underpinnings of melanoma vasculogenic mimicry are beginning to unfold. Recent experimental evidence has shown the importance of several key molecules in the formation of vasculogenic-like networks by melanoma cells, including VE (vascular endothelial)–cadherin (CD144 or cadherin 5), EphA2 (epithelial cell kinase), and laminin 5 $\gamma 2$ chain.[7–9] VE-cadherin is an adhesive protein, previously considered to be endothelial cell–specific, belonging to the cadherin family of transmembrane proteins promoting homotypic cell-to-cell interaction.[13–16] EphA2 is a receptor protein tyrosine kinase that is part of a large family of ephrin receptors.[17] Binding of EphA2 to its ligand ephrin-A1 results in the phosphorylation of EphA2; however, there is evidence indicating that EphA2 can also be constitutively phosphorylated in unstimulated cells.[18] Strong expression of EphA2 and ephrin-A1 has been associated with increased melanoma thickness and decreased survival, and the EphA2/ephrin-A1 pathway has been linked to tumor cell proliferation.[19] Laminins are major components of basement membranes and play an active role in neurite outgrowth, tumor metastasis, cell attachment and migration, and angiogenesis.[19–21] Proteolytic cleavage of laminin, particularly the laminin 5 $\gamma 2$ chain, can alter and regulate the integrin-mediated migratory behavior of certain cells,[22–24] illustrating its potential importance as a molecular trigger in the microenvironment.

This article will review the experimental strategy that is in progress to characterize the unique cellular and molecular aspects of distinctive aggressive and poorly aggressive melanoma tumor cells within heterogeneous tumors, with a special focus on the role of the microenvironment in maintaining and inducing an aggressive tumor cell phenotype.

METHODS

Cell Cultures

The primary choroidal or ciliary body human uveal melanoma cell lines (OCM-1A, C918), liver metastasis cell lines (MUM-2B, MUM-2C), metastatic cutaneous melanoma cell line (C8161), and a poorly invasive cutaneous melanoma cell line (C81-61) have been described previously[7,9,25–28] and were maintained in RPMI 1640 medium (OCM-1A, C918, MUM-2B, MUM-2C, and C8161; Invitrogen, Carlsbad, CA) or Hams F-10 (C81-61; Invitrogen) supplemented with 10% fetal bovine serum (15% for C81-61), 0.1% gentamycin sulfate (Gemini Bioproducts, Calabasas, CA), and 1× Mito+ (C81-61; Collaborative Biomedical, Bedford, MA). The cultures were determined to be free of *mycoplasma* contamination using a *mycoplasma* PCR ELISA detection kit (Roche, Indianapolis, IN).

Preconditioned Matrices

A critical aspect of this work was the careful production of 3-D type I collagen gels[7,9] as follows: Twenty-five µL of type I collagen (average 3–4 mg/mL; Collaborative Biomedical) was dropped onto 18-mm glass coverslips in 12-well culture dishes and polymerized with an application of 100% ethanol for 5 min at room temperature. After washing with PBS, tumor cells were seeded onto the 3-D gel in complete medium. The cultures were then observed after 4 days and images captured digitally using a Zeiss Televal inverted microscope (Carl Zeiss, Inc., Thornwood, NY) and Hitachi HV-C20 CCD camera (Hitachi Denshi America, Ltd., Woodbury, NY). For experiments designed to analyze the ability of the poorly aggressive melanoma cells to engage in vasculogenic mimicry when placed on a matrix preconditioned by the highly aggressive melanoma cells in the presence or absence of 3 µg/mL COL3 (a kind gift from CollaGenex, Newton, PA), the aggressive cells were removed after 3 days with 20 mM NH_4OH followed by three quick washes with water, PBS, and then complete medium. Poorly aggressive cells were then seeded onto the preconditioned matrix and images acquired after 3–4 days.

Western Blot Analysis

Expression of laminin 5 γ2 and its promigratory fragments by C8161 and C81-61 cells was examined as previously described.[9] Briefly, cells were harvested in sample buffer (plus β-mercaptoethanol), incubated at >95°C for 5 min, microfuged, and loaded onto 4–12% sodium dodecyl sulfate polyacrylamide gradient gels. The separated proteins were then transblotted onto an Immobilon-P membrane (Millipore, Bedford, MA). Laminin was detected using antibody D4B5 to the laminin 5 γ2 chain (Chemicon, Temecula, CA), a secondary antibody conjugated to horseradish peroxidase (Jackson Immunoresearch Laboratories, West Grove, PA), and enhanced chemiluminescence (ECL; Amersham, Piscataway, NJ).

SEMIQUANTITATIVE RT-PCR ANALYSIS

Total RNA from untreated and 3 µg/mL COL-3–treated melanoma cell lines was prepared using Trizol reagent (Invitrogen) as per the manufacturer's protocol and was

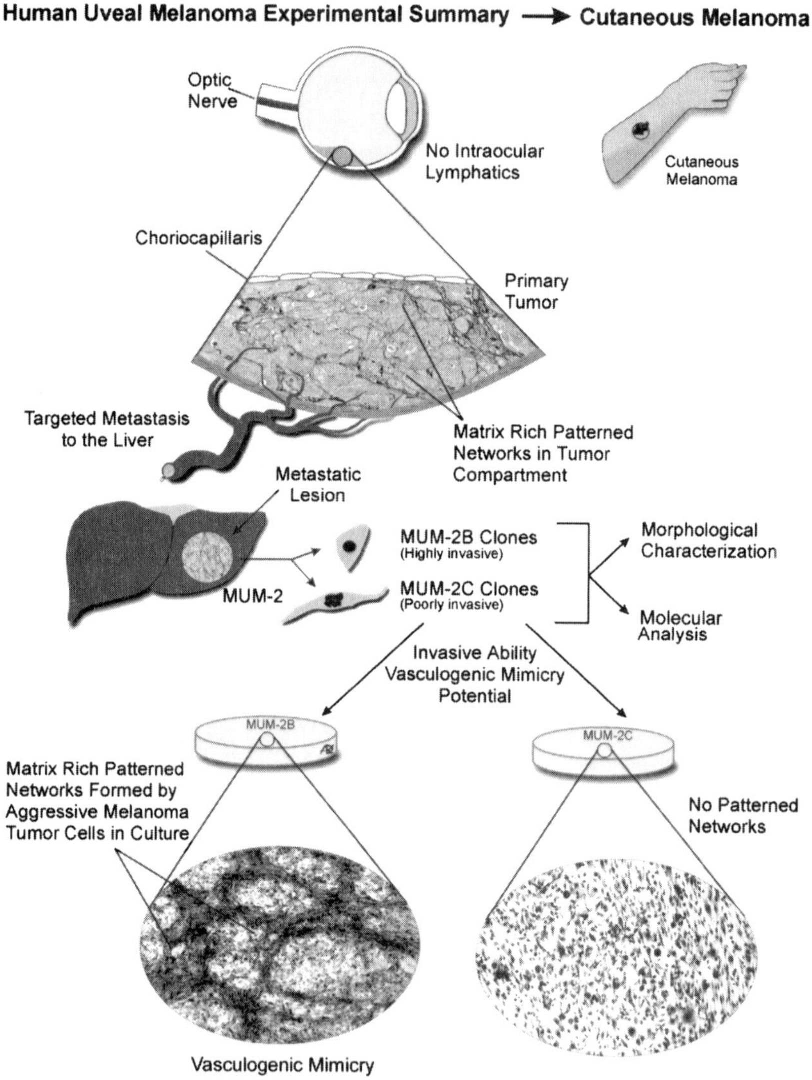

FIGURE 1. Overview of the experimental approach used to study the cellular and molecular aspects of highly aggressive vs. poorly aggressive human uveal and cutaneous melanoma. Within both primary tumors and metastases, subpopulations consisting of highly and poorly aggressive melanoma cells coexist. There are morphological and molecular distinctions between these subpopulations. In addition, aggressive melanoma cells form matrix-rich, patterned, vasculogenic-like networks (known as vasculogenic mimicry) when cultured on 3-D collagen I gels, while poorly aggressive (and nonaggressive) tumor cells are unable to do so.

reverse-transcribed using the Advantage PCR kit as per manufacturer's instructions (BD Clontech Laboratories, Palo Alto, CA). PCR amplifications were performed with gene-specific primers, and annealing temperature and number of amplification cycles were optimized using cDNA from untreated MUM-2B and MUM-2C cells. PCR amplification reactions were performed in a Robocycler Infinity Thermocycler (Stratagene, La Jolla, CA) as follows: 1 cycle, 94°C for 1 min; 25–30 cycles, 94°C for 1 min, 68°C for 2.5 min, 72°C for 1 min; and 1 cycle, 72°C for 5 min. GAPDH primers (BD Clontech) were used as controls for PCR amplification and PCR fragments were ligated into the pCR2.1 TOPO sequencing vector as per manufacturer's instructions (Invitrogen). Plasmid DNA was isolated and subjected to DNA sequencing analysis using fluorescent Sanger-based dideoxy sequencing on a PE-Applied Biosystems Model 3700 Automated Sequencer (University of Iowa DNA Facility). Two plasmids from each primer set were sequenced and shown to have 100% identity to the expected DNA sequence.

RESULTS

The overall experimental strategy for the current investigation is illustrated in FIGURE 1. The experimental study utilized both well-established and newly characterized uveal and cutaneous human melanoma cell lines, isolated from aggressive and poorly aggressive primary tumors and metastatic foci, along with distinctive subpopulations of highly invasive and poorly invasive uveal melanoma cells cloned from a heterogeneous liver metastasis. The aggressive and poorly aggressive melanoma cells were further characterized using morphological and molecular analyses (microarray and RT-PCR), in addition to assays testing the invasive ability and vasculogenic mimicry potential. As shown in FIGURE 1, the highly aggressive cells that are also highly invasive[28] form laminin-rich vasculogenic-like networks in 3-D collagen I gels (*in vitro*) that recapitulate the laminin-positive patterned networks seen in patients' aggressive tumors.[9] In contrast, the poorly aggressive tumor cells were unable to form network structures under similar culture conditions. Additional phenotypic analysis of these melanoma cell lines (FIG. 2) revealed that the poorly aggressive cells appeared spindle-shaped in monolayer culture and expressed vimentin intermediate filaments as their predominant mesenchymal marker. The aggressive melanoma cells, however, expressed both vimentin and keratin(s) 8/18 intermediate filaments and appeared epithelioid in monolayer culture. This latter phenotype has been associated with a dedifferentiated, interconverted cell type found in embryonic and metastatic cells.[29]

In order to better understand the interactions between melanoma tumor cells and their microenvironment in 3-D collagen I gels, an experimental strategy was followed that allowed a molecular analysis of the remodeling of the extracellular matrix (ECM) by aggressive melanoma cells. Based on the previous observations that aggressive melanoma cells overexpress vasculogenic mimicry–associated genes (VE-cadherin, EphA2, TIE-1, and VEGF-C), laminin 5 $\gamma 2$ chain, and MMPs,[9] the ability of these cells to precondition 3-D collagen I gels with factors that could induce poorly aggressive melanoma cells to assume an aggressive phenotype was examined.

As shown in FIGURE 3, aggressive melanoma cells capable of forming vasculogenic-like networks on collagen I gels (FIG. 3A) were removed from their matrix at

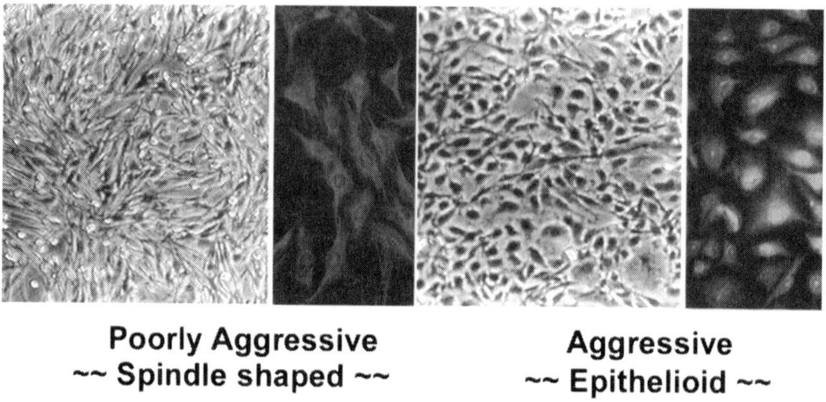

FIGURE 2. (Left to right) Phase-contrast microscopy of poorly aggressive human uveal melanoma cells that are spindle-shaped and express vimentin, a mesenchymal intermediate filament marker, as shown by rhodamine immunofluorescence microscopy; phase-contrast microscopy of aggressive human uveal melanoma cells that are epithelioid and coexpress vimentin and keratin(s) 8/18, by immunofluorescence microscopy.

2–3 days after preconditioning it. In parallel experiments, Western blot analysis (FIG. 3B) of the aggressive tumor cells (C8161) revealed the presence of laminin 5 γ2 chain and cleaved fragments (laminin γ2′; laminin γ2x) of the laminin 5 γ2 chain, which have been shown to elicit promigratory signals in cells activated by secreted MMPs.[22] Another noteworthy observation from this Western blot is that poorly aggressive melanoma cells (C81-61) did not express laminin 5 γ2 chain nor its cleaved, promigratory fragments. When the acellular matrix (after removal of the aggressive melanoma cells) was examined for laminin, a distinctive pattern of laminin-positive networks emerged within the collagen I gel (FIG. 3C). Whereas poorly aggressive melanoma cells are unable to form vasculogenic-like networks on collagen I gels (FIG. 3D), they were induced to form these patterned networks when seeded on the collagen I gels preconditioned by the aggressive melanoma cells (FIG. 3E). Furthermore, a molecular analysis of the differential expression of select genes (see FIG. 4) reveals the induction of EphA2, VE-cadherin, TIE-1, VEGF-C, laminin 5 γ2 chain, and keratin 7, for the first time, in the poorly aggressive melanoma cells grown on matrices preconditioned by aggressive melanoma cells, thus demonstrating a change in phenotype induced predominantly by the metastatic tumor cell microenvironment. In separate experiments, aggressive melanoma cells in 3-D culture were treated with an MMP inhibitor (COL-3, a chemically modified tetracycline), which resulted in a predominantly dramatic reduction in vasculogenic mimicry–associated genes (VE-cadherin, VEGF-C, and TIE-1), in addition to reduced levels of MMP-2, -9, and -14 (TABLE 1). Moreover, Western blot analysis revealed that COL-3 inhibited not only the generation of laminin 5 γ2 chain promigratory fragments in the aggressive cell preconditioned matrix, but also the induction of laminin 5 γ2 chain gene expression in poorly aggressive melanoma cells exposed to this microenvironment.

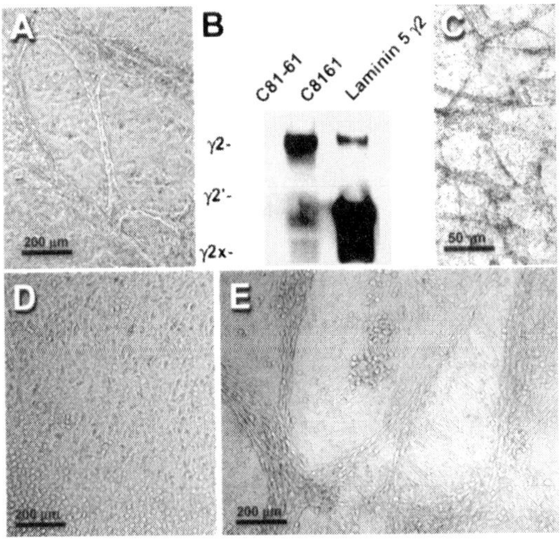

FIGURE 3. Phase-contrast images **(A, C, D, E)** and Western blot analyses **(B)** of human cutaneous melanoma tumor cells. **(A)** Aggressive cells (C8161) form vasculogenic-like networks in 4 days. **(B)** Western blot analysis of the aggressive (C8161) and poorly aggressive (C81-61) melanoma cells revealed the expression of laminin 5 γ2 chain (γ2) and its cleavage products (γ2′; γ2x) in cultures of C8161 cells only. **(C)** An anti-laminin antibody showed a laminin-rich meshwork secreted by the aggressive melanoma cells within the collagen I gel. **(D)** Poorly aggressive C81-61 cells cultured on a collagen I gel for 4 days showed no vasculogenic mimicry. **(E)** When the C81-61 cells were seeded onto collagen gels preconditioned by C8161 aggressive cells (which were subsequently removed), they formed vasculogenic-like networks.

TABLE 1. Semiquantitative RT-PCR changes in expression of vasculogenic mimicry–associated markers in response to COL-3

Gene marker	[C8161 cells + COL-3]/ [C8161 cells – COL-3] × 100%	[MUM-2B cells + COL-3]/ [MUM-2B cells – COL-3] × 100%
MMP-2	32%	64%
MMP-9	18%	13%
MMP-14	14%	76%
VE-cadherin	14%	19%
VEGF-C	27%	47%
TIE-1	89%	69%

NOTE: C8161 or MUM-2B metastatic human melanoma cells were cultured on collagen I 3-D matrices plus or minus 3 µg/mL COL-3. Vasculogenic mimicry was inhibited in those cells treated with COL-3. The cells were recovered and semiquantitative RT-PCR performed. The gels were digitized and the level of expression of the vasculogenic mimicry markers in the COL-3-treated cells determined relative to the untreated (control) cells × 100% after correction for loading using the values determined for GAPDH.

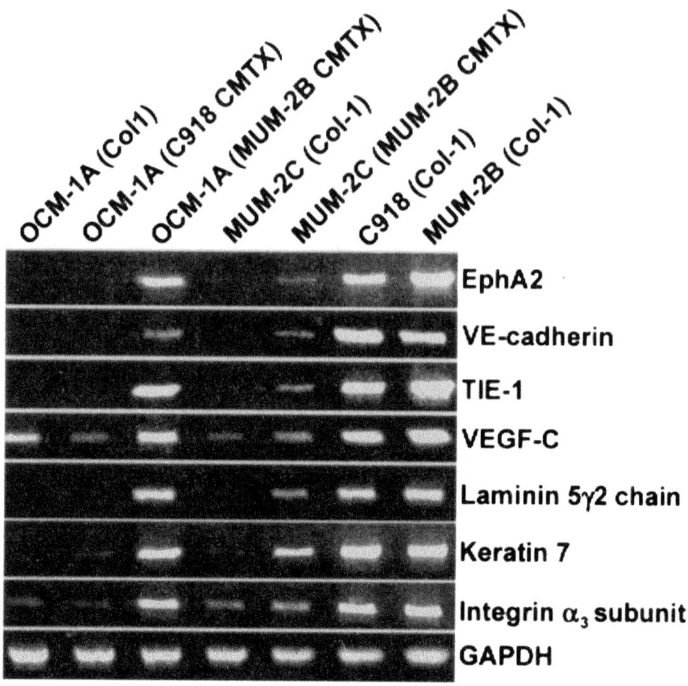

FIGURE 4. Semiquantitative RT-PCR analyses of vasculogenic mimicry–associated genes expressed by poorly aggressive melanoma cells (OCM-1A and MUM-2C), highly aggressive melanoma cells (C918 and MUM-2B), and poorly aggressive cells grown on matrices preconditioned (CMTX) by aggressive primary tumor–derived C918 cells vs. aggressive metastatic–derived MUM-2B cell CMTX. GAPDH was used to demonstrate equal loading of RNA samples.

Based on these collective observations, a model has been proposed regarding the molecular signals in the tumor microenvironment, as presented in FIGURE 5. Aggressive melanoma cells that escape a primary tumor site have the ability to deposit molecular signals in the tumor microenvironment that can provide migratory signals for other tumor cells to follow, possibly facilitating secondary sites of metastasis. For example, migratory signals can be formed through the cooperative interactions of MMPs and laminin 5 $\gamma 2$ chain, resulting in promigratory fragments that are involved in the induction of poorly aggressive tumor cells to acquire a more aggressive, vasculogenic phenotype, as demonstrated in FIGURE 4 by RT-PCR analysis.

DISCUSSION

The molecular profiling of tissues and cells has contributed significantly to our understanding of the cellular and molecular underpinnings of tumor progression. The data summarized in this paper focus on the molecular interactions of tumor cells

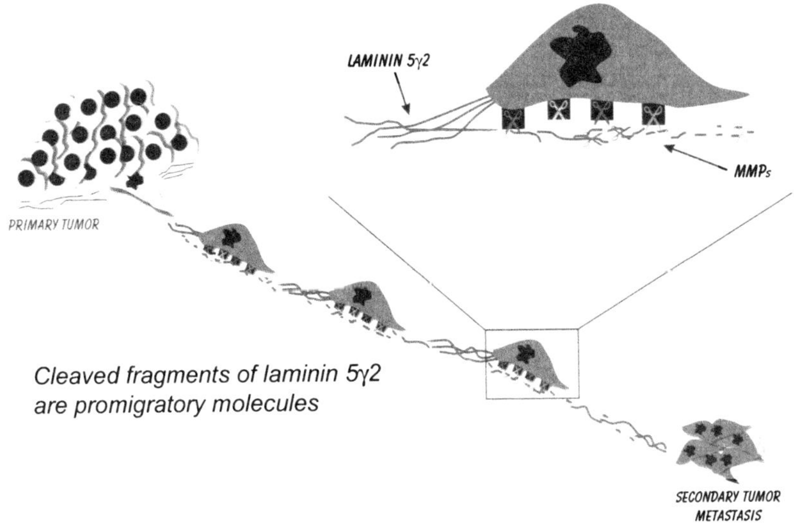

FIGURE 5. Hypothetical model for molecular signals deposited in the tumor microenvironment by aggressive melanoma cells through the molecular interactions of laminin 5 γ2 chain and MMPs, resulting in the production of promigratory molecules and thus facilitating tumor cell dissemination.

with their microenvironment, an understudied area in cancer biology, and provide a novel perspective on the plasticity and remodeling capabilities of aggressive melanoma cells.

The molecular analyses of various cancers, including melanoma, have generated more questions than answers about the phenotype of highly aggressive versus poorly aggressive tumor cells. As has been shown in the molecular profiling of other tumor types such as rhabdomyosarcoma, breast, prostatic, and renal cell carcinomas,[30–33] the molecular signature resembles that of multiple phenotypes and supports the notion of a deregulated genotype. In our molecular analyses of cutaneous and uveal melanoma,[6–9,28,34] we have tried to address the biological significance of several endothelial- and epithelial-related phenotypes as they relate to vasculogenic mimicry. Based on the expression of key angiogenesis/vasculogenesis genes, we have speculated that aggressive melanoma tumor cells have the potential to revert to an embryonic-like phenotype, as demonstrated by the *de novo* formation of vasculogenic-like networks reminiscent of embryologic events (summarized in FIG. 1). Furthermore, the significance of the overexpression of key extracellular matrix molecules, such as laminin 5 γ2 chain, and the cooperative interactions with MMPs resulting in promigratory and inductive signals, suggests that new treatment strategies focused on the tumor microenvironment might yield fruitful results.

In conclusion, the ability of aggressive melanoma tumor cells to mimic embryonic vasculogenesis when cultured on collagen I gels requires the cooperative interaction between the laminin 5 γ2 chain and specific MMPs.[9] Furthermore, blocking the function or activity of MMPs results in the inhibition of tumor cell vasculogenic mimicry. Remarkably, collagen gels preconditioned by aggressive melanoma cells can induce poorly aggressive melanoma cells to assume a vascular phenotype and form vasculogenic-like networks. Collectively, these observations may help identify specific molecular targets that function as templates for the coordinated migration of aggressive tumor cells and their proteolytic remodeling of the extracellular matrix, and may have profound implications for the development of novel therapies directed at the extracellular matrix to alter tumor progression.

ACKNOWLEDGMENTS

We gratefully acknowledge the scientific contributions of Naohiko Koshikawa (Scripps Research Institute) and Karla Daniels (University of Iowa), in addition to the kind gift of COL-3 from CollaGenex Pharmaceuticals, Inc. This research was supported by NIH Grant Nos. CA80318, CA59702, and CA88043-2S (to M. J. C. Hendrix); GM46902, CA47858-13S, and CA47858 (to V. Quaranta); and CA83137 (to R. E. B. Seftor).

REFERENCES

1. RISAU, W. 1997. Mechanisms of angiogenesis. Nature **386:** 671–674.
2. CARMELIET, P. 2000. Mechanisms of angiogenesis and arteriogenesis. Nat. Med. **6:** 389–395.
3. FOLKMAN, J. 1995. Clinical applications of research on angiogenesis: seminars in medicine of the Beth Israel Hospital, Boston. N. Engl. J. Med. **333:** 1757–1763.
4. RAK, J. & R.S. KERBEL. 1996. Treating cancer by inhibiting angiogenesis; new hopes and potential pitfalls. Cancer Metastasis Rev. **15:** 231–236.
5. KUMAR, R. & I.J. FIDLER. 1998. Angiogenic molecules and cancer metastasis. In Vivo **12:** 27–34.
6. MANIOTIS, A.J., R. FOLBERG, A.R. HESS *et al.* 1999. Vascular channel formation by human melanoma cells *in vivo* and *in vitro*: vasculogenic mimicry. Am. J. Pathol. **155**(3)**:** 739–752.
7. HESS, A.R., E.A. SEFTOR, L.M.G. GARDNER *et al.* 2001. Molecular regulation of tumor cell vasculogenic mimicry by tyrosine phosphorylation: role of epithelial cell kinase (Eck/EphA2). Cancer Res. **61:** 3250–3255.
8. HENDRIX, M.J.C., E.A. SEFTOR, P.S. MELTZER *et al.* 2001. Expression and functional significance of VE-cadherin in aggressive human melanoma cells: role in vasculogenic mimicry. Proc. Natl. Acad. Sci. USA **98:** 8018–8023.
9. SEFTOR, R.E.B., E.A. SEFTOR, N. KOSHIKAWA *et al.* 2001. Cooperative interactions of laminin 5 $\gamma 2$ chain, matrix metalloproteinase-2, and membrane type-1–matrix/metalloproteinase are required for mimicry of embryonic vasculogenesis by aggressive melanoma. Cancer Res. **61:** 6322–6327.
10. CLARIJS, R., I. OTTE-HÖLLER, D.J. RUITER *et al.* 2002. Presence of a fluid-conducting meshwork in xenografted cutaneous and primary human uveal melanoma. Invest. Ophthalmol. Visual Sci. **43**(4)**:** 912–918.
11. WARSO, M.A., A.J. MANIOTIS, X. CHEN *et al.* 2001. Prognostic significance of periodic acid-Schiff-positive patterns in primary cutaneous melanoma. Clin. Cancer Res. **7**(3)**:** 473–477.
12. HENDRIX, M.J.C., R.E.B. SEFTOR, E.A. SEFTOR *et al.* 2002. Transendothelial function of human metastatic melanoma cells: role of the microenvironment in cell-fate determination. Cancer Res. **62**(3)**:** 665–668.

13. HYNES, R.O. 1992. Specificity of cell adhesion in development: the cadherin superfamily. Curr. Opin. Genet. Dev. **2:** 621–624.
14. KEMLER, R. 1992. Classical cadherins. Semin. Cell Biol. **3:** 149–155.
15. LAMPUGNANI, M.G., M. RESNATI, M. RAITERI *et al.* 1992. A novel endothelial-specific membrane protein is a marker of cell-cell contacts. J. Cell Biol. **118:** 1511–1522.
16. GUMBINER, B.M. 1996. Cell adhesion: the molecular basis of tissue architecture and morphogenesis. Cell **4:** 345–357.
17. PASQUALE, E.B. 1997. The Eph family of receptors. Curr. Opin. Cell Biol. **9:** 608–615.
18. ROSENBURG, I.M., M. GOKE, M. KANAI *et al.* 1997. Epithelial cell kinase-B61: an autocrine loop modulating intestinal epithelial migration and barrier function. Am. J. Physiol. **273**(4, part 1)**:** G824–G832.
19. STRAUME, O. & L.A. AKSLEN. 2001. Expression of vascular endothelial growth factor, its receptors (FLT-1, KDR), and TSP-1 related to microvessel density and patient outcome in vertical growth phase melanomas. Am. J. Pathol. **159**(1)**:** 223–235.
20. MALINDA, K.M. & H.K. KLEINMAN. 1996. The laminins. Int. J. Biochem. Cell Biol. **28**(9)**:** 957–995.
21. COLOGNATO, H. & P.D. YURCHENCO. 2000. Form and function: the laminin family of heterotrimers. Dev. Dyn. **218:** 213–234.
22. GIANNELLI, G., J. FALK-MARZILLIER, O. SCHIRALDI *et al.* 1997. Induction of cell migration by matrix metalloprotease-2 cleavage of laminin-5. Science **277:** 225–228.
23. MALINDA, K.M., N. MOTOYOSHI, C. MELISSA *et al.* 1999. Identification of laminin α1 and β1 chain peptides active for endothelial cell adhesion, tube formation, and aortic sprouting. FASEB J. **13:** 53–62.
24. KOSHIKAWA, N., G. GIANNELLI, V. CIRULLI *et al.* 2000. Role of cell surface metalloprotease MT1-MMP in epithelial cell migration over laminin-5. J. Cell Biol. **148:** 615–624.
25. KAN-MITCHELL, J., M.S. MITCHELL, N. RAO *et al.* 1989. Characterization of uveal melanoma cell lines that grow as xenografts in rabbit eyes. Invest. Ophthalmol. Visual Sci. **30:** 829–834.
26. DANIELS, K., H.C. BOLDT, J.A. MARTIN *et al.* 1996. Expression of type VI collagen in uveal melanoma: role in pattern formation and tumor progression. Lab. Invest. **75:** 55–66.
27. HENDRIX, M.J.C., E.A. SEFTOR, Y-W. CHU *et al.* 1992. Coexpression of vimentin and keratins by human melanoma tumor cells: correlation with invasive and metastatic potential. J. Natl. Cancer Inst. **84:** 165–174.
28. SEFTOR, E.A., P.S. MELTZER, D.A. KIRSCHMANN *et al.* 2002. Molecular determinants of human uveal melanoma invasion and metastasis. Clin. Exp. Metastasis **19:** 233–246.
29. HENDRIX, M.J.C., E.A. SEFTOR, Y-W. CHU *et al.* 1996. Role of the intermediate filaments in migration, invasion, and metastasis. Cancer Metastasis Rev. **15:** 507–525.
30. KHAN, J., R. SIMON, M. BITTNER *et al.* 1998. Gene expression profiling of alveolar rhabdomyosarcoma with cDNA microarrays. Cancer Res. **58:** 5009–5013.
31. ROSS, D.T., U. SCHERF, M.B EISEN *et al.* 2000. Systematic variation in gene expression patterns in human cancer cell lines. Nat. Genet. **24:** 227–234.
32. PEROU, C.M., T. SORLIE, M.B. EISEN *et al.* 2000. Molecular portraits of human breast tumours. Nature **406:** 747–752.
33. LUO, J., D. DUGGAN, D. CHEN *et al.* 2001. Human prostate cancer and benign prostatic hyperplasia: molecular dissection by gene expression profiling. Cancer Res. **61:** 4683–4688.
34. BITTNER, M., P. MELTZER, Y. CHEN *et al.* 2000. Molecular classification of cutaneous malignant melanoma by gene expression profiling. Nature **406:** 536–540.

Notch Signaling in Primary Endothelial Cells

CARRIE J. SHAWBER, INDRANIL DAS, ESTHER FRANCISCO, AND JAN KITAJEWSKI

Department of Pathology and OB/GYN, Columbia University, College of Physicians and Surgeons, New York, New York 10032, USA

ABSTRACT: The Notch family of cell-surface receptors has been proposed to regulate cell-fate decisions by modulating the ability of each cell to respond to environmental cues. In vertebrates, gain-of-function and loss-of-function studies have demonstrated a requirement for Notch signaling for proper patterning of the vasculature during embryogenesis. To examine the molecular mechanisms by which Notch regulates vascular development, we analyzed changes in gene expression in response to Notch signaling. Notch signal transduction and function were assessed in primary human endothelial cells isolated from the dermal microvasculature of neonates, HMVECd. We demonstrate that HMVECd cells express a heterodimeric form of endogenous Notch4 on their cell surface. Using an *in vitro* coculture assay, we found that Delta4 can function as a ligand for Notch4 in HMVECd cells. Moreover, ectopic expression of an activated allele of Notch4 upregulated the expression of Delta4, suggesting that there may be a regulatory loop between Notch4 and its ligand, Delta4. Notch4 activation also induced the expression of the transcriptional repressors, HES1, HERP1, and HERP2, as well as ephrinB2, an angiogenic factor proposed to be involved in arterial/venous endothelial cell specification.

KEYWORDS: Notch; Delta; primary endothelial cells; angiogenesis

INTRODUCTION

Notch functions in an evolutionarily conserved signaling pathway that is required for numerous cell-type specifications. In mammals, the Notch family consists of four receptors (Notch 1–4) and has five ligands (Jagged 1–2 and Delta 1, 3, 4).[1] Depending on the environmental context, Notch signaling has been proposed to influence many different types of cell-fate decisions by providing inhibitory, inductive, or proliferative signals (reviewed in refs. 2–4). This pleiotropic function prompts the idea that Notch interfaces with multiple signaling pathways in a spatiotemporal manner in the embryo and adult.

Notch regulates cell-fate decisions by altering patterns of gene expression. Upon ligand activation, the cytoplasmic domain of Notch is proteolytically released and translocates to the nucleus where it interacts with CSL [CBF1, Su (H), Lag-2] transcriptional repressors and converts them to transcriptional activators. Both *in vivo*

Address for correspondence: Jan Kitajewski, Department of Pathology and OB/GYN, Columbia University, College of Physicians and Surgeons, New York, NY 10032. Fax: 212-305-3624. jkk9@columbia.edu

Ann. N.Y. Acad. Sci. 995: 162–170 (2003). © 2003 New York Academy of Sciences.

and *in vitro* studies indicate that the HES and HERP families of transcriptional repressors are the direct targets of Notch/CSL-dependent signaling.[5–8] Expression of HES1 or the *Drosophila* homologue E(Spl) has been shown to repress the expression of proneural bHLH transcription factors, preventing neuronal differentiation.[1,7] However, in most other cell types, the genes that HES and HERP gene products regulate have not been identified.

A number of observations suggest that the Notch signaling pathway may play a role in cell-fate determination and patterning of the vascular system. Notch1, Notch4, Jagged1, and Dll4 are all expressed in the developing vasculature, while Notch3 is expressed in the accessory smooth muscle cells.[9–12] Moreover, gain-of-function and loss-of-function studies in vertebrates have demonstrated that Notch is required for proper patterning and development of blood vessel networks during embryogenesis.[11,13,14] Mice nullizygous for Jagged1 or Notch1 and Notch4 die *in utero* due to severe hemorrhaging and vascular disorganization in the embryo, yolk sac, and intersomitic vessels.[11,13] Similarly, exogenous expression of an activated Notch4 in the developing endothelium also results in embryonic lethality due to vascular patterning defects.[14] These data suggest that appropriate levels of Notch signaling are crucial for the development of the embryonic vasculature. Recent experiments have also implicated Notch signaling in arterial/venous specification. Expression analysis of E13.5 mouse embryos found that Notch1, Notch3, Notch4, Delta4, Jagged1, and Jagged2 expression became restricted to the arteries and was absent from the veins.[12] Consistent with the expression data, disruption of Notch signaling in Zebrafish was associated with a loss of the arterial marker ephrinB2, while ectopic expression of an activated form of Notch led to a loss in the venous cell marker EphB4 within the dorsal aorta.[15] Taken together, these data suggest that Notch signaling may function at a number of critical steps during vascular development, including vasculogenesis, angiogenesis, and arterial/venous specification. However, the molecular mechanism(s) by which the Notch signaling pathways influence these different steps of vascular development has yet to be elucidated.

Here, we report our findings using the human primary endothelial cell line derived from neonatal dermal microvasculature (HMVECd). HMVECd cells endogenously express Notch4 and Delta4. In coculture experiments using ligand-expressing fibroblasts and Notch-expressing HMVECd cells, we demonstrate that Delta4 binds and activates Notch4. Finally, we have identified a number of genes whose expression is upregulated in response to Notch4 activation. These genes include Delta4, HES1, HERP1, HERP2, and ephrinB2.

MATERIALS AND METHODS

Constructs

The full-length Notch4 cDNA and the naturally occurring Notch4/int3 cDNA[14] were engineered in the adenovirus expression vector pAd-Lox. Adenoviral stocks were generated and titered as previously described.[16] The full-length Delta4 cDNA fused in frame with a myc/His tag was engineered in the mammalian expression vector pcDNA3.1 (Chiron). For detection with the anti-HA monoclonal 12CA5 antibody, Notch4 constructs have been tagged at their C-terminus with one copy of the HA epitope. The CBF1 luciferase reporter has been previously described.[17] The LacZ cDNA was engineered into both pAd-Lox and the expression vector pHyTc[18] (Genbank accession numbers for mouse Notch4: U43691; mouse Notch4/int3: M80456; and human Delta4: AF253468).

Cell Lines and Growth Conditions

Human primary endothelial cells derived from the dermal microvasculature of neonates (HMVECd) were purchased from Clonetics (Maryland) and maintained according to manufacturer's protocol. Bosc23[19] cells were maintained in Dulbecco's modified medium (DMEM) containing 10% fetal bovine serum (FBS) and penicillin-streptomycin.

Adenoviral Infections and Westerns

Adenoviruses encoding Notch4, Notch4/int3, or LacZ were used to infect HMVECd cells at an m.o.i. of 100 pfu/cell. At an m.o.i. of 100, Ad-LacZ infection led to greater than 95% of the cells being LacZ-positive at 24 hours postinfection. Twenty-four hours postinfection, LacZ and Notch4/int3HA-expressing HMVECd were lysed in RIPA buffer (20 mM Tris, pH 7.5, 2 mM EDTA, 150 mM NaCl, 1% NP40, 1% DOC, 0.1% SDS). Lysates were separated on an 8% SDS/PAGE gel and transferred to nitrocellulose (MSI). Western analysis was performed using the monoclonal anti-HA antibody 12CA5 (Babco) at a 1:1000 dilution in BLOTTO, and proteins were visualized using Enhanced Chemiluminescence (ECL; Amersham) in conjunction with a secondary antimouse IgG-HRP conjugate.

Cell Surface Biotinylation and Westerns

Twenty-four hours postinfection, parental and Notch4-expressing HMVECd cells were incubated with 0.5 mg/mL EZ-Link Sulfo-NHS-Biotin reagent (Pierce) in PBS supplemented with 1 mM $MgCl_2$, 0.1 mM $CaCl_2$, and 0.1% glucose (PBS-CMG) for 20 min at 4°C. Cells were then incubated with PBS–100 mM glycine for 15 min at 4°C and rinsed 2× with PBS-CMG to wash off excess biotin. Cells were lysed in 300 mM NaCl, 50 mM Tris, pH 8.0, 0.5% NP40, 0.5% DOC, 1 mM $CaCl_2$, and 1 mM $MgCl_2$. Lysates were incubated 20 min on ice and cleared. Supernatants were then incubated overnight with 30 μL streptavidin-conjugated beads (Pierce). Streptavidin beads were washed 3× with TENT buffer (50 mM Tris, pH 8.0, 2 mM EDTA, 150 mM NaCl, 1% Triton X-100). Proteins were eluted in boiling sample buffer, separated on a 7.5% SDS/PAGE gel, and transferred to PVDF membrane (Millipore). Membranes were probed with rabbit polyclonal antisera generated against the intracellular domain of Notch 4 (CoRb2-2) as previously described.[18]

Coculture Notch/CSL Reporter Assay

HMVECd cells were plated at 8×10^4 cells/well in 12-well plates. The following day, HMVECd cells were infected with adenoviruses encoding either Notch4 or GFP at an m.o.i. of 100 pfu/cell. Following the adenoviral infection, HMVECd cells were transfected using Lipofectamine 2000 (Invitrogen) containing 1.5 μg CBF1 luciferase reporter and 0.25 μg pHyTc-lacZ. The same day, Bosc23 cells were $CaPO_4$ transfected with 25 μg pcDNA3.1 Delta4 or dsRed (Clonetech) plasmids. Twenty-four hours posttransfection, 4×10^5 parental or Delta4-expressing Bosc23 cells were added to each well of Notch4 or GFP-expressing HMVECd cells. The following day, coculture lysates were collected and assayed for luciferase (Enhanced Luciferase Assay Kit, BD) and β-galactosidase (Galacto-light Plus, Tropix) activity following manufacturers' protocols.

Semiquantitative RT-PCR (sqRT-PCR)

Forty-eight hours postinfection, total RNAs were isolated by guanidium thiocyanate-phenol-chloroform extraction[20] from Notch4/int3 and LacZ-expressing HMVECd cells. One to two μg of total RNA was DNaseI (Ambion) digested followed by first-strand syntheses using random hexamers and Superscript II reverse transcriptase (Invitrogen). PCR primers were designed to recognize both human and murine transcripts for Jagged1, Delta4, HES1, HES5, HERP1, HERP2, ephrinB2, and EphB4. For each gene, human PCR amplicons were subcloned into the pDrive PCR cloning vector (Stratagene) for use as PCR reference standards. PCR reactions were prepared in quadruplicate for each unknown sample, as well as for a matching serially diluted reference standard. Reactions are removed at five cycle intervals, separated on 7% nondenaturing bis-acrylamide gels, and stained with Syber Green (Molecular Probes). Band intensities were determined using Kodak Digital Science: 1D Image Analysis Software. A standard curve for each gene was generated from the reference standard dilution series and the values for the unknown samples were extrapolated. To correct for sample-to-sample variations in RT-PCR efficiency and errors in sample quantitation, sqRT-PCR analysis of β-actin expression was used to normalize the different RNA samples.

RESULTS AND DISCUSSION

Expression of Full-Length and Truncated Forms of Notch4 in HMVECd Cells

We chose to introduce the full-length and truncated Notch4 constructs into human primary endothelial cells derived from the dermal microvasculature of neonates (HMVECd) by adenoviral infection. An advantage of adenoviral vectors is in the ability to infect primary endothelial cell cultures with 100% efficiency. Parental HMVECd cells or HMVECd cells infected with LacZ or full-length Notch4 adenoviruses were biotinylated to analyze the cell-surface expression of Notch4. Biotinylated cell-surface proteins were purified from protein lysates with streptavidin beads, and Westerns were performed. Both full-length and a heterodimeric, furin-processed form of Notch1 have been shown to be present at the cell surface.[21] In the HMVECd cells that ectopically expressed full-length Notch4, both the unprocessed (250-kDa) and processed (70-kDa) forms of Notch4 were detected with antibodies against the cytoplasmic domain of Notch4 (FIG. 1A, lane 2) as compared to the LacZ controls (FIG. 1A, lane 1). Cell surface expression of endogenous Notch4 could also be seen as an approximately 70-kDa band after longer exposure of the Western to film (FIG. 1A, lane 3). A slower migrating band running above the 85-kDa marker was also observed and may represent a phosphorylated or otherwise modified form

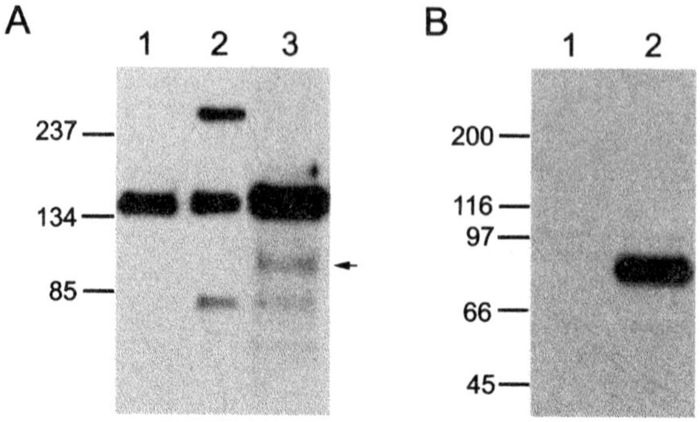

FIGURE 1. Western analysis of endogenous and ectopic Notch4. **(A)** HMVECd cells infected with adenoviruses encoding LacZ (lane 1), full-length Notch4 (lane 2), and parental HMVECd cells (lane 3) were cell-surface biotinylated. Biotinylated Notch4 proteins captured with streptavidin beads were detected by Western analysis using the cytoplasmic polyclonal Notch4 antibody (CoRb2-2). In lane 2, an unprocessed 250-kDa and a processed 70-kDa protein were detected on the surface of HMVECd ectopically expressing Notch4. A longer exposure was necessary to visualize the endogenous Notch4 (lane 3). Both the processed 70-kDa and a slower migrating 85-kDa band (*arrow*) were detected in the parental cells. A strong background band migrates above the 134-kDa marker that is present in both Notch4-expressing and nonexpressing parental cell lines (data not shown). **(B)** Total cell lysates from HMVECd cells infected with adenoviruses encoding either LacZ (lane 1) or Notch4/int3HA (lane 2) were analyzed by Western analysis using the anti-HA monoclonal 12CA5 antibody. Notch4/int3HA migrates as a 70-kDa protein.

of the processed Notch4 cytoplasmic domain.[22] To constitutively activate Notch4 signaling, HMVECd cells were infected with an adenovirus that encodes Notch4/int3HA, a naturally occurring truncated and active allele of Notch4 fused in frame with one copy of the HA epitope tag.[18,23] The expression of the Notch4/int3HA construct was confirmed by sqRT-PCR analysis (data not shown) and Western analysis using the monoclonal 12CA5 antibody that recognizes the C-terminal HA epitope (FIG. 1B, lane 2).

Delta4 Activates Notch4 in HMVECd Cells

The coexpression of Delta4 and Notch4 in the developing vascular endothelium suggests that Delta4 may be a ligand for Notch4. To address this hypothesis, we used an *in vitro* coculture assay to determine if Delta4 could activate Notch4 in HMVECd cells. In this assay, dsRed or Delta4-expressing 293 cells (Bosc23) were cocultured with HMVECd cells infected with adenoviruses encoding either full-length Notch4 or GFP, and then transiently transfected with a CBF1-responsive luciferase reporter and a LacZ expression construct. CBF1 is the mammalian member of the CSL family of transcription factors. Twenty-four hours after coculturing, lysates were collected and assayed for β-gal and luciferase activity. Notch4-expressing HMVECd cells cocultured with Delta4-expressing Bosc23 cells transactivated CBF1 3.8-fold relative to the cocultures with dsRed-expressing Bosc23 cells (FIG. 2). Interestingly, Notch4-expressing HMVECd cells alone or cocultured with the control dsRed-expressing Bosc23 cells also activated CBF1 as compared to the GFP-expressing HMVECd

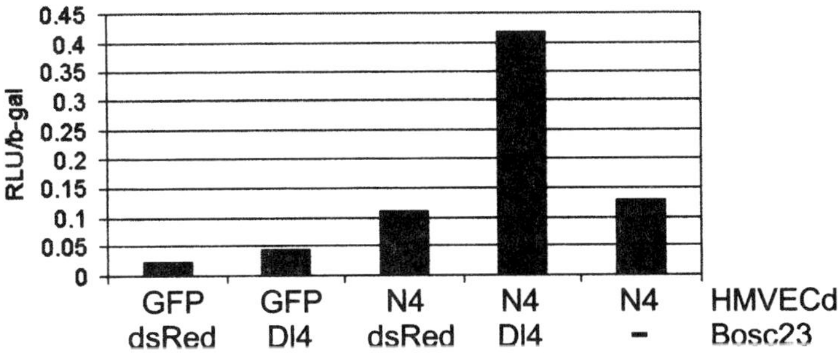

FIGURE 2. Delta4 activates Notch4 in HMVECd cells. To examine whether Delta4 (Dl4) could induce signaling through Notch4 (N4), an *in vitro* coculture assay was used. Dl4 or dsRed-expressing control Bosc23 cells were cocultured with either Notch4 or GFP-infected HMVECd cells that were also transfected with a CBF1 reporter and LacZ plasmid. The data points represent the relative luciferase units (RLU) normalized for transfection efficiency by dividing by the relative β-gal units. Data in this figure represent results from a single coculture experiment. These coculture results have been confirmed in three separate experiments.

cells (FIG. 2). Notch signaling in the absence of ectopic Delta4 expression suggests that Notch4 may have been activated by endogenous ligands. In fact, parental HMVECd cells express both Jagged1 and Delta4 (FIG. 3A). Consistent with our findings in endothelial cells, ectopic expression of both Delta4 and Notch4 in *Xenopus* embryos upregulated the expression of the HES family members ESR-1 and ESR-7 within the neural ectoderm.[9] This proneural phenotype was similar to that seen for activated forms of Notch1. Together with our results, these observations suggest that Delta4 is a ligand for Notch4 in the endothelium.

Identification of Genes Regulated in Response to Notch4 Activation

Studies in vertebrates indicate that Notch signaling is essential for proper vascular development in the embryo. However, the mechanism(s) by which Notch regulates this process has yet to be elucidated. During vascular development, Notch likely modulates cell-fate decisions by regulating the expression of genes involved in Notch signaling and vascular development. Using sqRT-PCR, we examined whether Notch4 signal activation altered the expression of three different categories of genes: Notch ligands (Jagged1 and Delta4), HES/HERP transcriptional repressors (HES1, HES5, HERP1/2), and ephrins (ephrinB2, EphB4).

To activate the Notch pathway in cultured endothelial cells, we have made an adenoviral vector that encodes the int3 allele of Notch4, Notch4/int3. The Notch4/int3 allele is a truncated cytoplasmic form of Notch4 that behaves as a gain-of-function of Notch4 and activates the CBF1-luciferase reporter in cultured endothelial cells (data not shown).[18] We chose to use the constitutively active Notch4/int3 construct for two reasons: (1) the magnitude of Notch4 signaling is greater than that for ligand-activated Notch signaling and (2) coculture assays include ligand-presenting 293 cells that will contribute to the total RNA and obfuscate the data. HMVECd cells were infected with adenoviruses encoding Notch4/int3HA and total RNA was isolated 48

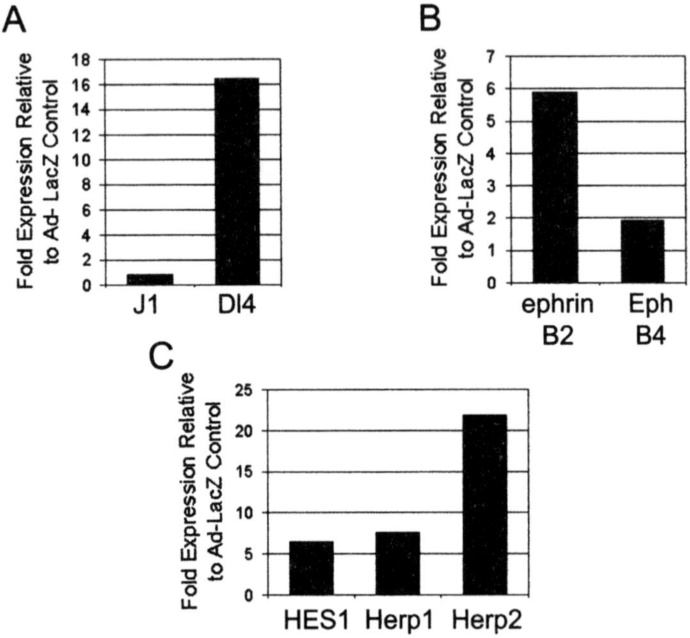

FIGURE 3. Expression analysis of Notch4/int3HA-expressing HMVECd. Semiquantitative (sq) RT-PCR analysis was performed using gene-specific PCR primer pairs for **(A)** the Notch ligands Jagged1 and Delta4, **(B)** the angiogenic factors ephrinB2 and EphB4, and **(C)** the bHLH transcriptional repressors HES1, HERP1, and HERP2. The bar graphs represent the fold-expression of each gene transcript in HMVECd cells expressing Notch4/int3 relative to the HMVECd LacZ control. For each PCR, a standard curve was generated from a serially diluted control plasmid. An R^2 value of greater than 0.93 was achieved for each sqRT-PCR standard curve. Samples were normalized with sqRT-PCR values for β-actin. Data represent an average of two separate sqRT-PCR experiments for each gene examined.

hours postinfection. Adenoviruses encoding LacZ were also used to infect HMVECd cells to control for alterations in gene expression due to the adenoviral infection.

The expression of both Notch4 and its ligand, Delta4, has been shown to be restricted to the developing vasculature in the embryo.[9–12] Consistent with their overlapping expression patterns, Notch4/int3HA expression resulted in a 16-fold increase in endogenous Delta4 expression as compared to the LacZ control (FIG. 3A). In contrast, activation of Notch4 did not significantly alter the expression of Jagged1 (FIG. 3A). Since Delta4 can activate Notch4 in cultured endothelial cells (FIG. 2), this result may suggest a regulatory loop between Notch4 and Delta4 in the vascular endothelium.

Both genetic and biochemical studies have demonstrated that the Notch/CSL complex directly upregulates the expression of the transcriptional repressors of the Hairy Enhancer of Split (HES) and HES-Related Protein (HERP) families in several cell types.[5–8,24] In HMVECd cells, Notch4/int3HA induced the expression of HES1, HERP1, and HERP2 to differing degrees, from 5-fold for HES1 to over 21-fold for HERP2 (FIG. 3C). We also examined if Notch4 signaling altered HES5 expression, but it was not expressed in HMVECd cells (data not shown). During neuronal differentiation, HES1 has been shown to inhibit the expression of proneuronal bHLH transcription factors.[1,7] Overexpression of HERP2 in human adipose tissue–derived capillary endothelial cells blocked proliferation, migration, and network formation in culture.[25] Thus, Notch4 may also regulate vascular development through a similar mechanism. However, proangiogenic bHLH transcription factors have yet to be identified.

Genetic studies in Zebrafish indicate that Notch signaling is required for the specification of the first embryonic artery by upregulating the expression of the cell-surface ligand ephrinB2 and downregulating the expression of its receptor, EphB4.[15] Consistent with these observations, expression of Notch4/int3HA in HMVECd cells resulted in an almost 6-fold induction in ephrinB2 expression relative to the LacZ controls (FIG. 3B). In contrast, EphB4 expression was relatively unaffected (FIG. 3B). EphrinB2 has been proposed to be critical for the remodeling of the primary vascular plexus and arterial endothelial cell specification.[26] In Zebrafish, the expression of ephrinB2 within arterial endothelium was shown to be dependent on the expression of gridlock, a homologue of HERP1.[27,28] In response to Notch4 activation, we found that HERP1 and HERP2 were induced over 7- and 22-fold, respectively (FIG. 3C). These results suggest that, in HMVECd cells, Notch4 signaling may regulate the expression of ephrinB2 and arterial specification through the induction of HERP1 and/or HERP2.

SUMMARY

Using an *in vitro* coculture assay, we demonstrated that Delta4 can function as a ligand for Notch4 in HMVECd cells. Moreover, expression of Notch4/int3 induced the expression of Delta4, suggesting a regulatory loop between Notch4 and its ligand, Delta4. We also found that Notch4 activation led to an increase in the expression of the transcriptional repressors HES1, HERP1, and HERP2, as well as ephrinB2, an angiogenic factor proposed to be involved in arterial/venous endothelial cell specification. With these analyses, we are beginning to identify the Notch ligand/receptor pairs that function in endothelial cells and some of the target genes regulated by Notch in the vasculature.

ACKNOWLEDGMENTS

We would like to thank Peter Canoll for critical reading of the manuscript. This work was supported by grants to J. Kitajewski from the NIH (No. RO1 HL62454) and the American Heart Association (No. AHA-x0150435N). C. J. Shawber was supported by NIH Training Grant No. 2T32CA-09503-16 and I. Das by NIH Training Grant No. 2T32 DK07328.

REFERENCES

1. WEINMASTER, G. 1997. The ins and outs of Notch signaling. Mol. Cell. Neurosci. **9:** 91–102.
2. ARTAVANIS-TSAKONAS, S. *et al.* 1995. Notch signaling. Science **268:** 225–232.

3. GREENWALD, I. 1998. LIN-12/Notch signaling: lessons from worms and flies. Genes Dev. **12:** 1751–1762.
4. ROBEY, E. 1997. Notch in vertebrates. Curr. Opin. Genet. Dev. **7:** 551–557.
5. BAILEY, A.M. & J.W. POSAKONY. 1995. Suppressor of hairless directly activates transcription of enhancer of split complex genes in response to Notch receptor activity. Genes Dev. **9:** 2609–2622.
6. EASTMAN, D.S. *et al.* 1997. Synergy between suppressor of Hairless and Notch in regulation of Enhancer of split m gamma and m delta expression. Mol. Cell. Biol. **17:** 5620–5634.
7. WETTSTEIN, D.A. *et al.* 1997. The Xenopus homolog of Drosophila Suppressor of Hairless mediates Notch signaling during primary neurogenesis. Development **124:** 693–702.
8. NAKAGAWA, O. *et al.* 2000. Members of the HRT family of basic helix-loop-helix proteins act as transcriptional repressors downstream of Notch signaling. Proc. Natl. Acad. Sci. USA **97:** 13655–13660.
9. SHUTTER, J.R. *et al.* 2000. Dll4, a novel Notch ligand expressed in arterial endothelium. Genes Dev. **14:** 1313–1318.
10. UYTTENDAELE, H. *et al.* 1996. Notch4/int-3, a mammary proto-oncogene, is an endothelial cell–specific mammalian Notch gene. Development **122:** 2251–2259.
11. KREBS, L.T. *et al.* 2000. Notch signaling is essential for vascular morphogenesis in mice. Genes Dev. **14:** 1343–1352.
12. VILLA, N. *et al.* 2001. Vascular expression of Notch pathway receptors and ligands is restricted to arterial vessels. Mech. Dev. **108:** 161–164.
13. HRABE DE ANGELIS, M. *et al.* 1997. Maintenance of somite borders in mice requires the *Delta* homologue *Dll1*. Nature **386:** 717–721.
14. UYTTENDAELE, H. *et al.* 2001. Vascular patterning defects associated with expression of activated Notch4 in embryonic endothelium. Proc. Natl. Acad. Sci. USA **98:** 5643–5648.
15. LAWSON, N.D. *et al.* 2001. Notch signaling is required for arterial-venous differentiation during embryonic vascular development. Development **128:** 3675–3683.
16. HARDY, S. *et al.* 1997. Construction of adenovirus vector through Cre-lox recombination. J. Virol. **71:** 1842–1849.
17. HONJO, T. 1996. The shortest path from the surface to the nucleus: RBP-J kappa/Su(H) transcription factor. Genes Cells **1:** 1–9.
18. UYTTENDAELE, H. *et al.* 2000. Notch4 and Jagged-1 induce microvessel differentiation of rat brain endothelial cells. Microvasc. Res. **60:** 91–103.
19. PEAR, W. *et al.* 1993. Production of high-titer helper-free retroviruses by transient transfection. Proc. Natl. Acad. Sci. USA **90:** 8392–8396.
20. CHOMCZYNSKI, P. & N. SACCHI. 1987. Single-step method of RNA isolation by acid guanidinium thiocyanate–phenol–chloroform extraction. Anal. Biochem. **162:** 156–159.
21. BUSH, G. *et al.* 2001. Ligand-induced signaling in the absence of furin processing of Notch1. Dev. Biol. **229:** 494–502.
22. WU, G. *et al.* 2001. SEL-10 is an inhibitor of notch signaling that targets notch for ubiquitin-mediated protein degradation. Mol. Cell. Biol. **21:** 7403–7415.
23. GALLAHAN, D. & R. CALLAHAN. 1997. The mouse mammary tumor associated gene INT3 is a unique member of the NOTCH gene family (NOTCH4). Oncogene **14:** 1883–1890.
24. ISO, T. *et al.* 2002. HERP1 is a cell type–specific primary target of Notch. J. Biol. Chem. **277:** 6598–6607.
25. HENDERSON, A.M. *et al.* 2001. The basic helix-loop-helix transcription factor HESR1 regulates endothelial cell tube formation. J. Biol. Chem. **276:** 6169–6176.
26. WANG, H.U. *et al.* 1998. Molecular distinction and angiogenic interaction between embryonic arteries and veins revealed by ephrin-B2 and its receptor Eph-B4. Cell **93:** 741–753.
27. ZHONG, T.P. *et al.* 2001. Gridlock signalling pathway fashions the first embryonic artery. Nature **414:** 216–220.
28. ZHONG, T.P. *et al.* 2000. Gridlock, an HLH gene required for assembly of the aorta in zebrafish. Science **287:** 1820–1824.

Macrophage Migration Inhibitory Factor (MIF)

Its Potential Role in Tumor Growth and Tumor-Associated Angiogenesis

JUN NISHIHIRA,[a] TERUO ISHIBASHI,[a] TSUYOSHI FUKUSHIMA,[b] BAILONG SUN,[b] YUJI SATO,[b] AND SATORU TODO[b]

[a]*Department of Molecular Biochemistry,* [b]*First Department of Surgery, Hokkaido University Graduate School of Medicine, Sapporo 060-8638, Japan*

ABSTRACT: Macrophage migration inhibitory factor (MIF) functions as a pluripotent cytokine involved in broad-spectrum pathophysiological events in association with inflammation and immune responses. Several reports, including ours, have suggested that MIF is also involved in tumorigenesis; however, its precise role has not been fully investigated. We examined the effectiveness of anti-MIF antibodies on tumor growth and tumor-associated angiogenesis using murine colon cancer cell line, colon 26. We observed a significant inhibition of growth of tumors embedded on the back of BALB/c mice by treatment with anti-MIF antibodies. Next, we implanted a Millipore chamber filled with colon cancer cells in the subcutaneous fascia of the flanks of mice and then treated them with anti-MIF antibodies. We found that angiogenesis was markedly suppressed within the region of the subcutaneous fascia that was in contact with the chamber. To further assess the role of MIF in tumorigenesis, we established MIF transgenic mice, which demonstrated that tumor growth and the associated angiogenesis were significantly enhanced in comparison with control mice.

KEYWORDS: angiogenesis; cytokine; inflammation; tumor; wound repair

INTRODUCTION

Macrophage migration inhibitory factor (MIF) plays a pivotal role in systemic as well as local inflammatory and immune responses. MIF was first identified as a T cell–derived lymphokine that inhibits the random migration of macrophages out of capillary tubes *in vitro*.[1,2] MIF exerts a variety of biological functions in macrophage activation, for example, enhancement of adherence, phagocytosis, and tumoricidal activity.[3–6] In addition, recent reports have demonstrated MIF expression in a variety of cells other than activated T cells, indicating its involvement within and beyond the

Address for correspondence: Jun Nishihira, Department of Molecular Biochemistry, Hokkaido University Graduate School of Medicine, Kita-15, Nishi-7, Kita-ku, Sapporo 060-8638, Japan. Voice: +81-11-706-5046.
j_nisihi@med.hokudai.ac.jp

Ann. N.Y. Acad. Sci. 995: 171–182 (2003). © 2003 New York Academy of Sciences.

immune system.[7,8] It is particularly intriguing that MIF functions as a cytokine, an anterior pituitary–derived hormone, and a glucocorticoid-induced immunomodulator.[9] Accordingly, it has been suggested that MIF is the initial inflammatory mediator to stimulate the expression of cytokines such as tumor necrosis factor (TNF)–α and interleukin (IL)–1β via suppression of the anti-inflammatory actions of glucocorticoids.

In addition to inflammatory and immunologic functions, MIF is considered to play an essential role in cell proliferation and differentiation according to an array of recent reports. With regard to tumorigenesis, the enhancement of MIF mRNA expression has been identified in prostatic lymph node metastases[10] and ductal breast carcinoma.[11] More recently, it has been shown that MIF plays a key role in tumor growth and neovascularization based on the results of *in vivo* and *in vitro* studies using lymphoma cells,[12] melanoma cells,[13] and colon cancer cells.[14,15] Consistent with these findings, it was revealed that MIF has the potential to suppress the action of the tumor suppressor gene p53, leading to promotion of cell growth as well as inhibition of apoptosis.[16] It also became evident that MIF stimulates the cell proliferation of fibroblasts.[17] Alternatively, we have shown that MIF mRNA is upregulated in parallel with the wound-healing process, in which an anti-MIF antibody delays wound repair.[18] In this context, it is conceivable that MIF might play an important role not only in tumor growth, but also in wound repair.

In this paper, we focus on the biological properties of MIF regarding tumorigenesis and wound repair by presenting recent data on the biological functions of MIF.

EVIDENCE FOR THE POTENTIAL INVOLVEMENT OF MIF IN CELL GROWTH

Wistow *et al.* reported that MIF mRNA expression is correlated with differentiation of lens cells, suggesting the involvement of MIF in cell proliferation and differentiation.[19] Following this report, we demonstrated evidence showing the potential regulatory role of MIF in pregnancy, development, and cell growth.[20–22] For example, high expression of MIF was noted in oocyte development, the early gestational period, and embryonal stage of the nervous system. Recently, Kobayashi *et al.* have revealed that the MIF expression pattern parallels tissue specification and organogenesis during embryogenesis.[23] These findings have indicated specific functional roles of MIF in cell growth and differentiation, particularly during development.

MIF has been reported to be one of the genes offering early responses to growth factors.[24] However, the biological significance of the increased MIF expression in the early phases has not been well investigated. We examined this biological phenomenon and demonstrated that MIF mRNA expression was significantly upregulated in colon cancer cells in response to various growth factors, including transforming growth factor (TGF)–β, basic fibroblast growth factor (bFGF), and platelet-derived growth factor (PDGF) (FIG. 1).[14] Following this, we further assessed the expression of MIF mRNA in mouse osteoblastic cell line MC3T3-E1 and showed that TGF-α, bFGF, insulin-like growth factor-II, and fetal calf serum (FCS) also markedly upregulate MIF mRNA expression.[25] These facts suggest the possibility that MIF is, regardless of the cell type, upregulated by growth factors and exerts biological functions in concert with growth factors.

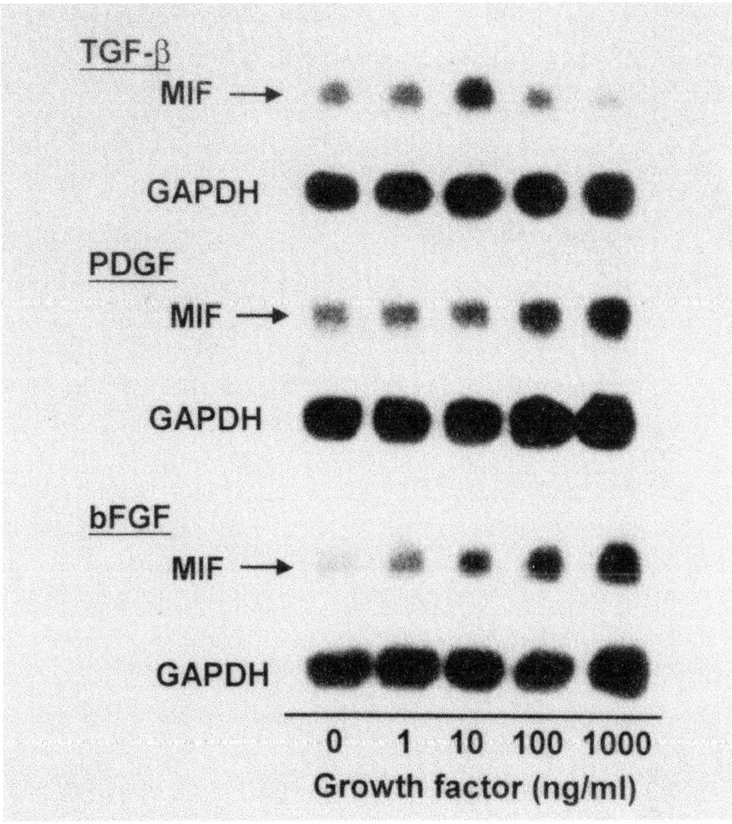

FIGURE 1. Northern blot analysis of MIF mRNA expression in colon 26 cells in response to growth factors. Total RNA was extracted from cultured cells after 24 h in RPMI-1640 containing 1% fetal calf serum in the presence of various doses of each growth factor. The profile of GAPDH is shown at the bottom of each lane.

THE ROLE OF MIF IN TUMOR GROWTH AND ANGIOGENESIS

We found a high level of MIF mRNA expression in cancer cells, including myelomonocytic leukemia cells.[26] For example, a squamous cell carcinoma showed a typical picture of increased production of MIF protein in the cancerous lesions shown by immunohistochemistry (FIG. 2). To further evaluate the clinical aspects of MIF in tumorigenesis, we examined the MIF expression in human lung cancer cells in relation to patient prognosis. We performed immunohistochemical staining of lung tissues obtained surgically from 38 patients with primary lung adenocarcinoma.[27] The results showed that, in most cases (73.8%), MIF is diffusely expressed in the cytoplasm. On the other hand, positive MIF staining was also observed in the nucleus in several cases (15.8%). It is of interest that the patients with negative MIF staining in the nuclei had a poorer prognosis than those with positive staining. Although the

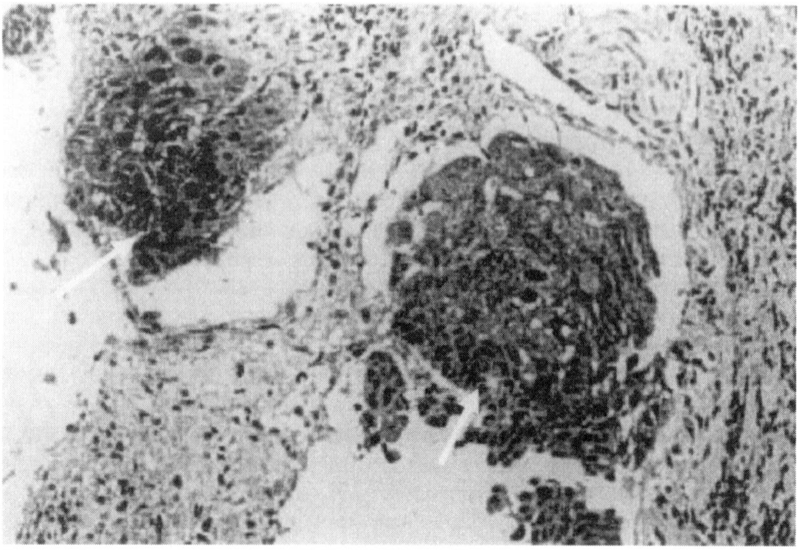

FIGURE 2. Immunohistochemical staining of MIF in squamous cell carcinoma. A surgical specimen obtained from a patient with squamous cell carcinoma in the neck. Formalin-fixed, paraffin-embedded tissue was stained with the DAKO CSA system (DAKO, Copenhagen, Denmark). MIF protein is abundantly observed in carcinomatous lesions (indicated by white arrows) in comparison with surrounding noncarcinomatous tissues.

role of MIF in the nucleus is not clear, these findings support the concept that MIF functions within the cytoplasm and nucleus in a different manner depending on the magnitude of malignancy, which could predict the prognosis of patients. These findings prompted us to examine whether MIF is involved in tumor cell proliferation.

Using an antisense MIF plasmid (pBK-MIFAS), we transfected it by electroporation into murine adenocarcinoma cell line colon 26.[14] The plasmids pBK-MIFAS and pBK vector were linearized with SspI before use. The proliferation rate of the transfected colon 26 cells with pBK-MIFAS was markedly suppressed ($61.1 \pm 10.5\%$) compared to the control cells (FIG. 3). These results strongly suggest that MIF has the potential to induce tumor growth.

Recently, the effectiveness of anti-MIF antibodies in reducing the growth of lymphoma and neovascularization was demonstrated.[12] Consistent with this finding, we have verified the effectiveness of anti-MIF antibodies in slowing tumor growth and angiogenesis of murine melanoma[13] and colon cancer cells.[15] As for murine colon adenocarcinoma cell line colon 26 cells, we inoculated the cancer cells subcutaneously in the flanks of BALB/c mice. After 9 days, we treated the tumor-bearing mice with anti-MIF antibodies ip on days 9, 11, 13, 15, 17, 19, and 21. We found significant tumor regression caused by this treatment from day 15 to day 22 (FIG. 4). Alternatively, we implanted a chamber filled with the cancer cells in the subcutaneous fascia of the flanks of mice, which were then treated with anti-MIF antibodies. The results showed that angiogenesis was markedly suppressed within the region of subcutaneous fascia that was in contact with the chamber (FIG. 5).

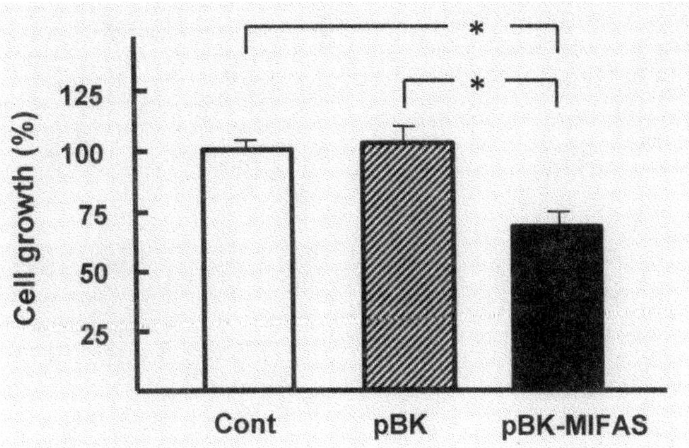

FIGURE 3. Effect of antisense MIF plasmid transfection on cell growth of colon cancer cells. Growth of colon 26 cells (5×10^3 cells) transfected with antisense MIF cDNA (pBK-MIFAS) was examined using a CellTiter MTS assay kit assay after 96-h incubation. Lane 1, control colon 26 cells; lane 2, cells transfected with pBK vector alone; lane 3, colon 26 cells transfected with pBK-MIFAS. The result is the mean ± SD of three different experiments ($n = 9$). *Statistically significant ($p < 0.01$).

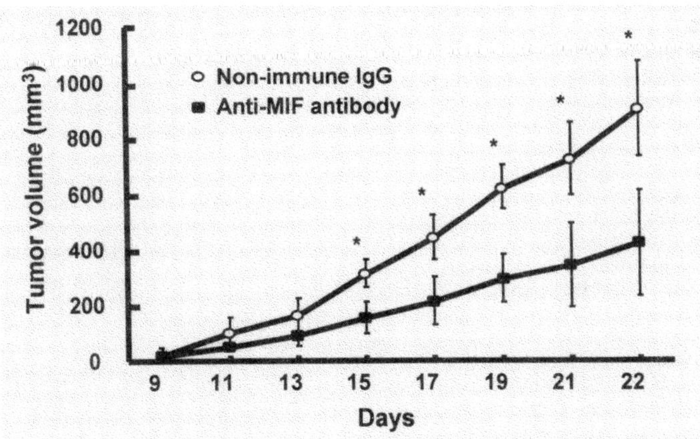

FIGURE 4. Effects of anti-MIF antibodies on tumor growth. Colon 26 cells were inoculated subcutaneously into the right flanks of BALB/c mice. When tumors became measurable on day 9, treatment with intraperitoneal injection of the antirat MIF antibody (IgG) (200 µg/mouse/day) was initiated, and the anti-MIF antibody was injected at days 9, 11, 13, 15, 17, 19, and 21. Tumor volume (V) was determined by the following formula: $V = 0.5 \times a \times b^2$ (mm^3), where a is the longest diameter and b is the shortest diameter. Values are expressed as mean ± SD ($n = 6$). *$p < 0.01$ vs. control.

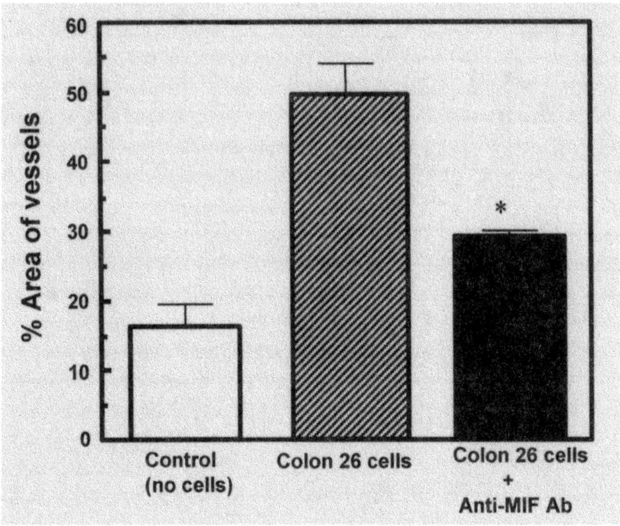

FIGURE 5. Effect of anti-MIF antibody on tumor-associated angiogenesis. Effectiveness of the antirat MIF antibody on tumor-induced angiogenesis was examined by the subcutaneous dorsal air sac technique using Millipore chambers. The chambers were filled with colon 26 cells (1×10^7 cells in 200 μL of medium). Anti-MIF antibodies were administered ip (200 μg IgG/mouse/day) at days 1, 3, and 5. Values are expressed as mean ± SD ($n = 5$). $^*p < 0.01$ vs. control.

These facts convincingly indicated that MIF had the potential to promote tumor growth and tumor-associated angiogenesis.

TUMOR GROWTH AND ANGIOGENESIS IN MIF TRANSGENIC MICE

To further assess the function of MIF in tumorigenesis, we generated transgenic (Tg) mice carrying MIF cDNA driven by a cytomegalovirus enhancer and a β-actin/β-globin promoter. Cells of murine sarcoma cell line S-180 were subcutaneously inoculated into the right back of Tg and non-Tg mice. When tumors became measurable on day 4, we started to measure tumor size every 2 days. We observed a significant increase of the tumor growth rate in MIF Tg mice in comparison with non-Tg mice up to day 9 after tumor inoculation (FIG. 6). This result was compatible with our previous result that tumor growth was suppressed by anti-MIF antibody treatment.[15]

Furthermore, an angiogenesis assay using the dorsal air sac method was performed to evaluate the influence of MIF on the angiogenesis induced by S-180 cells. A Millipore chamber, which allows only soluble factors to pass through, was used. The chamber was filled with S-180 cells and placed into the right back in MIF Tg and non-Tg mice. Mice were sacrificed on day 5 and the area of neovascularization was quantitatively assessed by the ratio of the area of blood vessels to the area in contact with the chamber. Angiogenesis within the subcutaneous fascia of MIF Tg

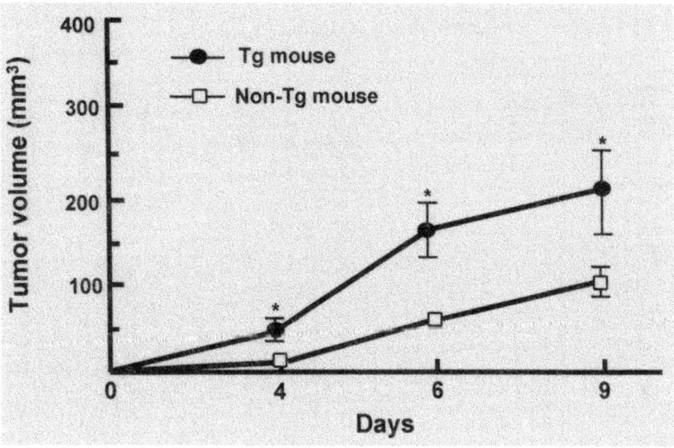

FIGURE 6. Tumor growth in MIF transgenic mice. S-180 cells were inoculated subcutaneously into the right backs of MIF Tg mice and non-Tg mice. When tumors became measurable on day 4, tumor size was measured every second day. Values are expressed as mean ± SD ($n = 10$). *$p < 0.05$ vs. non-Tg mice.

mice was compared with that of non-Tg mice. In the Tg mice embedded with chambers with S-180 cells, tumor-induced angiogenesis was significantly enhanced in comparison with that in non-Tg mice. The ratios of the areas of blood vessels to the areas in contact with a chamber were 12.4 ± 0.9%, 17.7 ± 2.3%, and 30.7 ± 8.5% in the control, non-Tg, and Tg mice, respectively. These results strongly suggested that MIF potentiated tumor growth and tumor-associated angiogenesis.

EXPRESSION OF MIF IN WOUND REPAIR

We previously reported upregulation of MIF mRNA in corneal epithelium concomitant with keratocyte infiltration during corneal wound healing after a penetration injury.[28] The MIF-positive staining of the corneal epithelium diminishes immediately after injury, and the positive MIF staining reappears on the basal cells of the injured area concomitant with wound healing. Similarly, we found upregulation of MIF mRNA in the event of peripheral nerve injury, and the expression was well correlated with the magnitude of inflammation within the wound healing process.[29] Based on time-course studies after the injury, it is conceivable that MIF contributes to the repair of damaged tissues.

During the healing of injured rat skin, we found similar expression of MIF mRNA.[18] In this event, the increase in MIF mRNA expression was biphasic with maximum peaks at 3 and 24 h after injury. Administration of anti-MIF antibodies induced a delay in wound healing *in vivo* (FIG. 7). In *in vitro* study, cultured fibroblasts from skin wound lesions secreted a higher level of MIF protein into culture medium in response to endotoxin than fibroblasts from normal skin. Interestingly, MIF showed a chemotactic effect on freshly prepared keratinocytes from injured rat skin in a Boyden chamber assay. These results indicate that the cells became more

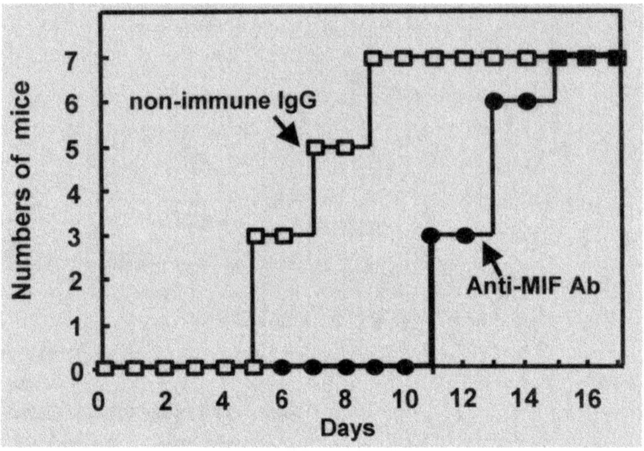

FIGURE 7. Effects of anti-MIF antibodies on healing of skin wound lesions. Wound healing appearance was assessed after treatment with anti-MIF antibodies in mice. The number of mice with complete healing of four skin lesions was plotted at different time points after penetrating wounds through both the epidermis and dermis.

sensitive with regard to the release of MIF into extracellular space, suggesting the potential role of MIF in wound healing.

MOLECULAR BASIS ASSESSMENT OF MIF FUNCTION

We found that the tertiary structure of MIF revealed by X-ray crystallography is similar to those of 5-carboxymethyl-2-hydroxymuconate isomerase and 4-oxalocrotonate tautomerase.[30] Following this discovery, MIF was found to have phenylpyruvate tautomerase and thiol protein oxidoreductase activities.[31,32] However, the physiological significance of these enzymatic activities remains to be investigated.

It is of great interest that MIF has the potential to inhibit the biological action of the tumor suppressor oncogene p53, which plays a key role in the negative regulation of cell proliferation through not only maintaining genomic stability, but also suppressing transformation and tumorigenesis.[16] Two functional screenings were performed to isolate genes that bypassed either p53-mediated growth arrest or apoptosis. Surprisingly, MIF cDNA was unexpectedly isolated in both cases as a factor suppressing p53 bioactivity in conjunction with the recombinant MIF overcoming p53 activity. Based on these facts, it is considered that MIF is capable of inactivating tumor suppressor p53, and this protein is a key molecule to provide a link between inflammation and tumorigenesis.

It should be noted that both endogenously secreted and exogenously added MIF stimulate the proliferation of NIH/3T3 fibroblasts.[12] It was demonstrated that the cellular response to MIF is associated with the activation of p44/p42 mitogen-activated protein kinases (MAPK). In addition, it was revealed that MIF regulates cytosolic phospholipase A2 activity via protein kinase A. Furthermore, the gluco-

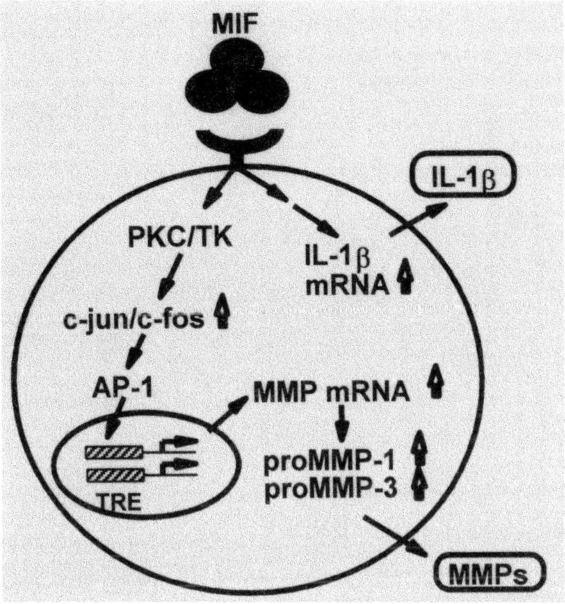

FIGURE 8. Signal transduction of matrix metalloproteinases (MMPs) in response to MIF. MIF stimulates the production of MMP-1 and MMP-3, in which MIF upregulates the mRNA via tyrosine kinase (TK)–, protein kinase C (PKC)–, and AP-1–dependent pathways. TRE, TPA-responsive element.

corticoid suppression of cytokine-induced cytoplasmic phospholipase A2 activity and arachidonic acid release were reversed by the addition of recombinant MIF. These data indicate that MIF possesses growth factor–like biological activity and, moreover, may cause inflammation, at least in part, via the arachidonic acid cascade.

We have shown the upregulation of neutral matrix metalloproteinase (MMP)–1 and MMP-3 mRNAs of cultured synovial fibroblasts retrieved from rheumatoid arthritis (RA) patients in response to MIF.[33] MMPs not only are responsible for pathological features of RA, such as the degradation of cartilage, but also play a critical role in tumor invasion and angiogenesis. Concerning the MMP upregulation, we investigated the signal transduction of MIF. The elevation of MMP-1 and MMP-3 mRNA is dose-dependent and begins 6 h poststimulation by MIF, reaching the maximum level at 24 h poststimulation. These events are preceded by the upregulation of c-jun and c-fos mRNAs. The mRNA upregulation of MMPs is inhibited by curcumin, an inhibitor of transcription factor AP-1. The mRNA upregulation by MIF is also inhibited by the tyrosine kinase inhibitor genistein as well as by the protein kinase C inhibitor H-7. On the other hand, this inhibition is not seen after the addition of the cyclic AMP–dependent kinase inhibitor H-8. Thus, it is likely that MIF may upregulate MMP-1 and MMP-3 via tyrosine kinase–, protein kinase C–, and AP-1–dependent pathways (FIG. 8).

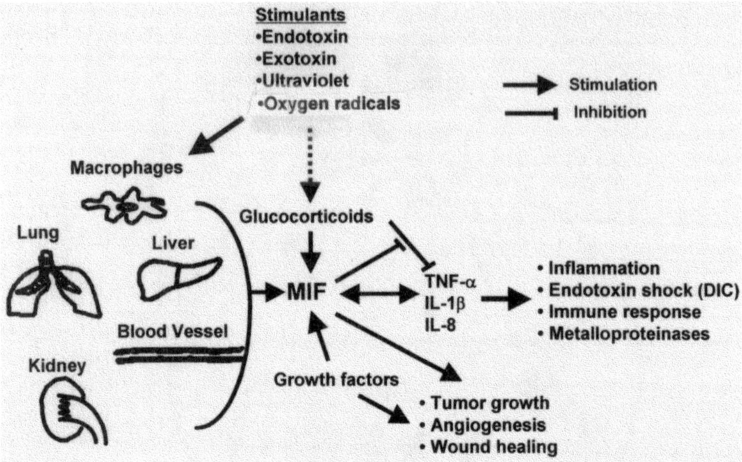

FIGURE 9. Schematic view of biological functions of MIF. MIF is readily released into extracellular space and circulation in response to various stimuli, such as endotoxins, and functions as a pluripotent cytokine involved in broad-spectrum pathophysiological events—not only inflammation and immune responses, but also tumorigenesis and wound healing.

PERSPECTIVES

Since the isolation of the cDNA of human MIF from T lymphocytes,[34] a number of novel biological properties of this protein have been discovered. The unique feature of MIF is that this protein is abundantly stored within various cells and appears to exert its biological actions as a cytokine, hormone, or enzyme. Based on extensive investigation, it is becoming clear that MIF contributes in innate and acquired fashions to promote cell growth and tumor-associated angiogenesis as well as the wound healing process (FIG. 9). However, a number of questions remain unanswered regarding the pathophysiological roles of MIF. In this context, MIF study should be focused on molecular-based investigations, particularly for isolation of its specific receptor. Last, it is expected that regulation of MIF bioactivity could potentially have clinical therapeutic use in various diseases, including autoimmune diseases and malignant tumors, in the near future.

ACKNOWLEDGMENTS

This research was supported by a Grant-in-Aid for Scientific Research (No. 13214004) from the Ministry of Education, Science, Sports, and Culture, Japan.

REFERENCES

1. DAVID, J.R. 1966. Delayed hypersensitivity *in vitro*: its mediation by cell-free substances formed by lymphoid cell–antigen interaction. Proc. Natl. Acad. Sci. USA **56:** 72–77.

2. BLOOM, B.R. & B. BENNETT. 1966. Mechanism of a reaction *in vitro* associated with delayed-type hypersensitivity. Science **153:** 80–82.
3. NATHAN, C.F., M.L. KARNOVSY & J.R. DAVID. 1971. Alterations of macrophage functions by mediators from lymphocytes J. Exp. Med. **133:** 1356–1376.
4. NATHAN, C.F., H.G. REMOLD & J.R. DAVID. 1973. Characterization of a lymphocyte factor which alters macrophage function. J. Exp. Med. **137:** 275–288.
5. CHURCHHILL, W.H., W.F. PIESSENS, C.A. SULIS *et al.* 1975. Macrophages activated as suspension cultures with lymphocyte mediators devoid of antigen become cytotoxic for tumor cells. J. Immunol. **115:** 781–786.
6. HERRIOTT, M.J., H. JIANG, C.A. STEWART *et al.* 1993. Mechanistic differences between migration inhibitory factor (MIF) and IFN-gamma for macrophage activation: MIF and IFN-gamma synergize with lipid A to mediate migration inhibition, but only IFN-gamma induces production of TNF-alpha and nitric oxide. J. Immunol. **150:** 4524–4531.
7. BUCALA, R. 1996. MIF rediscovered: pituitary hormone and glucocorticoid-induced regulator of cytokine production. FASEB J. **7:** 19–24.
8. NISHIHIRA, J. 2000. Macrophage migration inhibitory factor (MIF): its essential role in the immune system and cell growth. J. Interferon Cytokine Res. **20:** 751–762.
9. CALANDRA, T., J. BERNHAGEN, C.N. METZ *et al.* 1995. MIF as a glucocorticoid-induced modulator of cytokine production. Nature **377:** 68–71.
10. MEYER, S.K. & P.B. HUDSON. 1996. Enhanced expression of macrophage migration inhibitory factor in prostatic adenocarcinoma metastases. Urology **48:** 448–452.
11. BINI, L., B. MAGI, B. MARZOCCHI *et al.* 1997. Protein expression profiles in human breast ductal carcinoma and histologically normal tissue. Electrophoresis **18:** 2832–2841.
12. CHESNEY, J., C.N. METZ, M. BACHER *et al.* 1999. An essential role for macrophage migration inhibitory factor (MIF) in angiogenesis and the growth of a murine lymphoma. Mol. Med. **5:** 181–191.
13. SHIMIZU, T., R. ABE, H. NAKAMURA *et al.* 1999. High expression of macrophage migration inhibitory factor in human melanoma cells and its role in tumor cell growth and angiogenesis. Biochem. Biophys. Res. Commun. **264:** 751–758.
14. TAKAHASHI, N., J. NISHIHIRA, Y. SATO *et al.* 1998. Involvement of macrophage migration inhibitory factor (MIF) in the mechanism of tumor cell growth. Mol. Med. **4:** 707–714.
15. OGAWA, H., J. NISHIHIRA, Y. SATO *et al.* 2000. An antibody for macrophage migration inhibitory factor suppresses tumor growth and inhibits tumor-associated angiogenesis. Cytokine **12:** 309–314.
16. HUDSON, J.D., M.A. SHOAIBI, R. MAESTRO *et al.* 1999. A proinflammatory cytokine inhibits p53 tumor suppressor activity. J. Exp. Med. **190:** 1375–1382.
17. MITCHELL, R.A., C.N. METZ, T. PENG *et al.* 1999. Sustained mitogen-activated protein kinase (MAPK) and cytoplasmic phospholipase A2 activation by macrophage migration inhibitory factor (MIF): regulatory role in cell proliferation and glucocorticoid action. J. Biol. Chem. **274:** 18100–18106.
18. ABE, R., T. SHIMIZU, A. OHKAWARA *et al.* 2000. Enhancement of macrophage migration inhibitory factor (MIF) expression in injured epidermis and cultured fibroblasts. Biochim. Biophys. Acta **1500:** 1–9.
19. WISTOW, G.J., M.P. SHAUGHNESSY, D.C. LEE *et al.* 1993. A macrophage migration inhibitory factor is expressed in the differentiating cells of the eye lens. Proc. Natl. Acad. Sci. USA **90:** 1272–1275.
20. SUZUKI, H., H. KANAGAWA & J. NISHIHIRA. 1996. Evidence for the presence of macrophage migration inhibitory factor in murine reproductive organs and early embryos. Immunol. Lett. **51:** 141–147.
21. WADA, S., S. FUJIMOTO, Y. MIZUE *et al.* 1997. Macrophage migration inhibitory factor in the human ovary: presence in the follicular fluids and production by granulosa cells. Biochem. Mol. Biol. Int. **41:** 805–814.
22. SUZUKI, T., A. OGATA, K. TASHIRO *et al.* 1999. Augmented expression of macrophage migration inhibitory factor (MIF) in the telencephalon of the developing rat brain. Brain Res. **816:** 457–462.
23. KOBAYASHI, S., K. SATOMURA, J.M. LEVSKY *et al.* 1999. Expression pattern of macrophage migration inhibitory factor during embryogenesis. Mech. Dev. **84:** 153–156.

24. LANAHAN, A., J.B. WILLIAMS, L.K. SANDERS *et al.* 1992. Growth factor–induced delayed early response genes. Mol. Cell. Biol. **12:** 3919–3929.
25. ONODERA, S., K. SUZUKI, K. KANEDA *et al.* 1999. Growth factor–induced expression of macrophage migration inhibitory factor in osteoblasts: relevance to the plasminogen activator system. Semin. Thromb. Hemostasis **25:** 563–568.
26. NISHIHIRA, J., Y. KOYAMA & Y. MIZUE. 1996. Identification of macrophage migration inhibitory factor in human leukemia HL-60 cells and its induction by lipopolysaccharide. Biochem. Mol. Biol. Int. **40:** 861–869.
27. KAMIMURA, A., M. KAMACHI, J. NISHIHIRA *et al.* 2000. Intracellular distribution of macrophage migration inhibitory factor predicts the prognosis of patients with adenocarcinoma of the lung. Cancer **89:** 334–341.
28. MATSUDA, A., Y. TAGAWA, H. MATSUDA *et al.* 1997. Expression of macrophage migration inhibitory factor in corneal wound healing in rats. Invest. Ophthalmol. Visual Sci. **38:** 1555–1562.
29. NISHINO, Y., A. MINAMI, H. KATO *et al.* 1999. Identification of macrophage migration inhibitory factor (MIF) in rat peripheral nerves: its possible involvement in nerve regeneration. Biochim. Biophys. Acta **1453:** 74–82.
30. SUZUKI, M., H. SUGIMOTO, A. NAKAGAWA *et al.* 1996. Crystal structure of the macrophage migration inhibitory factor from rat liver. Nat. Struct. Biol. **3:** 259–266.
31. ROSENGREN, E., P. AMAN, S. THELIN *et al.* 1997. The macrophage migration inhibitory factor MIF is a phenylpyruvate tautomerase. FEBS Lett. **417:** 85–88.
32. KLEEMANN, R., A. KAPURNIOTU, R.W. FRANK *et al.* 1998. Disulfide analysis reveals a role for macrophage migration inhibitory factor (MIF) as thiolprotein oxidoreductase. J. Mol. Biol. **280:** 85–102.
33. ONODERA, S., K. KANEDA, Y. MIZUE *et al.* 2000. Macrophage migration inhibitory factor up-regulates expression of matrix metalloproteinases in synovial fibroblasts of rheumatoid arthritis. J. Biol. Chem. **275:** 444–450.
34. WEISER, W.Y., P.A. TEMPLE, J.S. WITEK-GIANNOTI *et al.* 1989. Molecular cloning of a cDNA encoding a human macrophage migration inhibitory factor. Proc. Natl. Acad. Sci. USA **86:** 7522–7526.

ADAMTS1: A Matrix Metalloprotease with Angioinhibitory Properties

M. LUISA IRUELA-ARISPE, DARREN CARPIZO, AND ALFONSO LUQUE

Department of Molecular, Cell, and Developmental Biology, Molecular Biology Institute and Jonsson Comprehensive Cancer Center, UCLA, Los Angeles, California 90095, USA

ABSTRACT: Neovascularization is a hallmark of cancer progression. Suppression of the angiogenic response in tumors has been associated with inhibition and even regression of total tumor mass. Therefore, the derivation of synthetic or natural products that could interfere with proangiogenic signaling pathways can greatly impact cancer therapy. Using the antiangiogenic motifs in thrombospondin-1, we have recently cloned METH1/ADAMTS1, a secreted metalloproteinase with three thrombospondin-1, and shown that the protein inhibits endothelial cell proliferation *in vitro* and blocks the neovascular response induced by growth factors *in vivo*. The mechanism of action responsible for these events has not been elucidated. In this report, we present evidence to support two effects of METH1/ADAMTS1 that impact proangiogenic signaling events. ADAMTS1 binds to VEGF and dampens VEGFR2 phosphorylation. The ability of ADAMTS1 to bind VEGF and functionally inactivate VEGFR2 is reversible as dissociation of the complex results in active growth factor. A second activity of ADAMTS1 requires the catalytic domain as a single point mutation in the metalloproteinase domain renders the protein inactive in tumor xenograft assays. The emerging theme is that both domains are likely required for the antiangiogenic/antitumor activities of ADAMTS1.

KEYWORDS: angiogenesis; extracellular matrix; tumor suppression

INTRODUCTION

The field of angiogenesis has been propelled by the concept that suppression of angiogenesis in tumors results in regression of tumor burden.[1] The overall consensus in the field is that understanding the mechanisms that control the neovascular response may make it possible to design therapeutic strategies to selectively prevent or halt pathological vascular growth and restrain cancer progression.[2,3] An added advantage for targeting the vasculature is that, unlike tumor cells, the cellular components of blood vessels, namely endothelial and smooth muscle, are not transformed and consequently are not prone to genetic instability and development of drug resistance.[4] The search for angiogenesis inhibitors has followed two general paths: (a) identification of small molecules that block recognized angiogenic signaling

Address for correspondence: M. Luisa Iruela-Arispe, Ph.D., Molecular Biology Institute, UCLA, 611 Charles Young Drive East, Los Angeles, CA 90024. Voice: 310-794-5763; fax: 310-794-5766.

arispe@mbi.ucla.edu

Ann. N.Y. Acad. Sci. 995: 183–190 (2003).

pathways and (b) search for endogenous inhibitors of physiological angiogenesis that by targeted delivery could display a similar inhibitory function in tumors. Our work has focused on the latter strategy.[5]

Extracellular matrix proteins have been shown to regulate migration, proliferation, and morphogenesis, mostly (but not exclusively) through modulation of integrin signaling.[6] Interestingly, some of the angiogenesis inhibitors currently in clinical trials include either antibodies or small peptides shown to block integrin engagement or fragments of extracellular matrix proteins.[7–9] Thrombospondin (TSP) is one example of an extracellular matrix molecule that has been shown to suppress capillary formation both *in vivo* and *in vitro*.[10–12] The antiangiogenic effect of TSP1 has been mapped to the type I or TSP motifs.[13,14] Interestingly, other molecules in addition to TSP1 harbor these repeats, some with reported antiangiogenic activity, such as TSP2 and ADAMTS1; however, others such as properdin do not appear to exhibit antiangiogenic properties.[15–17] Much of our recent work has centered on ADAMTS1 as a potential candidate for suppression of the neovascular response.[17] Because ADAMTS1 contains three repeats with high homology to the TSP1 antiangiogenic motifs, we were prompted to ask whether these domains, in the context of ADAMTS1, are responsible for its angiostatic properties.

METH1/ADAMTS1 belongs to a growing family of matrix metalloproteases (MMPs) characterized by the presence of metalloproteinase, disintegrin-like, and TSP repeats.[18] Currently, there are 20 members of this family.[19,20] Unlike the closely related, frequently membrane-bound ADAM proteins, ADAMTSs are secreted proteases with a rather restricted spectrum of catalytic specificity. The variable number of TSP repeats at the carboxy-terminal end appears to confer binding properties and stability to the catalytic domain.[21,22] In fact, removal of the carboxy-terminal end of ADAMTS1, ADAMTS4, and ADAMTS5 results in a catalytically inactive protease.[5,23,24] Thus far, only ADAMTS1 and ADAMTS8 display antiangiogenic activities, albeit the other molecules are yet to be evaluated.[17]

RESULTS AND DISCUSSION

We have previously demonstrated that ADAMTS1 is able to suppress endothelial cell proliferation mediated by a cocktail of growth factors, including vascular endothelial growth factor 165 ($VEGF_{165}$).[17] VEGF confers signals to endothelial cells in a paracrine manner via two receptor tyrosine kinases: VEGFR1 and VEGFR2. Proliferative signals, in particular, appear to be mediated by VEGFR2. Thus, we sought to determine the effect of ADAMTS1 on the activation of VEGFR2. FIGURE 1 shows that phosphorylation of VEGFR2 is significantly decreased in the presence of conditioned media from cells overexpressing ADAMTS1, but is not affected by conditioned media from cells transfected with vector alone. This effect required the presence of the active protein as heat denaturation resulted in no effect.[25] The experiments were also performed with purified protein with identical findings.[25] These results are likely to explain the effect of ADAMTS1 on endothelial cell proliferation, at least with respect to VEGF-mediated responses.

Next, we sought to ascertain the reason for such an effect; possibilities included (1) ADAMTS1 affected VEGFR2 directly, by either cleavage or direct blockade of the ligand binding site, and (2) ADAMTS1 inactivated VEGF through degradation,

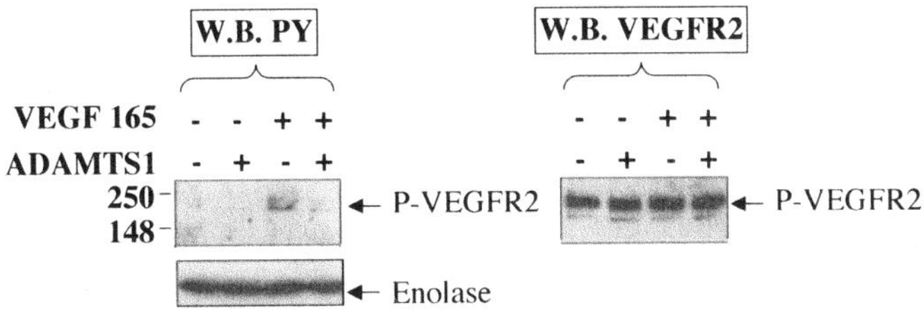

FIGURE 1. ADAMTS1 dampens VEGFR2 phosphorylation. **(Left)** ADAMTS1 partially blocks VEGFR2 phosphorylation in $VEGF_{165}$-activated human aortic endothelial cells (HAECs). HAECs were stimulated with VEGF (25 ng/mL) for 6 min in the presence of conditioned media from 293T cells expressing ADAMTS1 or vector alone. Phosphorylated proteins were visualized using the antiphosphotyrosine polyclonal Ab. **(Right)** The same membrane was stripped and reprobed with anti-VEGFR2 to evaluate total levels of the receptor.

binding, or inactivation. Neither VEGF nor VEGFR2 is a substrate for ADAMTS1, and we also determined that ADAMTS1 does not interact with VEGFR2 (unpublished observations). Instead, we have been able to determine that the suppression in VEGFR2 phosphorylation results from the binding and inactivation of VEGF by ADAMTS1.[25] The interaction between ADAMTS1 and VEGF has been confirmed by coimmunoprecipitation and cross-linking analysis. When in complex with ADAMTS1, $VEGF_{165}$ was unable to phosphorylate VEGFR2; however, the activity of the growth factor can be regained if ADAMTS1 is removed using gel filtration chromatography.[25] This functional reversibility could imply that ADAMTS1 is in fact a modulator of VEGFR2 function with the potential to sequester and subsequently release the molecule, albeit this hypothesis remains to be tested *in vivo*.

The interaction of ADAMTS1 and VEGF is not entirely surprising. TSP repeats have been shown to bind and modulate VEGF activity, at least within the context of TSP and connective tissue growth factor.[26,27] Structural-deletion mutants of ADAMTS1 were generated to determine if, in fact, the TSP1 repeats present in ADAMTS1 were responsible for binding to $VEGF_{165}$.[25] As predicted, the three terminal TSP repeats, but not the metalloprotease or disintegrin-like domains, bound to $VEGF_{165}$.[25] Furthermore, we were able to determine that the spacer region, a motif located between the first and second TSP1 repeats in ADAMTS1, was indeed an important component of such an interaction.

These findings were interesting because ADAMTS1 is processed *in vivo* to release the three-TSP cassette.[22] Based on the fact that several other matrix molecules have been shown to harbor antiangiogenic motifs that are subsequently released from the intact protein by proteolytic events, it was tempting to speculate that the same could be true for ADAMTS1. The carboxy-terminal processing event can be mediated *in vitro* by MMP2, MMP8, and MMP15,[22] but it is unclear whether one or all of these proteases perform the same effect *in vivo*.

To determine the effect of ADAMTS1 on tumor growth, we initiated a series of xenograft assays.[5] T47D human carcinoma cells were selected because of their low expression of ADAMTS1 and their rapid growth in xenograft assays. Cultures were cotransfected by standard calcium phosphate precipitation with pcDNA3.1-

ADAMTS1myc-his or empty vector (CON), and a hygromycin resistance plasmid. Resistant clones were expanded and characterized for ADAMTS1 expression by Western blot analysis of conditioned media. Generation of the expression constructs has been previously described.[28] After selection, 5×10^6 cells were injected subcutaneously in 6-week-old male nude mice. Both single- (1 tumor per mouse) and dual-tumor (1 CON, 1 ADAMTS1) assays were performed. In addition, separate xenograft tumor assays were done in which ADAMTS1 tumors were generated using either 1 clone or multiple clones (mixtures of 3 independent clones). Caliper measurements were made every 3–4 days and tumor volumes were calculated using width and length measurements. Mice were euthanized when either control or ADAMTS1 tumors reached 1500 mm^3. Mean tumor volumes were determined for each CON and ADAMTS1 group per time point and analyzed using the two-sided unpaired *t* test. Histological evaluation of the tumors followed standard procedures, including fixation in 4% paraformaldehyde and embedding in paraffin. Sections of 5 microns were obtained and stained with hematoxylin-eosin.

ADAMTS1 tumors grew much more slowly than controls and histologically displayed a sacular-like structure with a rim of viable tumor cells surrounding a hemorrhagic and necrotic center (FIG. 2d). This was in sharp contrast to control tumors, which expanded equally in three dimensions and showed a dense consistency and virtual lack of necrotic areas (FIG. 2c). These differences were consistent in more than 35 independent tumors evaluated thus far.

Tumor vasculature was evaluated by confocal microscopy after perfusion with FITC-conjugated *Lycopersicum esculentum* lectin followed by perfusion with 1% PFA as described previously.[29] Specimens were embedded in 7% agar and sectioned with a vibratome. Sections were mounted in 10% glycerol containing propidium iodide for visualization of cell nuclei. Control tumors were well vascularized in comparison to ADAMTS1-overexpressing tumors (FIGS. 2f and 2g). Both the number of vessels as well as their distribution were affected in ADAMTS1 tumors. The poor vascularization in these tumors is likely to explain the poor cellularity and the high degree of edema and necrosis in these tumors when compared to the highly cellular control tumors (FIGS. 2c and 2d).

Since ADAMTS1 is an MMP, we felt compelled to determine whether the catalytic domain contributed to the antitumor and antiangiogenic effects. We thus generated a catalytically inactive construct with a single amino acid substitution in the zinc-binding pocket of the metalloprotease domain ($\text{ADAMTS1}^{\text{E385A}}$). The generation of this construct and its catalytic activity have been described in a previous report.[28] Surprisingly, tumors with this construct were indistinguishable from control tumors in growth kinetics, histology, and vascular density (FIGS. 2b, 2e, and 2h). More notably, $\text{ADAMTS1}^{\text{E385A}}$ tumors were highly cellular with relatively little to no tumor necrosis, edema, or inflammation. These findings were surprising as they indicated that the catalytic domain was essential for the angioinhibitory effect displayed by ADAMTS1, and contrasted our previous findings that focused importance on the TSP repeats. These results prompted us to ask whether the metalloproteinase region was sufficient for its effect and whether the TSP cassette was relevant for *in vivo* suppression of blood vessel growth.

A series of deletion mutants were further generated with major focus on the catalytic domain. Unfortunately, removal of the first TSP1 repeat or the spacer region also renders the protease inactive (unpublished observations). Additional constructs substituting the TSP cassettes with similar motifs from related proteins that do not display antiangiogenic effects (such as those in the properdin protein or other members of the ADAMTS family) could help elucidate the contribution of the metalloprotease domain in the context of a different TSP cassette.

To ascertain the effect of the ADAMTS1-TSP cassette in the absence of the catalytic domain, an expression construct containing this region (aa 549–951; FIG. 3a)

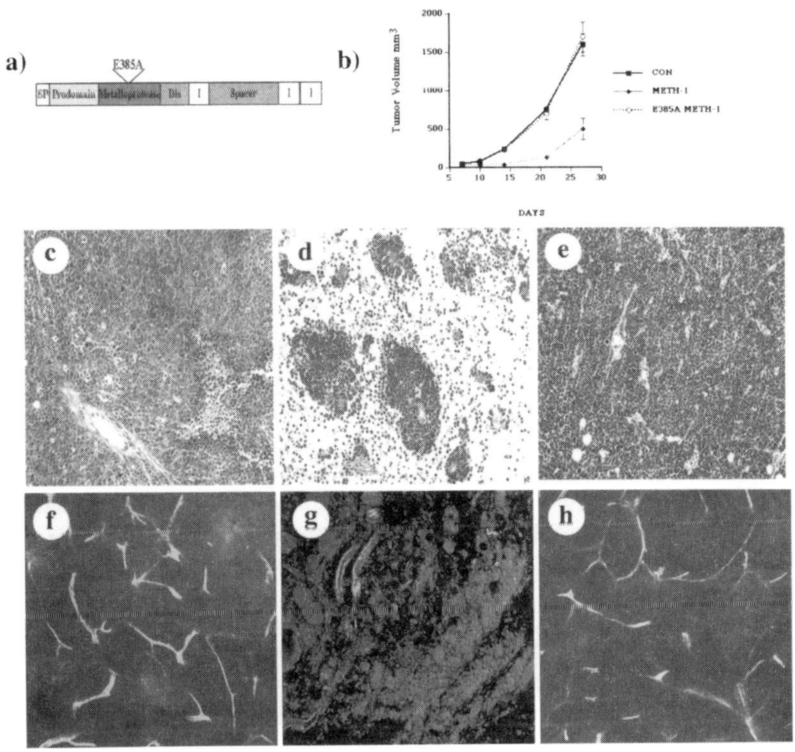

FIGURE 2. ADAMTS1 suppresses tumor angiogenesis and tumor growth. This effect requires intact catalytic activity. **(a)** Schematic representation of METH1/ADAMTS1 and location of the point mutation to generate the catalytically inactive mutant (E385A). **(b)** Tumor kinetics of control, METH1/ADAMTS1, and E385A METH1/ADAMTS1 tumors. Tumor growth was monitored by caliper measurement of length and thickness of the tumors every 3 days. Deduced volume versus time in a cohort of 4–6 animals per time point is presented (±SD). **(c–e)** Representative histological image of tumors. **(c)** Control tumors showed a high cellularity with little evidence of necrosis. **(d)** METH1/ADAMTS1 tumors showed profuse edema and necrosis. Viable tumor cells organized cylinders around viable blood vessels. **(e)** E385A METH1/ADAMTS1 tumors were indistinguishable from control tumors. **(f–h)** Representative confocal images of preparations perfused with *Lycopersicum esculentum* lectin. **(f)** Control tumor showed profuse vascularization and an even sheet of epithelial cells. **(g)** METH1/ADAMTS1-expressing tumors showed low vascular density as well as poor tumor cellularity. **(h)** E385A METH1/ADAMTS1 tumors displayed vascular density identical to control tumors.

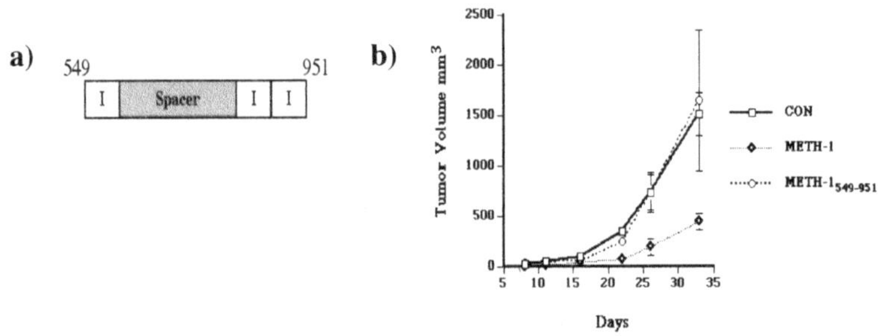

FIGURE 3. Effect of the carboxy-terminal half of METH1/ADAMTS1 on tumor growth. **(a)** Schematic representation of the fragment utilized in the xenograft assays. The numbers above the scheme denote the aa number (549–951). The letter "I" denotes the position of the TSP1 repeats. **(b)** Tumor kinetics of control and tumors expressing METH1/ADAMTS1 and a carboxy-terminal truncated construct (METH1$_{549-951}$). Tumor growth was monitored by caliper measurement of length and thickness of the tumors every 3 days. Deduced volume versus time in a cohort of 4–6 animals per time point is presented (±SD).

was transfected into T47D tumor cells. Clones were selected and xenograft assays were performed with several clones. FIGURE 3b shows the findings from these experiments. To our surprise, this construct had no effect on tumor growth. In contrast, a construct containing the first TSP1 repeat alone had a noticeable effect on tumor mass as well as tumor neovascularization.[16,30] Possible explanations for these contrasting results are likely to be linked to the relative solubility of each recombinant protein. We noted that the recombinant carboxy-terminal half of ADAMTS1 (FIG. 3b) is highly insoluble and deposited avidly into the matrix. High concentrations of heparin are required to detect the truncated protein in the supernatant, unlike the full-length enzyme. Clearly, this region confers the heparin affinity to ADAMTS1 as has been previously show;[21,22] however, its avidity for heparin/heparan sulfate proteoglycans is highly increased in the absence of the amino-terminal half of the protein. In contrast, the recombinant protein comprising the first TSP1 repeat of ADAMTS alone is highly soluble and more likely to interact and inactivate VEGF. This possibility remains to be scrutinized experimentally.

In conclusion, these experiments revealed that (1) the catalytic activity of ADAMTS1 is required for its antitumor/antiangiogenic effects and (2) the TSP1 repeats, at least *in vitro*, are able to bind and inactivate VEGF function. The interdependency of these two effects is yet to be determined; however, these results have opened the possibility for at least two potential mechanisms of action of ADAMTS1. Are these mechanisms related/interdependent? It is possible that the catalytic domain is required for the release of the TSP1 repeats in ADAMTS1 either directly (autocatalytic effect) or indirectly. It is also possible that they act in an additive manner. Clearly, experiments aimed at the identification of all potential ADAMTS1 substrates are an obvious direction to follow. Unlike many MMPs, the substrate spectrum of ADAMTS proteins appears to be restricted. From 17 purified extracellular matrix proteins tested *in vitro*, ADAMTS1 only cleaves aggrecan and versican.[28,31,32] These two proteoglycans, however, are unlikely to play any role in our tumor xenograft assays considering their low to undetectable expression levels. Unpublished data from our laboratory indicate that ADAMTS1 is able to shed proteins from the surface of the endothelium; this finding could shed some light into the significance of the catalytic activity of ADAMTS1 for its angioinhibitory properties.

CONCLUDING REMARKS

Angiogenesis requires a number of cell physiological events, such as migration, proliferation, and cell-cell association, that are dependent on a wide variety of molecules, including growth factors, receptors, extracellular matrix proteins, and proteases. ADAMTS1 exhibits a number of domains that have the potential to affect several tangents of the angiogenic process by modification of the extracellular microenvironment through its proteolytic activity or via its multiple TSP domains. While it is unclear whether ADAMTS1 plays a role in nonpathological angiogenesis,[33] its effects in a variety of assays and in tumor xenografts underscore its therapeutic potential. Clearly, a deeper understanding of the process by which this protease suppresses angiogenesis will likely reveal potential areas for intervention for suppression of vascular growth. In addition, our findings contradicted the general dogma that MMPs are "always" proangiogenic and prometastatic, adding caution to the use of broad MMP inhibitors for tumor growth suppression.

ACKNOWLEDGMENTS

This work was supported by National Institutes of Health Grant No. RO1CA77420-01 (to M. L. Iruela-Arispe) and fellowships from the Giannini Foundation (to D. Carpizo) and the Department of Defense (to A. Luque).

REFERENCES

1. HANAHAN, D. & R.A. WEINBERG. 2000. The hallmarks of cancer. Cell **100:** 57–70.
2. KERBEL, R. & J. FOLKMAN. 2002. Clinical translation of angiogenesis inhibitors. Nat. Rev. Cancer **2:** 727–739.
3. ROSEN, L. 2000. Antiangiogenic strategies and agents in clinical trials. Oncologist **5**(suppl. 1)**:** 20–27.
4. O'REILLY, M.S. *et al.* 1996. Angiostatin induces and sustains dormancy of human primary tumors in mice. Nat. Med. **2:** 689–692.
5. CARPIZO, D. & M.L. IRUELA-ARISPE. 2000. Endogenous regulators of angiogenesis—emphasis on proteins with thrombospondin–type I motifs. Cancer Metastasis Rev. **19:** 159–165.
6. STUPACK, D.G. & D.A. CHERESH. 2002. ECM remodeling regulates angiogenesis: endothelial integrins look for new ligands. Sci. STKE **119:** PE7.
7. GUTHEIL, J.C. *et al.* 2000. Targeted antiangiogenic therapy for cancer using vitaxin: a humanized monoclonal antibody to the integrin alphaV beta3. Clin. Cancer Res. **6:** 3056–3061.
8. O'REILLY, M.S. *et al.* 1997. Endostatin: an endogenous inhibitor of angiogenesis and tumor growth. Cell **88:** 277–285.
9. O'REILLY, M.S. *et al.* 1994. Angiostatin: a novel angiogenesis inhibitor that mediates the suppression of metastases by a Lewis lung carcinoma. Cell **79:** 315–328.
10. GOOD, D.J. *et al.* 1990. A tumor suppressor–dependent inhibitor of angiogenesis is immunologically and functionally indistinguishable from a fragment of thrombospondin. Proc. Natl. Acad. Sci. USA **87:** 6624–6628.

11. IRUELA-ARISPE, M.L. *et al.* 1991. Thrombospondin exerts an antiangiogenic effect on tube formation by endothelial cells *in vitro*. Proc. Natl. Acad. Sci. USA **88:** 5026–5030.
12. STREIT, M. *et al.* 2000. Thrombospondin-1 suppresses wound healing and granulation tissue formation in the skin of transgenic mice. EMBO J. **19:** 3272–3282.
13. TOLSMA, S.S. *et al.* 1993. Peptides derived from two separate domains of the matrix protein thrombospondin-1 have anti-angiogenic activity. J. Cell Biol. **122:** 497–511.
14. IRUELA-ARISPE, M.L. *et al.* 1999. Inhibition of angiogenesis by thrombospondin-1 is mediated by two independent regions within the type 1 repeats. Circulation **100:** 1423–1431.
15. HAWIGHORST, T. *et al.* 2001. Thrombospondin-2 plays a protective role in multistep carcinogenesis: a novel host anti-tumor defense mechanism. EMBO J. **20:** 2631–2640.
16. IRUELA-ARISPE, M.L. *et al.* 1999. Antiangiogenic domains shared by thrombospondins and metallospondins, a new family of angiogenic inhibitors. Ann. N.Y. Acad. Sci. **886:** 58–66.
17. VAZQUEZ, F. *et al.* 1999. METH1, a human ortholog of ADAMTS, and METH2 are members of a new family of proteins with angio-inhibitory activity. J. Biol. Chem. **274:** 23349–23357.
18. KUNO, K. *et al.* 1997. Molecular cloning of a gene encoding a new type of metalloproteinase-disintegrin family protein with thrombospondin motifs as an inflammation associated gene. J. Biol. Chem. **272:** 556–562.
19. CAL, S. *et al.* 2002. Cloning, expression analysis, and structural characterization of seven novel human ADAMTSs, a family of metalloproteinases with disintegrin and thrombospondin-1 domains. Gene **283:** 49–62.
20. LLAMAZARES, M. *et al.* 2003. Identification and characterization of ADAMTS-20 defines a novel subfamily of metalloproteinases-disintegrins with multiple thrombospondin-1 repeats and a unique GON-domain. J. Biol. Chem. In press.
21. KUNO, K. *et al.* 1998. ADAMTS-I protein anchors at the extracellular matrix through the thrombospondin type I motifs and its spacing region. J. Biol. Chem. **273:** 13912–13917.
22. RODRIGUEZ-MANZANEQUE, J.C. *et al.* 2000. Characterization of Meth-1/Adamts-1 processing reveals two distinct active forms. J. Biol. Chem. **275:** 33471–33479.
23. TORTORELLA, M. *et al.* 2000. The thrombospondin motif of aggrecanase-1 (ADAMTS-4) is critical for aggrecan substrate recognition and cleavage. J. Biol. Chem. **275:** 25791–25797.
24. FLANNERY, C.R. *et al.* 2002. Autocatalytic cleavage of ADAMTS-4 (aggrecanase-1) reveals multiple glycosaminoglycan-binding sites. J. Biol. Chem. **277:** 42775–42780.
25. LUQUE, A. *et al.* 2003. ADAMTS1/METH1 inhibits endothelial cell proliferation by direct binding and sequestration of VEGF165. J. Biol. Chem. In press.
26. GUPTA, K. *et al.* 1999. Binding and displacement of vascular endothelial growth factor (VEGF) by thrombospondin: effect on human microvascular endothelial cell proliferation and angiogenesis. Angiogenesis **3:** 147–158.
27. HASHIMOTO, G. *et al.* 2002. Matrix metalloproteinases cleave connective tissue growth factor and reactivate angiogenic activity of vascular endothelial growth factor 165. J. Biol. Chem. **277:** 36288–36295.
28. RODRIGUEZ-MANZANEQUE, J.C. *et al.* 2002. ADAMTS1 cleaves aggrecan at multiple sites and is differentially inhibited by metalloproteinase inhibitors. Biochem. Biophys. Res. Commun. **293:** 501–508.
29. RODRIGUEZ-MANZANEQUE, J.C. *et al.* 2001. Thrombospondin-1 suppresses tumor growth by a novel mechanism that includes blockade of matrix metalloproteinase-9 activation. Proc. Natl. Acad. Sci. USA **98:** 12485–12490.
30. IRUELA-ARISPE, M.L. *et al.* 1999. Anti-angiogenic domain of thrombospondin. *In* Angiogenesis in Health and Disease: Basic Mechanisms and Clinical Applications.
31. KUNO, K. *et al.* 2000. ADAMTS-1 cleaves a cartilage proteoglycan, aggrecan. FEBS Lett. **478:** 241–245.
32. SANDY, J.D. *et al.* 2001. Versican V1 proteolysis in human aorta *in vivo* occurs at the Glu441–Ala442 bond, a site which is cleaved by recombinant ADAMTS-1 and ADAMTS-4. J. Biol. Chem. **276:** 13372–13378.
33. SHINDO, T. *et al.* 2000. ADAMTS-1: a metalloproteinase-disintegrin essential for normal growth, fertility, and organ morphology and function. J. Clin. Invest. **105:** 1345–1352.

Genetic Analysis of Hypoxia Signaling and Response in *C. elegans*

CHUAN SHEN AND JO ANNE POWELL-COFFMAN

Department of Zoology and Genetics, Iowa State University, Iowa 50010-3260, USA

ABSTRACT: **During normal development and homeostasis, animals use cellular and systemic strategies to adapt to changing oxygen levels. In mammals, hypoxic tissues secrete growth factors to induce angiogenesis, and individual cells increase anaerobic metabolism in order to sustain basic cellular functions. Many of these critical responses to decreased oxygen availability are regulated by the hypoxia-inducible factors, dimeric transcriptional complexes consisting of α and β subunits. HIFα proteins are specialized for hypoxia response, and oxygen levels regulate their stability and activity. The *C. elegans hif-1* gene is orthologous to mammalian HIFα genes, and *C. elegans* has proven to be a powerful system for the study of hypoxia-inducible factor regulation and function. Mutants lacking *hif-1* function are viable in normoxic or anoxic conditions, but they cannot adapt to hypoxia. Recent genetic analyses in *C. elegans* led to the identification of the evolutionarily conserved enzyme that hydroxylates HIFα in an oxygen-dependent manner. Once modified, HIFα binds the von Hippel–Lindau tumor suppressor protein and is targeted for proteasomal degradation. Here, we briefly review the characterization of *C. elegans hif-1* and interacting genes, and discuss genetic strategies for studying hypoxia signaling and response.**

KEYWORDS: **bHLH-PAS proteins; *egl-9*; oxygen; transcription; von Hippel–Lindau tumor suppressor**

INTRODUCTION

In a multicellular organism, cells and tissues have varying access to oxygen, and cellular adaptation to hypoxia (low oxygen) plays a central role in many important developmental processes. Large animals require circulatory systems to distribute oxygen and nutrients, and angiogenesis is guided by the vascular endothelial growth factor, which is secreted by hypoxic cells.[1] Coronary occlusions, ischemic heart disease, and other pathologies that disrupt the circulatory system can cause immediate and disastrous effects, including hypoxic cell death and irreversible damage to critical tissues. These pathologies can be partially alleviated by hypoxic preconditioning,[2,3] suggesting that specific changes in gene expression can protect cells from short-term oxygen deprivation. Cellular response to hypoxia also plays a central role in tumor growth.[4–6]

Address for correspondence: Jo Anne Powell-Coffman, Iowa State University, 2108 Molecular Biology Building, Ames, IA 50010-3260. Voice: 515-294-3906; fax: 515-294-6755.
japc@iastate.edu

Ann. N.Y. Acad. Sci. 995: 191–199 (2003). © 2003 New York Academy of Sciences.

HYPOXIA-INDUCIBLE FACTORS

Many of the physiological responses to decreased oxygen availability are mediated by the sequence-specific DNA-binding hypoxia-inducible factors (HIF).[1] HIF complexes are dimeric, consisting of alpha and beta subunits. The HIFβ subunit is also termed ARNT (aryl hydrocarbon receptor nuclear translocator).[7,8] While ARNT can dimerize with other transcription factors, HIFα is apparently dedicated to hypoxia response. Researchers have isolated three murine HIFα genes (HIF-1α, HIF-2α, and HIF-3α). The HIFα proteins have striking sequence similarities, and the proteins have distinct, but overlapping, expression patterns.[7,9–11] HIF-1α has an essential role in angiogenesis. Mice lacking a functional HIF-1α gene die early in embryogenesis with severe vascular defects.[12,13] Mice that are deficient for HIF-2α (EPAS-1) suffer from bradycardia, which is associated with decreased catecholamine synthesis.[14]

Regulation of HIFα

The stability of HIFα is regulated by oxygen. When cellular oxygen levels are high, a conserved proline residue in the HIFα subunit is hydroxylated. This increases the affinity of the oxygen-dependent degradation domain (ODD) of HIFα for the von Hippel–Lindau tumor suppressor protein (VHL).[15–17] VHL targets HIFα for polyubiquitination and proteasomal degradation.[18,19] The prolyl hydroxylase that modifies HIFα was first identified in *Caenorhabditis elegans* (see below).[20] In hypoxic conditions, prolyl hydroxylation and subsequent degradation of HIFα are inhibited. HIFα translocates to the nucleus, dimerizes with ARNT, and activates the expression of target genes, which act to increase oxygen delivery or implement metabolic adaptation to hypoxia. While oxygen-dependent degradation is an important aspect of HIFα regulation, several studies suggest that other cell-type-specific factors or signaling pathways modulate HIFα activity. These have been reviewed recently elsewhere.[21]

A genetic model system for the study of hypoxia-inducible factor regulation and function has been developed in *C. elegans*. Genetic studies will complement the wealth of biochemical and pharmacological analyses in mammalian cells and will advance our understanding of hypoxia signaling and response.

HYPOXIA RESPONSE IN *C. ELEGANS*

The nematode *C. elegans* is a powerful animal model system for the dissection of evolutionarily conserved signaling pathways. *C. elegans* are small (~1 mm in length as adults), and they can be cultured on agar plates with a bacterial food source. There are two sexes, self-fertilizing hermaphrodites and males, and the generation time is less than 4 days at 20°C. The wild-type embryonic and postembryonic lineages have been extensively documented and are essentially invariant.[22,23] This allows mutant phenotypes to be diagnosed at the level of individual cells. The simplicity and experimental convenience of *C. elegans* make it an attractive model organism for genetic analyses, and a wealth of recent genome-wide studies has increased the power of this system. The sequence of the ~10^8 bp genome was completed in 1998, revealing approximately 19,000 genes.[24] Large-scale studies are under way to characterize the function of every gene by double-stranded RNA interference, full-genome micro-

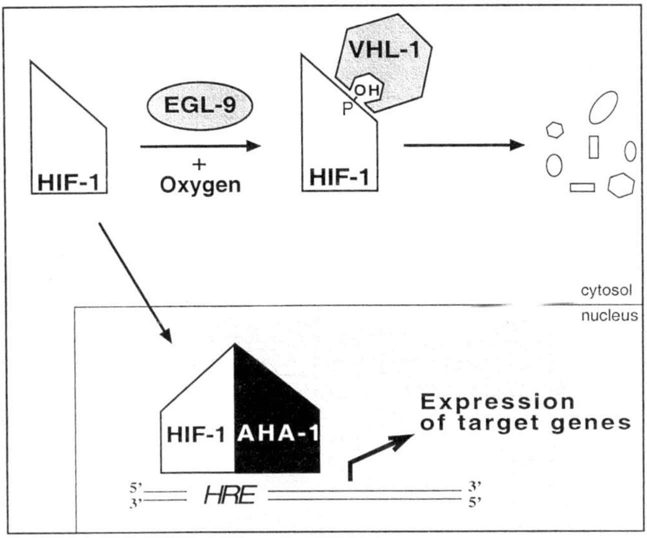

FIGURE 1. Regulation of *C. elegans* HIF-1. HIF-1 protein levels are regulated by oxygen availability. The EGL-9 protein is a member of the 2-oxoglutarate-dependent oxygenase superfamily, and oxygen is a required cofactor for EGL-9 function. EGL-9 acts directly on HIF-1 to hydroxylate the proline residue in the conserved LXXLAP motif. This modification is required for binding of HIF-1 to VHL-1. VHL-1 targets HIF-1 for proteasomal degradation. Under hypoxic conditions, EGL-9 does not efficiently modify HIF-1, and HIF-1 dimerizes with AHA-1 to form a transcriptional complex.[20] The HIF-1 heterodimer binds to the hypoxia response elements (HRE) in target genes to implement changes in gene expression (unpublished data).

arrays, and semiautomated analyses of protein protein interactions.[25–28] Much of these data are available to the scientific community through dedicated internet sites.[29]

In the wild, *C. elegans* inhabits the soil, where it can encounter hypoxic microenvironments. *C. elegans* is able to maintain a near-normal metabolic rate at environmental oxygen concentrations as low as 2%. In 0.5% or 1% oxygen, the animals must decrease oxygen consumption, but they continue to develop and reproduce.[30] *C. elegans* does not have a complex circulatory system. Any cell in the organism is only a few cell widths from the outer surface of the worm or the intestinal lumen,[22] and oxygen delivery is thought to be accomplished by diffusion. Thus, individual cells must sense and adapt to local environmental oxygen levels. Recent discoveries have revealed that the molecular mechanisms that govern transcriptional responses to hypoxia are, at least in part, conserved between *C. elegans* and humans (see text below and FIG. 1).

C. ELEGANS HIF-1

The *C. elegans hif-1* gene is orthologous to mammalian HIF alpha subunits.[20,31] Like its mammalian cognates, *C. elegans* HIF-1 protein is induced by hypoxia and is rapidly degraded upon reoxygenation. *hif-1* mRNA levels are not dramatically affected by oxygen concentration.[20,31] Under conditions in which *C. elegans* HIF-1 is stable, it binds AHA-1, the orthologue of mammalian ARNT/HIFβ, to form a complex that can bind DNA sequences containing the hypoxic regulatory element (FIG. 1).[31] Both HIF-1 and AHA-1 are expressed in most, if not all, cells, as assayed by a *hif-1*:GFP reporter gene and AHA-1-specific antibodies. In the intestine, which does not express any bHLH-PAS proteins other than HIF-1 and AHA-1, nuclear localization of AHA-1 is dependent upon HIF-1.[31]

VHL-1 and EGL-9 Mediate the Degradation of HIF-1

Recent studies have identified an evolutionarily conserved enzyme that directly couples cellular oxygen levels to the stability of HIF α subunits. In 2001, two groups demonstrated that a conserved proline in HIF-1α and HIF-2α was hydroxylated in an oxygen-dependent manner and that this modification increased binding of the HIF α subunits to VHL.[15,16] VHL targets the protein for proteasomal degradation.[18,19] In a tour de force published later in the year, Ratcliffe, Maxwell, Pugh, and colleagues identified the enzyme that hydroxylates HIFα. First, they demonstrated that oxygen-dependent hydroxylation of the LXXLAP motif in HIFα proteins and the subsequent interaction with VHL are conserved in nematodes and humans. Modification of *C. elegans* HIF-1 allows direct binding of VHL-1, the orthologue of VHL.[20] The importance of VHL-1 for HIF-1 degradation was further confirmed by isolation and analysis of mutants carrying a deletion in the *vhl-1* gene. These VHL-1-deficient animals express HIF-1 protein at high levels in both normoxic and hypoxic conditions. Second, the authors identified the enzyme that hydroxylates the oxygen-dependent degradation domain of HIF-1 by assaying HIF-1 protein levels in a series of mutant strains, each defective in a different candidate gene. They found that animals carrying loss-of-function mutations in the *egl-9* gene failed to downregulate HIF-1 expression in normoxic conditions. Additional experiments demonstrated that the EGL-9 protein acted directly on HIF-1 to hydroxylate the proline in the LXXLAP motif *in vitro*. Finally, the authors identified mammalian homologues of EGL-9 (termed PHD 1, 2, and 3) and demonstrated that the mammalian PHDs hydroxylate human HIF-1α or HIF-2α *in vitro*.[20] Thus, HIF-1, VHL-1, and EGL-9 appear to be part of an evolutionarily conserved regulatory network that senses hypoxia and implements appropriate transcriptional changes (FIG. 1).

The *egl-9* gene was originally isolated as a gene required for normal egg-laying,[32] and *egl-9* encodes a member of the 2-oxoglutarate-dependent oxygenase superfamily.[33,34] In an independent study, *egl-9* mutants were shown to have decreased sensitivity to cyanide exposure.[35] The cyanide sensitivity of *vhl-1*-deficient animals or *egl-9;hif-1* double-mutants has not been reported, but phenotypic and molecular studies indicate that *egl-9* and *vhl-1* are likely to have functions that are independent of their critical role in the oxygen-dependent degradation of HIF-1.

The **hif-1** *Gene Is Required for Adaptation to Hypoxia*

The *hif-1* (*ia04*) mutation is predicted to be a strong loss-of-function allele. It deletes exons 2–4 of *hif-1*, and it introduces an early translational stop codon to the most abundant forms of *hif-1* mRNA.[31] Animals that are homozygous for the *hif-1* mutation exhibit no visible defects under standard laboratory conditions. However,

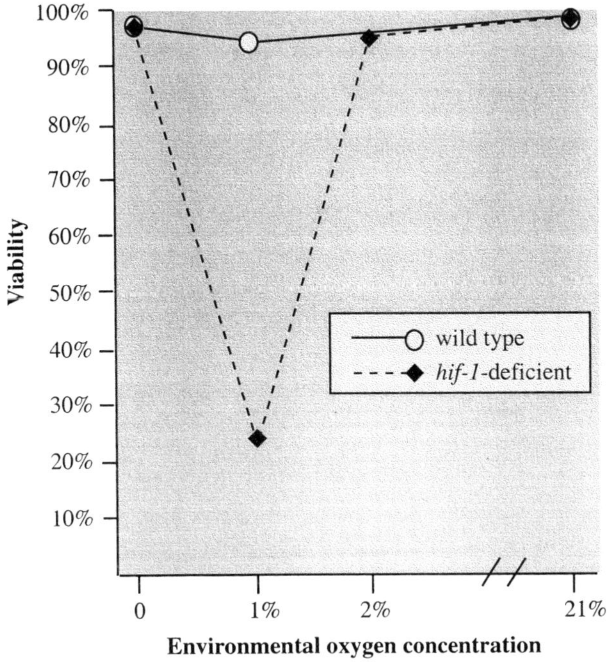

FIGURE 2. Viability of wild-type *C. elegans* and *hif-1* (*ia04*) mutants at varying oxygen levels. This graph summarizes published studies in which the survival of wild-type animals and *hif-1* (*ia04*) mutants was assayed at 21%, 2%, 1%, and 0% oxygen at 20°C.[31,36] The viability of *hif-1*-defective *C. elegans* at oxygen concentrations between 0% and 0.5% has not been reported. The survival curve is unlikely to be linear, and there may be a threshold oxygen concentration for survival. As discussed in the text, wild-type animals do not survive prolonged oxygen deprivation at 28°C.[37]

hif-1-defective animals are unable to adapt to 0.5% or 1% oxygen. While wild-type animals survive and reproduce in 1% oxygen, 66% of *hif-1*-defective animals do not survive embryogenesis in these conditions, and an additional 9% die during larval development (FIG. 2).[31]

Anoxia-Induced Suspended Animation versus Hypoxia

Analyses of *C. elegans hif-1* have revealed a fundamental difference in the molecular mechanisms by which *C. elegans* respond to hypoxia (1% or 0.5% O_2) versus anoxia (0% O_2) (FIG. 2). In anoxic conditions at 20°C, *C. elegans* at all stages arrest development and cell-cycle progression, and they enter a reversible state of "suspended animation". After 24 hours of incubation in anoxia, ~90% of *C. elegans* recover and continue development.[30,36] Padilla, Roth, and colleagues have shown that there is no detectable difference between *hif-1*-defective and wild-type animals in their ability to survive these conditions (FIG. 2).[36] Thus, *hif-1* function is essential in hypoxic conditions in which animals continue to grow and develop, but *hif-1* does not appear to have an important role in anoxia-induced suspended animation or in subsequent recovery upon reoxygenation.

The ability of *C. elegans* to survive oxygen deprivation decreases at high temperatures.[37] *C. elegans* are commonly cultured at temperatures between 15°C and 25°C. At 27°C, they are much more likely to form dauers, a stress-resistant alternative larval stage.[38] Crowder and colleagues have shown that when adult *C. elegans* are incubated for 20 h at 28°C in an anoxic chamber (5% CO_2, 10% H_2, 85% N_2 gas; oxygen levels measured at below 0.3%), fewer than 5% of adult animals survive.[37] Remarkably, this "hypoxic death" can be suppressed by certain loss-of-function mutations in the *daf-2* insulin-like growth factor receptor gene.[37] These *daf-2* alleles have also been shown to prolong the life of *C. elegans* grown in optimal conditions and to induce dauer formation at 25°C.[39] It will be interesting to learn whether mutation of *hif-1* influences the ability of animals to survive the hypoxic death assay.

Multiple Species of C. elegans hif-1 *mRNAs*

Over 45 *hif-1* cDNAs have been completely or partially sequenced by the *C. elegans* genome project, and 4 forms of *hif-1* mRNAs have been identified.[27] These are diagrammed in FIGURE 3. The three most abundant forms (*a*, *b*, and *c*) contain all of the previously defined domains in HIF-1, including the bHLH, PAS, and oxygen-dependent degradation domains. The *b* form utilizes an alternative splice acceptor site for exon 7, and this adds 6 bases and 2 codons. Both the *c* and *d* forms are rare. The *c* form includes an additional ~600 bp exon (exon 5.5 in FIG. 3). The smaller *d* species of *hif-1* appears to utilize an alternative promoter inside intron 4, and it does

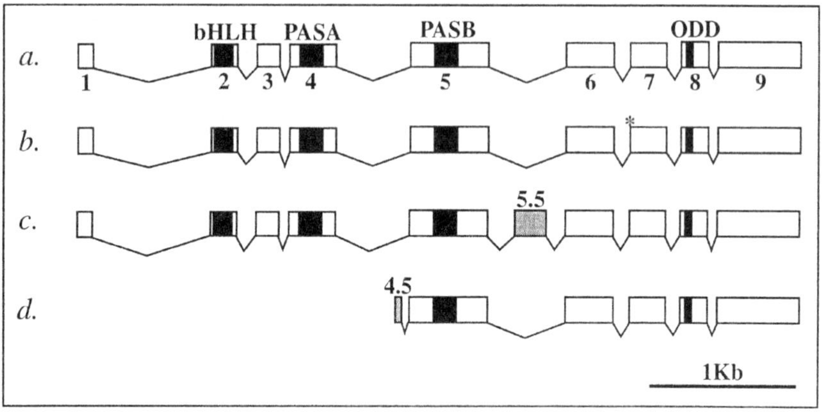

FIGURE 3. Multiple species of *hif-1* mRNAs. Exons are drawn as boxes, and the basic helix-loop-helix (bHLH), PER-ARNT-SIM (PAS), and oxygen-dependent degradation (ODD) domains are indicated in black. The *a* form has been characterized in previous biochemical and molecular studies.[20,31] The *b* form utilizes a different splice acceptor site for exon 7, and it encodes 2 additional amino acids (labeled *). Both the *c* and *d hif-1* mRNAs are rare. The *c* form includes an additional exon (labeled as 5.5). The *d* form appears to utilize an alternative promoter in intron 4, and it includes exon 4.5. Neither the bHLH nor PASA motifs are encoded by the *d* mRNA. This diagram summarizes data from The *C. elegans* Genome Consortium (available at http://www.wormbase.org/).

not encode the bHLH or PASA motifs. The protein encoded by the *a* class of mRNAs is well characterized,[20,31] but the functional and biochemical properties of the alternative gene products have not yet been examined.

Strategies for Genetic Analysis of Hypoxia Signaling and Response in C. elegans

C. elegans lacking *hif-1*, *vhl-1*, or *egl-9* are viable, and this enables full-genome microarrays to characterize hypoxia-induced changes in gene expression. In preliminary studies, we used this approach to clarify the role of HIF-1 in hypoxia response. We identified the following classes: (1) genes that are regulated by hypoxia in wild-type animals, but not in *hif-1* mutants; (2) genes that are regulated by hypoxia, independent of *hif-1* function; and (3) genes that require *hif-1* for basal levels of expression, but are not induced by a 4-h exposure to hypoxia. These data confirm that HIF-1 induces the expression of many genes in response to hypoxia. Further, they demonstrate that HIF-1 does not mediate all transcriptional responses to low oxygen levels. It is not immediately clear which *hif-1*-dependent changes in expression represent direct targets of HIF-1. As a first step towards identifying functionally important HIF-1:AHA-1 binding sites, we are assaying the binding specificity of this complex by gel mobility shift assays. Initial results suggest that the HIF-1:AHA-1 binding site is similar to the mammalian hypoxia response element (unpublished data).

Once direct targets of HIF-1 have been identified, it will be possible to construct GFP reporter genes for use in genetic screens. We have generated a *C. elegans* strain carrying a hypoxia-induced reporter gene that has a restricted expression pattern in wild-type animals in normoxic conditions. In environmental or genetic conditions that increase HIF-1 stability, the expression of the GFP reporter expands to other tissues (unpublished data). This and other reporters will be used to screen for mutations that result in increased or decreased HIF-1 activity. Such genetic analyses may identify additional evolutionarily conserved regulators of the hypoxic response.

ACKNOWLEDGMENTS

C. elegans strains were obtained from the *Caenorhabditis* Genetic Stock Center, which is supported by the National Institutes of Health National Center for Research Resources. The *hif-1* cDNAs were isolated and characterized by the *C. elegans* Genome Consortium. We are thankful to Danielle Thierry-Mieg for drawing our attention to the rare *hif-1* mRNAs. We are grateful to Min Jiang, Stuart Kim, and colleagues at Stanford University, who performed the microarray hybridizations. The hypoxia project in the Powell-Coffman lab is supported by grants from the National Science Foundation (No. 9874456) and from the American Heart Association (Established Investigator Award to J. A. Powell-Coffman).

REFERENCES

1. SEMENZA, G.L. 1999. Regulation of mammalian O_2 homeostasis by hypoxia-inducible factor 1. Annu. Rev. Cell Dev. Biol. **15:** 551–578.
2. TAJIMA, M., D. KATAYOSE, M. BESSHO *et al.* 1994. Acute ischaemic preconditioning and chronic hypoxia independently increase myocardial tolerance to ischaemia. Cardiovasc. Res. **28:** 312–319.

3. GIDDAY, J.M., J.C. FITZGIBBONS, A.R. SHAH *et al.* 1994. Neuroprotection from ischemic brain injury by hypoxic preconditioning in the neonatal rat. Neurosci. Lett. **168:** 221–224.
4. MAXWELL, P.H., C.W. PUGH & P.J. RATCLIFFE. 2001. Activation of the HIF pathway in cancer. Curr. Opin. Genet. Dev. **11:** 293–299.
5. HOCKEL, M. & P. VAUPEL. 2001. Tumor hypoxia: definitions and current clinical, biologic, and molecular aspects. J. Natl. Cancer Inst. **93:** 266–276.
6. HARRIS, A.L. 2002. Hypoxia—a key regulatory factor in tumour growth. Nat. Rev. Cancer **2:** 38–47.
7. WANG, G.L., B.H. JIANG, E.A. RUE *et al.* 1995. Hypoxia-inducible factor 1 is a basic-helix-loop-helix-PAS heterodimer regulated by cellular O_2 tension. Proc. Natl. Acad. Sci. USA **92:** 5510–5514.
8. JIANG, B.H., E. RUE, G.L. WANG *et al.* 1996. Dimerization, DNA binding, and transactivation properties of hypoxia-inducible factor 1. J. Biol. Chem. **271:** 17771–17778.
9. TIAN, H., S.L. MCKNIGHT & D.W. RUSSELL. 1997. Endothelial PAS domain protein 1 (EPAS1), a transcription factor selectively expressed in endothelial cells. Genes Dev. **11:** 72–82.
10. HOGENESCH, J.B., W.K. CHAN, V.H. JACKIW *et al.* 1997. Characterization of a subset of the basic-helix-loop-helix-PAS superfamily that interacts with components of the dioxin signaling pathway. J. Biol. Chem. **272:** 8581–8593.
11. FLAMME, I., T. FROHLICH, M. VON REUTERN *et al.* 1997. HRF, a putative basic helix-loop-helix-PAS-domain transcription factor is closely related to hypoxia-inducible factor-1 alpha and developmentally expressed in blood vessels. Mech. Dev. **63:** 51–60.
12. IYER, N.V., L.E. KOTCH, F. AGANI *et al.* 1998. Cellular and developmental control of O_2 homeostasis by hypoxia-inducible factor 1 alpha. Genes Dev. **12:** 149–162.
13. RYAN, H.E., J. LO & R.S. JOHNSON. 1998. HIF-1 alpha is required for solid tumor formation and embryonic vascularization. EMBO J. **17:** 3005–3015.
14. TIAN, H., R.E. HAMMER, A.M. MATSUMOTO *et al.* 1998. The hypoxia-responsive transcription factor EPAS1 is essential for catecholamine homeostasis and protection against heart failure during embryonic development. Genes Dev. **12:** 3320–3324.
15. JAAKKOLA, P., D.R. MOLE, Y.M. TIAN *et al.* 2001. Targeting of HIF-alpha to the von Hippel–Lindau ubiquitylation complex by O_2-regulated prolyl hydroxylation. Science **292:** 468–472.
16. IVAN, M., K. KONDO, H. YANG *et al.* 2001. HIFalpha targeted for VHL-mediated destruction by proline hydroxylation: implications for O_2 sensing. Science **292:** 464–468.
17. MIN, J.H., H. YANG, M. IVAN *et al.* 2002. Structure of an HIF-1alpha–pVHL complex: hydroxyproline recognition in signaling. Science **296:** 1886–1889.
18. MAXWELL, P.H., M.S. WIESENER, G.W. CHANG *et al.* 1999. The tumour suppressor protein VHL targets hypoxia-inducible factors for oxygen-dependent proteolysis. Nature **399:** 271–275.
19. OHH, M., C.W. PARK, M. IVAN *et al.* 2000. Ubiquitination of hypoxia-inducible factor requires direct binding to the beta-domain of the von Hippel–Lindau protein. Nat. Cell Biol. **2:** 423–427.
20. EPSTEIN, A.C., J.M. GLEADLE, L.A. MCNEILL *et al.* 2001. *C. elegans* EGL-9 and mammalian homologs define a family of dioxygenases that regulate HIF by prolyl hydroxylation. Cell **107:** 43–54.
21. WENGER, R.H. 2002. Cellular adaptation to hypoxia: O_2-sensing protein hydroxylases, hypoxia-inducible transcription factors, and O_2-regulated gene expression. FASEB J. **16:** 1151–1162.
22. SULSTON, J.E., E. SCHIERENBERG, J.G. WHITE *et al.* 1983. The embryonic cell lineage of the nematode *Caenorhabditis elegans*. Dev. Biol. **100:** 64–119.
23. SULSTON, J.E. & H.R. HORVITZ. 1977. Post-embryonic cell lineages of the nematode, *Caenorhabditis elegans*. Dev. Biol. **56:** 110–156.
24. THE *C. ELEGANS* SEQUENCING CONSORTIUM. 1998. Genome sequence of the nematode *C. elegans*: a platform for investigating biology. Science **282:** 2012–2018.
25. HILL, A.A., C.P. HUNTER, B.T. TSUNG *et al.* 2000. Genomic analysis of gene expression in *C. elegans*. Science **290:** 809–812.

26. WALHOUT, A.J., R. SORDELLA, X. LU *et al.* 2000. Protein interaction mapping in *C. elegans* using proteins involved in vulval development. Science **287:** 116–122.
27. REBOUL, J., P. VAGLIO, N. TZELLAS *et al.* 2001. Open-reading-frame sequence tags (OSTs) support the existence of at least 17,300 genes in *C. elegans*. Nat. Genet. **27:** 332–336.
28. JIANG, M., J. RYU, M. KIRALY *et al.* 2001. Genome-wide analysis of developmental and sex-regulated gene expression profiles in *Caenorhabditis elegans*. Proc. Natl. Acad. Sci. USA **98:** 218–223.
29. STEIN, L., P. STERNBERG, R. DURBIN *et al.* 2001. WormBase: network access to the genome and biology of *Caenorhabditis elegans*. Nucleic Acids Res. **29:** 82–86.
30. VAN VOORHIES, W.A. & S. WARD. 2000. Broad oxygen tolerance in the nematode *Caenorhabditis elegans*. J. Exp. Biol. **203**(part 16)**:** 2467–2478.
31. JIANG, H., R. GUO & J.A. POWELL-COFFMAN. 2001. The *Caenorhabditis elegans hif-1* gene encodes a bHLH-PAS protein that is required for adaptation to hypoxia. Proc. Natl. Acad. Sci. USA **98:** 7916–7921.
32. TRENT, C., N. TSUING & H.R. HORVITZ. 1983. Egg-laying defective mutants of the nematode *Caenorhabditis elegans*. Genetics **104:** 619–647.
33. DARBY, C., C.L. COSMA, J.H. THOMAS *et al.* 1999. Lethal paralysis of *Caenorhabditis elegans* by *Pseudomonas aeruginosa*. Proc. Natl. Acad. Sci. USA **96:** 15202–15207.
34. ARAVIND, L. & E.V. KOONIN. 2001. The DNA-repair protein AlkB, EGL-9, and leprecan define new families of 2-oxoglutarate- and iron-dependent dioxygenases. Genome Biol. **2:** research 0007.1–0007.8.
35. GALLAGHER, L.A. & C. MANOIL. 2001. *Pseudomonas aeruginosa* PAO1 kills *Caenorhabditis elegans* by cyanide poisoning. J. Bacteriol. **183:** 6207–6214.
36. PADILLA, P.A., T.G. NYSTUL, R.A. ZAGER *et al.* 2002. Dephosphorylation of cell cycle–regulated proteins correlates with anoxia-induced suspended animation in *Caenorhabditis elegans*. Mol. Biol. Cell **13:** 1473–1483.
37. SCOTT, B.A., M.S. AVIDAN & C.M. CROWDER. 2002. Regulation of hypoxic death in *C. elegans* by the insulin/IGF receptor homolog DAF-2. Science **296:** 2388–2391.
38. AILION, M. & J.H. THOMAS. 2000. Dauer formation induced by high temperatures in *Caenorhabditis elegans*. Genetics **156:** 1047–1067.
39. GEMS, D., A.J. SUTTON, M.L. SUNDERMEYER *et al.* 1998. Two pleiotropic classes of *daf-2* mutation affect larval arrest, adult behavior, reproduction, and longevity in *Caenorhabditis elegans*. Genetics **150:** 129–155.

Comparative Structure-Function Analysis of VEGFR-1 and VEGFR-2

What Have We Learned from Chimeric Systems?

ROSANA D. MEYER AND NADER RAHIMI

Departments of Ophthalmology and Biochemistry, Boston University, School of Medicine, Boston, Massachusetts 02118, USA

ABSTRACT: Activation of vascular endothelial growth factor receptor-1 and -2 (VEGFR-1 and VEGFR-2) plays a critical role in vasculogenesis and angiogenesis. However, the mechanism by which activation of VEGFRs elicits these cellular events is not fully understood. We recently constructed chimeric receptors containing the extracellular domain of human CSF-1R/c-*fms* fused with the entire transmembrane and cytoplasmic domains of murine VEGFR-1 and VEGFR-2. Selective activation of chimeric VEGFR-2, but not chimeric VEGFR-1, stimulated endothelial cell growth, migration, and differentiation. Stimulation of cells coexpressing chimeric VEGFR-1 and VEGFR-2 suppressed VEGFR-2-mediated endothelial cell growth. Site-directed mutagenesis demonstrated that tyrosines 799 and 1173 are required for VEGFR-2-mediated endothelial cell growth and activation of PI3 kinase. Further site-directed mutagenesis demonstrated that tyrosine 1212, located in the carboxyl tail of VEGFR-2, is required for the ligand-dependent autophosphorylation of the receptor and its ability to activate signaling proteins. Collectively, our results suggest that activation of VEGFR-1 and VEGFR-2 differentially regulates endothelial cell function and angiogenesis. Second, activation of VEGFR-2 is associated with many endothelial cell functions, including cell proliferation, migration, and differentiation. Third, activation of PI3 kinase by VEGFR-2 regulates endothelial cell proliferation.

KEYWORDS: VEGFR-1; VEGFR-2; angiogenesis; endothelial cells

INTRODUCTION

Vascular endothelial growth factor (VEGF) receptors and their ligands initiate signaling pathways that play a pivotal role in vasculogenesis during development and pathological angiogenesis. VEGF action is largely restricted to endothelial cells and it stimulates a broad spectrum of biological responses in endothelial cells, including cell proliferation, migration, survival, differentiation, and permeability to macromolecules.[1,2] VEGF exerts its pleiotropic effects on endothelial cells by binding

Address for correspondence: Nader Rahimi, Departments of Ophthalmology and Biochemistry, Boston University, School of Medicine, 715 Albany Street, Room L921, Boston, MA 02118. Voice: 617-638-5011; fax: 617-638-5337.
nrahimi@bu.edu

Ann. N.Y. Acad. Sci. 995: 200–207 (2003). © 2003 New York Academy of Sciences.

to and activating VEGFR-1/Flt-1 and VEGFR-2/FLK-1. VEGFR-1 and VEGFR-2 belong to a family of receptor tyrosine kinases (RTKs) whose expression is mainly restricted to endothelial cells.[1,2] In addition, VEGF also binds two nonreceptor tyrosine kinases, neuropilin-1 and neuropilin-2.[3] Initial gene targeting approaches demonstrated that activation of VEGFR-1 and VEGFR-2 might play a distinct role in endothelial cell functions, as illustrated by the nonidentical phenotypes of mice expressing homozygous knockouts of each of the genes. The VEGFR-2 knockout mice failed to develop endothelial cells, whereas knockout mice for VEGFR-1 gene contained endothelial cells, but they were poorly organized.[4,5] Further analysis revealed that disorganization of endothelial cells of the knockout mice was associated with an increase in the number of endothelial cells.[6,7]

Despite intensive research in angiogenesis in general and VEGFRs in particular, to date the molecular mechanisms responsible for different functions of VEGFR-1 and VEGFR-2 in angiogenesis and their signal transduction relays are not fully understood. Our study is aimed to address the contribution of VEGFR-1 and VEGFR-2 activation and their signal transduction relays to endothelial cell functions, including endothelial cell growth, migration, and differentiation.

RESULTS AND DISCUSSION

Chimeric Models of VEGFR-1 and VEGFR-2: A Model to Dissect the Biological Importance of VEGFR-1 and VEGFR-2 and Their Signal Transduction Relay in Angiogenesis

VEGF provokes angiogenesis by engaging multiple endothelial cell surface receptors, including VEGFR-1, VEGFR-2, neuropilin-1, and neuropilin-2, by generating potentially homologous and heterologous signaling networks during angiogenesis. Since these receptors are very often expressed in endothelial cells, it is difficult to address the role of individual receptors in angiogenesis. Understanding how activation of individual VEGF receptors generates unique cellular responses during angiogenesis is a major challenge in identifying signaling elements involved in angiogenesis. We have recently constructed a chimeric receptor containing the extracellular domain of human CSF-1R/c-*fms*, fused with the transmembrane and cytoplasmic domains of murine VEGFR-2. This model permitted us to dissect the function of VEGFR-1 and VEGFR-2 in endothelial cells by selectively stimulating the receptor with CSF-1. The resultant chimeric receptors are herein called CTR (CSF-1R-FLT-1 Receptor) and CKR (CSFR-1-FLK-1 Receptor), respectively.[8] They contain the five immunoglobulin repeat regions derived from the extracellular domain of the CSF-1R, which are essential for ligand binding. The VEGFR-1 and VEGFR-2 regions contain the entire cytoplasmic and transmembrane domains of VEGFR-1 and VEGFR-2.

Ligand-Dependent Autophosphorylation of VEGFR-1 and VEGFR-2

Ligand-dependent autophosphorylation of receptor tyrosine kinases is considered to be an essential step in receptor activation and recruitment of signaling molecules to the receptor. Stimulation of PAE cells expressing individual chimeric VEGFR-1

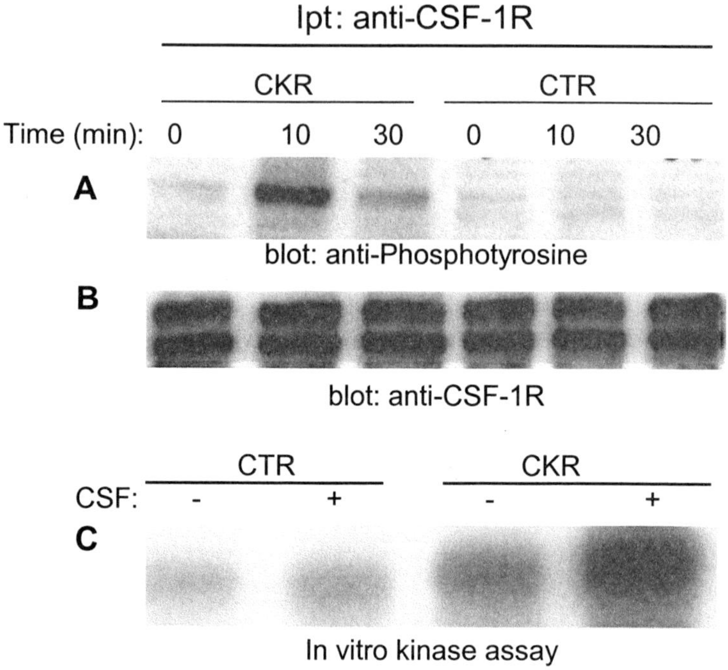

FIGURE 1. Ligand-dependent autophosphorylation of VEGFR-1 and VEGFR-2: Serum-starved PAE cells expressing chimeric VEGFR-1 (CTR) or chimeric VEGFR-2 (CKR) were stimulated with CSF-1 (40 ng/mL) for the indicated times **(A)** or for 10 min **(C)**. Cells were lysed, immunoprecipitated with anti-CSF-1R antibody, and subjected either to Western blotting using an antiphosphotyrosine antibody **(A)** or to *in vitro* kinase assay **(C)**. To confirm that an equal amount of protein is present in each group, the same membrane was reprobed with anti-CSF-1R antibody **(B)**.

(CTR) and VEGFR-2 (CKR) with CSF-1 demonstrated that, under comparable experimental conditions of expression and activation, VEGFR-1 and VEGFR-2 are strikingly different in their ability to undergo autophosphorylation and kinase activation. As demonstrated in FIGURE 1, stimulation of PAE cells expressing CKR resulted in robust tyrosine phosphorylation of CKR, but not CTR. Similarly, CKR, but not CTR, underwent autophosphorylation in an *in vitro* kinase assay (FIG. 1). These results suggest that VEGFR-1, unlike VEGFR-2, is unable to undergo ligand-dependent tyrosine phosphorylation and kinase activation.[8] Whether VEGFR-1 naturally is a poor kinase or its ligand-dependent tyrosine phosphorylation is rapidly dephosphorylated by tyrosine phosphatases is not clear.

Selective Activation of VEGFR-1 and VEGFR-2 Stimulates Distinct Biological Responses in Endothelial Cells

Although activation of VEGFR-2 is correlated with growth and migration of endothelial cells, it is not clear whether activation of VEGFR-2 alone is sufficient for induction of these biological responses. Most of the studies to date have used

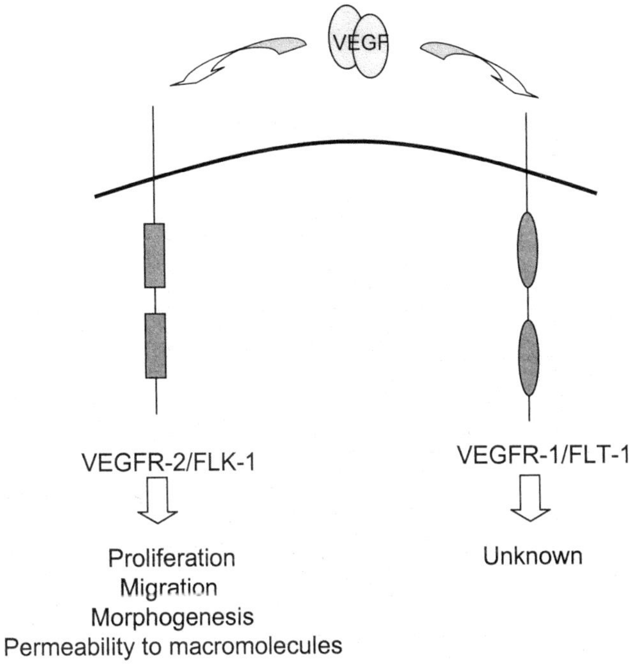

FIGURE 2. Contribution of VEGFR-1 and VEGFR-2 activity to endothelial cell functions: Selective activation of chimeric systems of VEGFR-1 and VEGFR-2 expressed in PAE cells shows that activation of VEGFR-2 is able to stimulate cell proliferation, migration, differentiation, and permeability. Unlike VEGFR-2, selective activation of VEGFR-1 in PAE cells results in no significant cell proliferation, migration, and differentiation.[8,14,16]

VEGF to activate VEGFR-2, which binds not only to VEGFR-1, but to VEGFR-2 as well.[1,2] To establish whether stimulation of VEGFR-2 and VEGFR-1 alone is sufficient to mediate the mitogenic responses in PAE cells, we subjected PAE cells expressing chimeric VEGFR-2 (CKR) or chimeric VEGFR-1 (CTR) to cell proliferation, migration, and differentiation assays. Stimulation of cells expressing CKR with CSF-1 stimulated growth, migration, differentiation, and *in vitro* angiogenesis of PAE cells in a dose-dependent manner[8,16] (FIG. 2).

Higher concentrations of CSF-1 (5–20 ng/mL) resulted in a profound effect on the morphology of PAE cells expressing CKR after a 24-h stimulation. The CSF-1-induced morphological changes were associated with elongation and acquisition of a dendritic shape. After 2–3 days of stimulation with CSF-1, these cells organized into short branching cords that formed a discontinuous network. However, stimulation of PAE cells expressing CTR with CSF-1 showed that CSF-1-dependent activation of CTR results in no significant growth and migration. Stimulation of PAE cells expressing CTR with high doses of CSF-1 also did not result in any obvious morphological changes.[8] Activation of CTR also did not stimulate tube formation of PAE cells in an *in vitro* angiogenesis assay (Meyer *et al.*, unpublished data).

Effect of Costimulation of VEGFR-1 and VEGFR-2 in Endothelial Cell Proliferation

The observation that activation of chimeric VEGFR-1 (CTR) alone is not sufficient to stimulate growth of endothelial cells raises the question of whether VEGFR-1 activation requires the presence of VEGFR-2 as a partner/coreceptor for its activation. For this purpose, we generated PAE cells coexpressing CKR and CTR. To gain insight into the biological consequences of CTR's effect on the CKR activation, PAE cells coexpressing CKR and CTR were subjected to a proliferation assay. Costimulation of CTR with CKR greatly reduces CKR-mediated PAE cell growth. To demonstrate that the observed effect of CTR modulation of CKR activity is not due to the chimera system, we also coexpressed VEGFR-1 and VEGFR-2 in PAE cells and subjected them to a proliferation assay in a similar manner. Essentially the same results were obtained when VEGFR-1 was costimulated with VEGFR-2 with VEGF. Thus, it appears that the VEGFR-2-mediated mitogenic signal in PAE cells is regulated by VEGFR-1 activation.[8]

Although VEGFR-1 was initially suggested to be a decoy receptor, this model of VEGFR-1 function is not fully able to illustrate the negative function of VEGFR-1 in angiogenesis for the following reasons: First, VEGFR-1, like VEGFR-2, has all

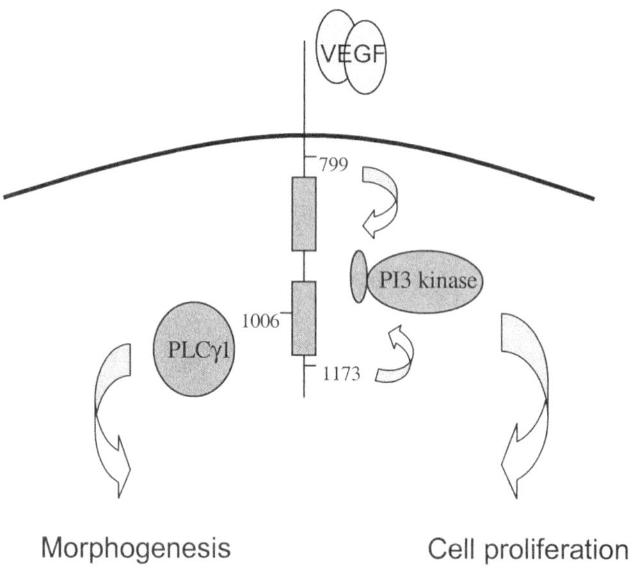

FIGURE 3. Activation of PI3 kinase and PLCγl by VEGFR-2 and their role in endothelial cell functions: Selective activation of VEGFR-2 in PAE cells shows that both PLCγl and PI3 kinase are activated by VEGFR-2. Site-directed mutagenesis on VEGFR-2 demonstrates that tyrosines 799 and 1173 are binding sites for p85 of PI3 kinase. Activation of PI3 kinase in these cells appears to be involved in cell proliferation. Activation of PLCγl by VEGFR-2 is mediated by tyrosine 1006, located in the kinase domain of VEGFR-2. Activation of PLCγl in this system is not involved in cell proliferation. PLCγl activation is involved with morphological changes in PAE cells associated with their differentiation.[8,16]

the features of RTKs, including ligand binding domain, autophosphorylation sites, and catalytic domain. This suggests that VEGFR-1 has the capacity to convey biological responses.[9] Second, activation of VEGFR-1 in nonendothelial cells such as monocytes and trophoblasts stimulates cell growth and cell migration.[10,11] If VEGFR-1 is a decoy receptor, one would expect it not to elicit any biological responses in any given cell type. Stimulation of cellular responses by VEGFR-1 in nonendothelial cells strongly suggests that VEGFR-1 is a functional receptor, and its activation and function are potentially regulated by cell type. Third, if VEGFR-1 elicits its negative effects by absorbing/trapping VEGF and preventing VEGF from binding to and activating VEGFR-2, one expects that replacing the ligand binding region of VEGFR-1 should abolish its inhibitory function. The chimeric system of VEGFR-1 and VEGFR-2 in which the extracellular domain of each receptor is replaced with the extracellular domain of human CSF-1R demonstrates that chimeric VEGFR-1 is still able to attenuate VEGFR-2-mediated endothelial cell growth.[8] This suggests that the antiangiogenic function of VEGFR-1 is not necessarily attributed to its ability to bind VEGF and prevent it from binding to and activating VEGFR-2. Fourth, a recent study by Dunk and Ahmed shows that VEGFR-1 may elicit its negative function in angiogenesis by nitric oxide production.[12] In addition, a study by Zeng *et al.* suggests that pertussis toxin–sensitive G proteins may mediate the negative function of VEGFR-1 in endothelial cells.[13] All together, these results strongly suggest that VEGFR-1 conveys its negative function in endothelial cells by actively participating in signal transduction and is involved in activation of unique signaling proteins in endothelial cells.

Role of Tyrosines 799 and 1173 of VEGFR-2 in Activation of PLCγ and PI3 Kinase

To determine the molecular determinant that influences the qualitative nature of cellular responses of VEGFRs, we mutated tyrosines 799 and 1173 to phenylalanine, these tyrosines are located in the juxtamembrane and C-terminus of VEGFR-2, respectively. We demonstrated that presence of tyrosines 799 and 1173 is not required for autophosphorylation and kinase activity of VEGFR-2; however, these tyrosine residues are required for mitogenic activity of VEGFR-2 and activation of PI3 kinase (FIG. 3).[14] Thus, mitogenic activity of VEGFR-2 in endothelial cells is established by phosphorylation of tyrosines 799 and 1173, leading to association with and activation of PI3 kinase and subsequently to endothelial cell growth. Interestingly, although these two tyrosine residues are also present on VEGFR-1, VEGFR-1 is unable to stimulate endothelial cell proliferation. Based on these findings, it appears that either these tyrosine sites are not phosphorylated in the VEGFR-1 or prerequisite requirements such as the presence of critical tyrosine residues on the VEGFR-1 are lacking.

Role of Tyrosine 1212 of VEGFR-2 in Autophosphorylation and Activation of Signaling Molecules

To analyze the contribution of other tyrosine sites to VEGFR-2 function, we have made additional mutations on VEGFR-2, including tyrosines 820, 949, 994, 1212, and 1221 (Meyer *et al.*, unpublished data). Our initial evaluation of these tyrosine mutant receptors showed that mutation of tyrosine 1212 unexpectedly impairs the ligand-dependent activation of VEGFR-2. Mutation of tyrosine 1212 to phenylalanine severely impaired the ligand-dependent autophosphorylation of VEGFR-2 and its ability to associate with and activate Src. This mutation also reduced the VEGFR-2's ability to phosphorylate PLCγ1 and MAPK. Unlike mutation of tyrosine 1212 to phenylalanine, replacement of tyrosine 1212 with glutamic acid preserved the ligand-dependent activation of VEGFR-2 and activation of VEGFR-2-associated signaling proteins, including Src, PLCγ1, and MAPK. To test whether Src activation is required for the ligand-dependent autophosphorylation of VEGFR-2, we have co-expressed chimeric VEGFR-2 with kinase dead Src (Src/KD). Our analysis showed that Src activation is not essential for ligand-dependent activation of VEGFR-2 since cells coexpressing wild-type receptor with kinase dead Src or wild-type Src displayed no apparent effect in the ligand-dependent autophosphorylation of VEGFR-2. Similarly, expression of wild-type VEGFR-2 in fibroblast (SYF) cells obtained from the triple-knockout Src family kinases showed normal ligand-dependent autophosphorylation. Collectively, these results suggest that phosphorylation of tyrosine 1212 of VEGFR-2 plays a crucial role in the activation of VEGFR-2 and subsequent VEGFR-2-mediated angiogenesis. However, Src kinases are not involved in the ligand-dependent autophosphorylation of VEGFR-2.[15]

Regulation of angiogenesis is the most critical step in the development of tumor and ocular neovascularization and in inflammation. The results presented in this work identify the key biological function of VEGFR-1 and VEGFR-2 by using unique and selective activation systems of VEGFR-1 and VEGFR-2. This approach offers an advantage over the use of wild-type models. This model can be exploited to define and characterize the function of these receptors in different systems. Furthermore, our work identifies the tyrosine sites involved in activation of PI3 kinase and PLCγ1 by VEGFR-2 and their contribution to endothelial cell functions. Further study will define the more detailed signal transduction relays employed by these receptors.

ACKNOWLEDGMENTS

This work was supported in part by NIH Grant Nos. EY012997 and EY0137061 to Nader Rahimi.

REFERENCES

1. Ferrara, N. 2002. Role of vascular endothelial growth factor in regulation of physiological angiogenesis. Am. J. Physiol. Cell Physiol. **280:** C1358–C1366.
2. Neufeld, G., T. Cohen, S. Gengrinovitch *et al.* 1999. Vascular endothelial growth factor (VEGF) and its receptors. FASEB J. **1:** 9–22.
3. Miao, H.Q. & M. Klagsbrun. 2000. Neuropilin is a mediator of angiogenesis. Cancer Metastasis Rev. **19**(1–2)**:** 29–37.
4. Shalaby, F., J. Rossant, T. Yamaguchi *et al.* 1995. Nature **376:** 62–66.
5. Fong, G.H., J. Rossant, M. Gertsenstein *et al.* 1995. Nature **376:** 66–70.
6. Fong, G.H., L. Zhang, D.M. Bryce *et al.* 1999. Increased hemangioblast commitment, not vascular disorganization, is the primary defect in flt-1 knock-out mice. Development **126:** 3015–3025.
7. Kearney, J.B., C.A. Ambler, K.A. Monaco *et al.* 2002. Vascular endothelial growth factor receptor Flt-1 negatively regulates developmental blood vessel formation by modulating endothelial cell division. Blood **99:** 2397–2407.

8. Rahimi, N., V. Dayanir & K. Lashkari. 2000. Receptor chimeras indicate that the vascular endothelial growth factor receptor-1 (VEGFR-1) modulates mitogenic activity of VEGFR-2 in endothelial cells. J. Biol. Chem. **275:** 16986–16992.
9. De Vries, C., J.A. Escobedo, H. Ueno *et al.* 1992. The fms-like tyrosine kinase, a receptor for vascular endothelial growth factor. Science **255:** 989–991.
10. Clauss, M., H. Weich, G. Breier *et al.* 1996. The vascular endothelial growth factor receptor Flt-1 mediates biological activities: implications for a functional role of placenta growth factor in monocyte activation and chemotaxis. J. Biol. Chem. **271:** 17629–17634.
11. Athanassiades, A. & P.K. Lala. 1998. Role of placenta growth factor (PIGF) in human extravillous trophoblast proliferation, migration, and invasiveness. Placenta **7:** 465–473.
12. Bussolati, B., C. Dunk, M. Grohman *et al.* 2001. Vascular endothelial growth factor receptor-1 modulates vascular endothelial growth factor mediated angiogenesis via nitric oxide. Am. J. Pathol. **159:** 993–1008.
13. Zeng, H., H.F. Dvorak & D. Mukhopadhyay. 2001. Vascular permeability factor (VPF)/vascular endothelial growth factor (VEGF) receptor-1 down-modulates VPF/VEGF receptor-2–mediated endothelial cell proliferation, but not migration, through phosphatidylinositol 3-kinase–dependent pathways. J. Biol. Chem. **276:** 26969–26979.
14. Dayanir, V., R.D. Meyer, K. Lashkari *et al.* 2001. Identification of tyrosine residues in vascular endothelial growth factor receptor-2/FLK-1 involved in activation of phosphatidylinositol 3-kinase and cell proliferation. J. Biol. Chem. **276:** 17686–17692.
15. Meyer, R.D., V. Dayanir, F. Majnoun *et al.* 2002. The presence of a single tyrosine residue at the carboxyl domain of vascular endothelial growth factor receptor-2/FLK-1 regulates its autophosphorylation and activation of signaling molecules. J. Biol. Chem. **277:** 27081–27087.
16. Meyer, R.D., C. Latz & N. Rahimi. 2003. Recruitment and activation of PLCγ1 by VEGFR-2 are required for tubulogenesis and differentiation of endothelial cells. J. Biol. Chem. **278:** 16347–16355.

Induction of Vascular Networks in Adult Organs: Implications to Proangiogenic Therapy

YUVAL DOR,[a] VALENTIN DJONOV,[b] AND ELI KESHET[a]

[a]*Department of Molecular Biology, Hebrew University–Hadassah Medical School, Jerusalem 91120, Israel*

[b]*Department of Anatomy, Bern University, Bern, Switzerland*

ABSTRACT: Most therapeutic proangiogenic approaches rely on the premise that a single factor (e.g., VEGF) can induce new vascular loops properly connected to the existing vasculature of an ill-perfused organ (e.g., the ischemic myocardium). To gain insights into VEGF-driven neovascularization in the adult, in general, and to uncover inherent obstacles in VEGF-based therapeutic angiogenesis, we have used a transgenic system for conditional "on" and "off" switching of VEGF expression in selected adult organs and analytical methods for three-dimensional visualization of the induced vascular network. VEGF induces massive accumulation of endothelial progenitor cells and their subsequent coalescence into new vascular loops. Continual ramification of the vasculature and fusion of adjacent vessels culminate in formation of hemangioma-like vessels. In the absence of guiding cues, abnormal branching patterns and random connections with the existing network are formed. Following cessation of the VEGF stimulus, the acquired network undergoes extensive hemodynamic remodeling acting to "normalize" its architecture. Premature withdrawal of VEGF leads to regression of most acquired vessels, thus challenging the utility of therapeutic approaches relying on short stimulus duration. Overproduction of VEGF leads also to highly disruptive edema, thereby emphasizing the need to uncouple VEGF-induced permeability from its angiogenic activity. This genetic switch system thus highlights remaining problems critical for the implementation of proangiogenic therapy.

KEYWORDS: angiogenesis; VEGF; therapy; conditional transgenic

INTRODUCTION

The gross vascular anatomy is characterized by a reproducible branching pattern and, at least for the major vessels, there are fixed branch angles, curvature, and size gradation along the vascular tree. There are also designated sites for branching for secondary branches, while microvessels formed by sprouting angiogenesis are mostly nonstereotyped. Cellular and molecular mechanisms underlying vascular patterning

Address for correspondence: Dr. Eli Keshet, Department of Molecular Biology, Hebrew University–Hadassah Medical School, Jerusalem 91120, Israel. Voice: +972-2-67584996; fax: +972-2-6757195.
keshet@cc.huji.ac.il

Ann. N.Y. Acad. Sci. 995: 208–216 (2003). © 2003 New York Academy of Sciences.

and, in particular, the cues for pathway guiding are not well understood, but have been shown to include both attractive and repulsive guidance signals and to be regulated by both positive and negative regulators.[1,2] Unlike the highly choreographed manner of vascular branching during embryonic vasculogenesis and angiogenesis, adult neovascularization is significantly less ordered. In the adult, neovascularization is mostly confined to the female reproductive system, but is also induced in response to perturbations in oxygen homeostasis, primarily following occlusion of major blood vessels. Ischemia-driven neovascularization mediated by VEGF,[3] acting together with a vessel remodeling process termed arteriogenesis,[4] operates to improve the perfusion of ischemic tissues. Tumor angiogenesis represents yet another example of extensive branching morphogenesis in the adult. Here, an elaborate network of arteries and veins is produced anew within an expanding tumor mass.

An even more complex situation is represented by a contemplated proangiogenic therapy. The latter is based on the premise that exogenous administration of angiogenic factor to the ischemic tissue will not only produce new vessels, but also that the new vessels will properly differentiate and make the right connections with the existing coronary network. It thus appears that in most, if not all, cases of adult neovascularization, whether physiological, pathological, or artificially induced, the process is instructed by proangiogenic activities elaborated in a spatially non-fashioned manner and in a dose solely determined by the severity of the ischemic insult or by the extent of deregulated expression in tumors.

With therapeutic angiogenesis, two factors, in particular, need to be considered: First, most contemplated proangiogenic therapies rely on the administration of a single factor (primarily VEGF), whereas the natural process is multifactorial. It is assumed, but not proven, that all complementary activities, including those essential for engagement with periendothelial cells and those required for specifying an arterial or venous fate, will be induced downstream of the initiating factor. Second, it remains to be determined how the complex morphogenetic process is executed in the absence of guiding cues and under circumstances where neither the normal dosage nor the spatial or temporal controls of the initiating morphogen are maintained. Improper connectivity might perturb rather than improve the coronary circulation.

Addressing these fundamental issues in preclinical studies has been hampered by several problems. Depending on the methodology used for growth factor delivery, complications may include insufficient or variable levels of VEGF in the target organ, "off target" effects, insufficient duration of the stimulus, an accompanying inflammatory response, and an immunological reaction. To distinguish between inherent problems and complications resulting from methodological shortcomings, we have developed a transgenic system for conditional "on" and "off" switching of VEGF in the organ of choice, thereby allowing evaluation of performance of this factor under better controlled conditions.

RESULTS AND DISCUSSION

A Transgenic Mouse System for Conditional Expression of VEGF in Selected Adult Organs

Considering previous findings that deviation from the normal dosage of VEGF during development leads to embryonic lethality,[5,6] we resorted to a conditional transgenic system for VEGF induction. An animal model system in which a source of VEGF can be specifically induced in the target organ, steadily maintained for the desired duration, and subsequently switched off was constructed. Transgenic mice expressing a tetracycline-regulated transactivator protein (tTA) exclusively in the myocardium, liver, or brain were employed as "driver lines". Another transgenic line harboring a VEGF164-encoding transgene (driven by a tetracycline-responsive promoter) was used as the "responder line". The respective "driver" mouse was mated with the VEGF responder animal, and pups that inherited both transgenes were selected. The onset of VEGF induction in these animals and the duration of expression were tightly controlled by the inclusion or omission of tetracycline from the drinking water ("off" and "on" modes, respectively). Littermates that inherited only one of the two transgenes and were concomitantly subjected to the same repression/induction regime served as controls. Heart-specific induction was achieved using a driver line in which tTA expression is driven by a myosin heavy chain-α (MHCα) heart-specific promoter, previously shown to be expressed specifically by cardiomyocytes.[7] Liver-specific induction was obtained by using a driver line in which tTA expression is driven by a C/EBPβ (CCAAT/enhancer binding protein β) promoter (also known as liver activator protein or PLAP), expressed primarily in hepatocytes.[8] Brain-specific induction is obtained by using a driver line in which tTA expression is driven by a calmodulin kinase II (CamKII) brain-specific promoter.[9] As anticipated, if kept unrepressed, misexpression of VEGF in either organ led to multiple developmental defects and embryonic lethality.[10] In the presence of tetracycline, however, tTA^{+}/$VEGF^{+}$ animals were born with the expected frequency, and VEGF expression was strongly induced, exclusively in the respective organ, upon withdrawal of tetracycline (see ref. 11 for profiles of induced expression in different target organs). Notably, VEGF induction was fully reversible and VEGF levels returned to baseline levels within a day after adding tetracycline, thereby allowing examination of remodeling and regression following termination of the inductive signal.

Formation of New Vessels Is Preceded by Accumulation of Endothelial Progenitor Cells (EPCs)

A major advantage of this system is a unique opportunity to initiate a massive angiogenic response in a synchronous manner, thereby discerning sequential steps of the process in its native organ environment. A striking observation was that at early time points after the onset of a VEGF switch (i.e., at stages preceding emergence of new vascular loops) foreign cells massively infiltrated the organ. Using a "heart switch", these cells appeared as clusters interspersed between the muscle fibers (FIG. 1A); using a "liver switch", these cells appeared as blood-borne or loosely associated with vessel walls; using a "brain switch", these cells appeared mostly in perivascular compacted clusters (data not shown). To determine that these cell masses indeed represent vascular progenitors and not merely components of an inflammatory infiltrate, a comprehensive *in situ* analysis was carried out using a panel of endothelial-specific markers. Results clearly showed that these cells indeed express multiple endothelial markers, including CD31, VEGFR2, VEGFR1, Tie1, Tie2, and others (data not shown). Histological examination of sections retrieved at progressive time points after the onset of the VEGF switch has shown that EPC clusters are remodeled and incorporated into patent vessels (FIG. 2B). It has been recently shown that circulating

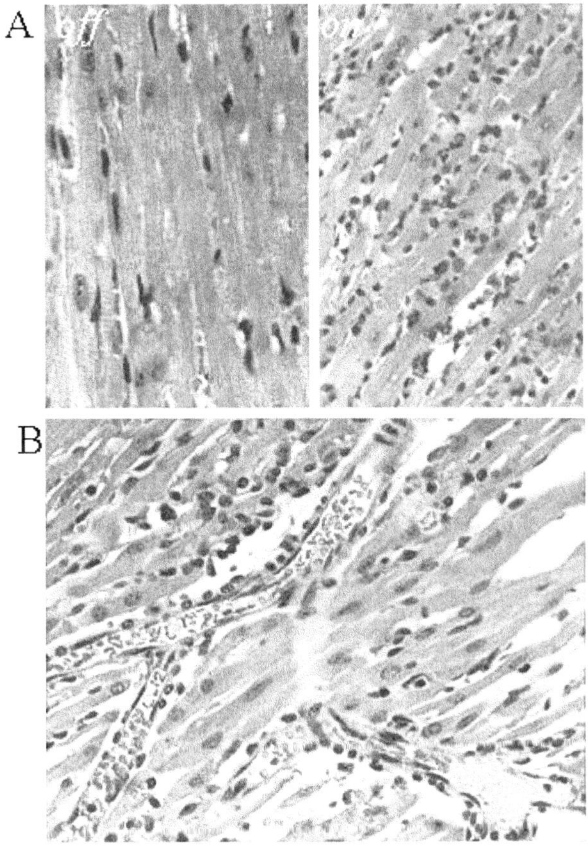

FIGURE 1. VEGF induces accumulation of EPCs in the myocardium. Myocardial VEGF expression was induced in 5-week-old mice and the myocardium was inspected at different times following induction. **(A)** Two weeks after switching on VEGF, the myocardium is loaded with presumptive EPCs that are interspersed among the muscle fibers [compare the H & E-stained tissue of littermates at the "off" (*left*) and "on" (*right*) modes]. **(B)** Two weeks later in an ongoing VEGF switch, EPCs are detectable alongside multiple, blood-filled vascular channels that have been formed in the myocardium.

EPCs mobilized from the bone marrow may significantly contribute to adult neovasculatures (for a recent review, see ref. 12). The relative contributions of BM-derived EPCs, proximal recruitment of vessel wall–derived circulating precursors, and *in situ* proliferation of activated endothelium to the overall process need to be determined.

Abnormal Architecture and Connections in the Induced Myocardial Network

For optimal functioning, new vessels induced by VEGF must not only form a well-graded vascular tree, but also appropriately connect with the existing coronary network. To visualize the three-dimensional architecture of the induced coronary network, corrosion casts were prepared and inspected by scanning EM. This methodology

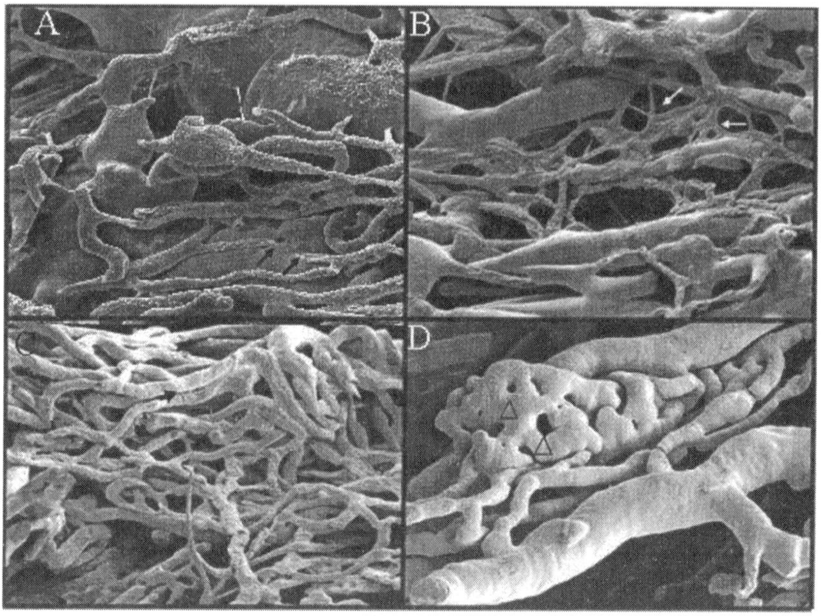

FIGURE 2. Vascular casts of VEGF-induced myocardial networks. Representative examples of Mercox-filled networks induced by VEGF are shown in images **A–D**. Note many abnormal features of the induced vasculature, including hemangioma-like vessels, abnormal branching patterns (*white arrows*), and circular vascular loops (*open triangles*). Irregularly shaped vessels are likely the result of ongoing vascular merging (*black arrows*) and parallel fusion (*black arrowhead*).

visualizes the entire vascular tree at a resolution that includes even the smallest capillaries. Since it detects only perfused vessels accessible to the casting resin, it highlights only functional vascular loops connected to the systemic circulation. Switching "on" VEGF expression and maintaining the stimulus for several weeks led to progressive ramification of the vasculature. Initially, this was evident mostly by reduced intercapillary distance and by structures suggestive of ongoing fusion events between contacting small vessels (arrowhead in FIG. 2). Further progression of the process during later stages resulted in the generation of very large vessels and eventually in the production of hemangioma-like vascular sacs (FIG. 2). Some of the vessels shown in FIGURE 2 seem to be in the process of flattening on top and merging their lumens with those of existing larger vessels (black arrows). Abnormal structural features of these vessels included a variable diameter in consecutive segments within a single vessel, a large proportion of vessels with multiple rather than bimodal branching points (white arrows), and abnormal branching angularities. Thus, it appears that, in the absence of guiding cues, vessels induced by VEGF in the adult myocardium connect with the existing coronary network in an arbitrary manner, creating a highly disorganized network.

VEGF-driven neovascularization was accompanied by excessive edema throughout the period in which VEGF was overexpressed. Switching "off" VEGF expression resulted in a rapid restoration of the normal water content.

Vessel Regression and Remodeling following Termination of the Inductive Stimulus

The ability to switch off VEGF production at any time after its initiation provided a good opportunity to address two important issues in adult neovascularization. First, will newly acquired vessels be maintained after withdrawal of the inductive stimulus? This issue is of particular concern for proangiogenic therapy in light of earlier studies showing that premature withdrawal of VEGF may lead to regression of immature vessels in settings of developmental and tumor angiogenesis.[13,14] Second, will the disorganized network produced initially be hemodynamically remodeled to normalize

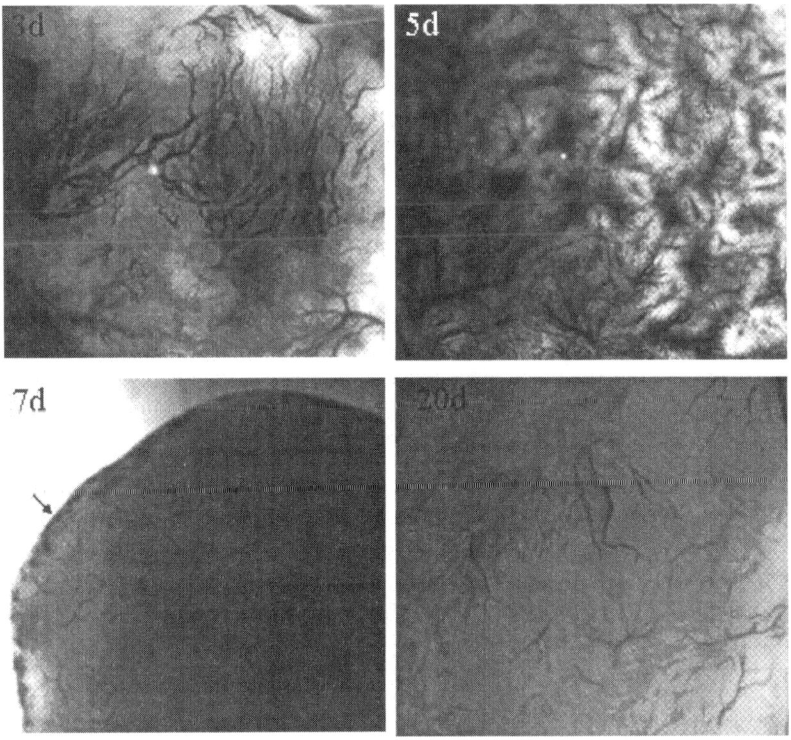

FIGURE 3. Kinetics of vessel regression and remodeling following termination of the VEGF stimulus. Liver-specific expression of VEGF was induced in 4-week-old mice, maintained for 4 weeks, and then switched off. Livers of transgenic littermates were retrieved at the indicated times after switching off VEGF expression and were photographed as whole mounts without staining (i.e., highlighting blood flowing in superficial vessels and blood hemorrhages). Note that, after 3 days, the elaborate newly formed network has begun to leak in many places. Five days after VEGF withdrawal, extensive hemorrhages are seen, presumably reflecting ongoing regression. By 7 days, the initial dense network has been considerably thinned, and hemorrhages are still detectable in peripheral regressing vessels (*arrow*). Twenty days after VEGF withdrawal, the remaining vessels have been remodeled, forming evenly spaced vascular trees.

its architecture? If not, it is unlikely that the VEGF-induced network will indeed improve the organ perfusion.

In a series of experiments, VEGF expression was induced in the heart or liver and was maintained for different durations before switching off. Changes in the newly formed networks were then monitored through inspection of littermates withdrawn for analysis at different time points after terminating VEGF expression. Remarkably, if VEGF was prematurely withdrawn, most if not all acquired vessels regressed. In a marked contrast, delaying the "off" switch until a few weeks after onset resulted in vessels that have become refractory to VEGF withdrawal and persisted for many months after cessation of the VEGF stimulus.[11] The molecular and cellular determinants of the transition from a VEGF-dependent to a VEGF-refractory state remain to be determined, but (extrapolating from other systems) may be associated with the stabilizing effect of recruited periendothelial cells.[15,16] These findings indicate that there is a critical transition point beyond which the initially unstable vessels no longer need VEGF for survival.

When sufficient time for vessel stabilization was allowed prior to withdrawing VEGF, a significant fraction of vessels persisted, thus allowing us to follow progressive changes in the vascular tree architecture. Importantly, the highly disorganized network was remodeled into orderly spaced and more refined treelike structures (FIG. 3). Hemodynamic remodeling is probably a major factor acting to rectify the abnormal network so that it may confer a significantly improved organ perfusion.[11]

CONCLUDING REMARKS

Findings outlined above demonstrate the remarkable degree of self-organization of VEGF-activated endothelium into a treelike network, as well as the key role of hemodynamic forces in rectifying abnormal branching and connections made initially. Thus, these findings suggest that, even in the absence of guiding cues, VEGF alone might be sufficient to produce a functional vascular network. The study also highlights, however, some major problems pertaining to VEGF-based therapy, including the problem of regression of newly formed vessels and the need to counteract VEGF-induced permeability during the critical period of vessel formation.

REFERENCES

1. CHILDS, S. *et al.* 2002. Patterning angiogenesis in the zebrafish embryo. Development **129:** 973–982.
2. LEE, S.H. *et al.* 2001. Inhibition of angiogenesis by mouse sprouty protein. J. Biol. Chem. **276:** 4123–4133.
3. DOR, Y. & E. KESHET. 1997. Ischemia-driven angiogenesis. Trends Cardiovasc. Med. **7:** 289–294.
4. VAN ROYEN, N. *et al.* 2001. Stimulation of arteriogenesis; a new concept for the treatment of arterial occlusive disease. Cardiovasc. Res. **49:** 543–553.
5. CARMELIET, P. *et al.* 1996. Abnormal blood vessel development and lethality in embryos lacking a single VEGF allele. Nature **380:** 435–439.
6. FERRARA, N. *et al.* 1996. Heterozygous embryonic lethality induced by targeted inactivation of the VEGF gene. Nature **380:** 439–442.
7. YU, Z., C.S. REDFERN & G.I. FISHMAN. 1996. Conditional transgene expression in the heart. Circ. Res. **79:** 691–697.

8. KISTNER, A. *et al.* 1996. Doxycycline-mediated quantitative and tissue-specific control of gene expression in transgenic mice. Proc. Natl. Acad. Sci. USA **93:** 10933–10938.
9. MAYFORD, M. *et al.* 1996. Control of memory formation through regulated expression of a CaMKII transgene. Science **274:** 1678–1683.
10. DOR, Y. *et al.* 2001. A novel role for VEGF in endocardial cushion formation and its potential contribution to congenital heart defects. Development **128:** 1531–1538.
11. DOR, Y. *et al.* 2002. Conditional switching of VEGF provides new insights into adult neovascularization and pro-angiogenic therapy. EMBO J. **21:** 1939–1947.
12. RAFII, S. *et al.* 2002. Vascular and hematopoietic stem cells: novel targets for anti-angiogenic therapy? Nat. Rev. Cancer **2:** 826–835.
13. ALON, T. *et al.* 1995. Vascular endothelial growth factor acts as a survival factor for newly formed retinal vessels and has implications for retinopathy of prematurity. Nat. Med. **1:** 1024–1028.
14. BENJAMIN, L.E. & E. KESHET. 1997. Conditional switching of vascular endothelial growth factor (VEGF) expression in tumors: induction of endothelial cell shedding and regression of hemangioblastoma-like vessels by VEGF withdrawal. Proc. Natl. Acad. Sci. USA **94:** 8761–8766.
15. LINDAHL, P. *et al.* 1997. Pericyte loss and microaneurism formation in PDGF-B-deficient mice. Science **277:** 242–245.
16. BENJAMIN, L. *et al.* 1999. Selective ablation of immature blood vessels in established human tumors follows vascular endothelial growth factor withdrawal. J. Clin. Invest. **103:** 159–165.

Index of Contributors